MASS TRANSFER AND SEPARATION PROCESSES

Principles and Applications

Second Edition

MASS TRANSFER AND SEPARATION PROCESSES

PRINCIPLES AND APPLICATIONS

Second Edition

Diran Basmadjian

CRC Press
Taylor & Francis Group
Boca Raton London New York

CRC Press is an imprint of the
Taylor & Francis Group, an **informa** business

CRC Press
Taylor & Francis Group
6000 Broken Sound Parkway NW, Suite 300
Boca Raton, FL 33487-2742

Printed in the United States of America on acid-free paper
10 9 8 7 6 5 4 3 2 1

International Standard Book Number-10: 1-4200-5159-8 (Hardcover)
International Standard Book Number-13: 978-1-4200-5159-9 (Hardcover)

Library of Congress Cataloging-in-Publication Data

Basmadjian, Diran.
Mass transfer and separation processes : principles and applications / author, Diran Basmadjian. -- [2nd ed.]
p. cm.
Rev. ed. of: Mass transfer.
Includes bibliographical references and index.
ISBN-13: 978-1-4200-5159-9 (hardcover : acid-free paper)
ISBN-10: 1-4200-5159-8 (hardcover : acid-free paper)
1. Mass transfer--Textbooks. 2. Separation (Technology)--Textbooks. I. Basmadjian, Diran. Mass transfer. II. Title.

QC318.M3B37 2007
660'.28423--dc22 2007008242

Visit the Taylor & Francis Web site at
http://www.taylorandfrancis.com

and the CRC Press Web site at
http://www.crcpress.com

Occam's Razor (Principle of Parsimony)

"Entia non sunt multiplicanda praeter necessitatem"

"Entities should not be multiplied beyond necessity"

William of Ockham (1280–1349)

Foreword

Professor Diran Basmadjian passed away on February 28, 2007, just days after finalizing this textbook. In fact, he checked that the publisher had received the last portion of the corrected galleys just hours before he died. He will be greatly missed by his family, his colleagues, his students, the readers of this and his many other texts, and by all who benefited from his scholarly contributions. He also will be missed by the kindergarten children at a local public school, whom he taught a hands-on science program after he retired as a full-time faculty member. One of us was asked annually to prepare a polyvinyl alcohol solution to his specification so that he could illustrate polymer rheology through "slime."

This playful spirit suffused all his teaching. He had a gift for developing creative physical insights into complex problems and he cultivated this in his students. He introduced the concept of a "surprise experiment" into our Unit Operations lab where students were confronted with a new problem complete with equipment first thing in the morning, and were then expected to build the pilot scale equipment [an important lesson for modern students who had rarely handled a screwdriver or wrench in their lives] and devise a series of meaningful experiments by the end of the day. His courses were in high demand whenever they were offered, even his graduate applied mathematics course that began at 7:30 a.m.!

Diran had a spectacular knack for simplifying complex systems in a way that provided physical insight into the situation along with useful analytical and/or graphical solutions. The essence of this gift was his "art of assumption" and his ability to find a mathematical solution to a problem derived from widely different fields; he believed that most differential equations that could be solved had been solved and it was just a matter of looking in the right place for the solutions. For example, Diran became interested in problems of blood compatibility of biomaterials, including the drag forces on a developing blood clot that would cause it to embolize. He recognized that the problem was physically similar to that of deposition on a river bed and sought out the literature on the latter to "solve" the mathematics of the embolization problem. He only got frustrated when he could not find an analytical solution and had to resort to numerical simulation, something he felt was a lesser approach. He was the one who showed us the importance of visualizing what was going on by "creative doodling" and plotting the equation, and also by looking to groupings of variables to give insight into the key underlying phenomena — something that was harder to do when the first approach to a solution was computer modeling. This talent for simplifying the complex to provide useful insights is well illustrated in his first text, *The Little Adsorption Book: A Practical Guide for Engineers and Scientists*, a text that summarizes this complex subject in less than 130 pages and includes much of his seminal work in the area.

He also consistently emphasized the value of "bracketing the solution," in which one makes *simplifying* assumptions that allow for *simple* solutions that represent the best and worse cases for a problem. Bracketing the limits of the solution is often all one needs to make a decision or solve a problem in practice, and these limiting cases frequently reveal important phenomena for further mathematical and/or experimental exploration. One of his favorite quotations was from Wolfgang Pauli, who upon hearing of a new theory said, "That's not right, that is not even wrong." It was important to Diran to be wrong, so that he understood enough to get it right.

Diran's dedication to teaching went beyond the classroom. He loved to discuss his craft with students and colleagues over a good cigar, particularly if they brought it along to class for him to enjoy. Fluent in five languages and coming to Toronto as a foreign student, he had a special affinity for international students. His office door was always open (if you didn't mind the cigar smoke) or he would wander into your office and begin a discussion centered around politics and current events but inevitably circling back to a research question or some other academic issue. Many of the faculty in our department were mentored through these life discussions.

Diran delighted in his students. But mostly, he delighted in his family — his wife, his daughters, and his granddaughters — who exceeded them all.

We are comforted knowing that Professor Basmadjian's legacy lives on in his scholarship and in the mark he made on so many people. As the Talmud says "a scholar is a builder, a builder of the world." Diran was a spectacular builder.

Michael V. Sefton
Michael E. Charles Professor of Chemical Engineering
University of Toronto

D. Grant Allen
Professor of Chemical Engineering and Applied Chemistry
University of Toronto

Preface

The philosophy and goals of this text remain the same as those of the first edition. Its principal aim is to convey, *at an introductory level,* the essential notions of mass transfer theory, and to reinforce them with classical and contemporary illustrations drawn from the engineering, environmental, and biosciences (Chapters 1 through 5). This provides the groundwork for an introduction to the sister topic of separation processes, which is taken up in Chapters 6 through 9. The treatment is designed to permit coverage of each topic in a *single* term, typically at the second- and third-year level of a regular engineering curriculum. The mathematics is kept at a simple, but not simplistic, level.

It was thought best to leave the more theoretical, and in some ways more restrictive, treatments of these topics, enshrined in the concept of "transport phenomena," for graduate-level study. For similar reasons, we excluded the more complex and exotic separation processes, which call for extensive computer simulations (multicomponent distillation, pressure swing adsorption, etc.). These are again best taken up in a follow-up course, typically an elective at the fourth-year level. The treatment is, nevertheless, detailed enough, and profound enough, for the student to enter the engineering world, or to proceed to graduate work, with some degree of confidence.

The theories of heat conduction and transfer are utilized not so much to draw analogies, but rather to make fruitful use of existing solutions in ways not seen in other texts. We use shape factors to present simple solutions to Laplace's equation (Chapter 2), the Lévêque solution for entry region mass transfer (most membrane processes operate in this region), and solutions to heat source problems to describe mass emissions (Chapter 4). We also borrow the Biot and Fourier numbers to analyze transient diffusion. In spite of these repeated intrusions of heat transfer theory, the author does not subscribe to the view that the two topics should be taught in unison (a currently fashionable trend). Beyond a certain communality of rate and conservation laws (and hence identical forms of their solutions), the interests, goals, and applications of the two disciplines diverge dramatically and irreversibly. Heat transfer theory rarely intrudes in the environmental and biosciences except to provide ready-made solutions for certain problems, and plays only a marginal role in separation processes (e.g., in distillation). If mass transfer is to have a companion, it makes more sense to twin it with the environmental or separation sciences. This is, at least, the approach taken here.

Although large sections have been rewritten and expanded, the organizational structure of the text is the same as that of the first edition. Chapter 1 provides an early introduction to basic notions of mass transfer theory and enables the reader, by the time it ends, to perform some simple but meaningful calculations. The alternative and usual practice of starting with elaborate dissertations on the molecular theories of diffusivity is a surefire recipe to bore an audience. (The reader will be bored in Chapter 3, but not terribly.) Readers who question the large amount of space devoted to modeling (Chapter 2) are reminded of the exasperated cry uttered by almost every instructor at least once: "They can't even do a simple mass balance!"

In the treatment of separation processes, we have again shunned the "Unit Operations" approach and have organized the material instead into the four unified topics of "Phase Equilibria," "Staged Operations," "Continuous-Contact Operations," and "Simultaneous Heat and Mass Transfer" (the latter not strictly a separation process). Some sections, and one chapter in particular (Phase Equilibria), contain material that is known, or should be known, from previous courses. It is surprising, however, how quickly such material is forgotten, or its relevance fully grasped. There have been no complaints about its inclusion, but the lagging instructor may wish to omit some portions to make up for lost time. Individuals who wish to use the text for a course in separation processes only will find that Chapters 1 through 5 are an *indispensable* adjunct. All too often, this material, typically taught in an earlier course, has either not been put in its proper context or, if taught as part of a "transport phenomena" package, been rendered irrelevant.

Even more than in the first edition, emphasis is placed on developing the art of making simplifying assumptions and conveying to the student a *sense of scale*, in part through the inclusion of numerous photographs of actual installations and the use of many "real-world" problems. Mere words do not seem to have the same effect, and some horrific errors in judgment have been made as a result.

The author's gratitude goes out, first and foremost, to his many colleagues who fielded countless queries and phone calls over the past 3 years. It is a source of wonder to him that they did not make greater use of their caller ID. Here are the victims:

Dr. Graeme Norval who never wavered in his faith and helped restore the F-words (both of them) to their rightful place.

Professor Vladimir Papangelakis, between excursions to ancient Greece, educated the author in the fine points of hydrometallurgy and mineral processing.

Professor Levente Diosady patiently dipped into his vast knowledge of food engineering for the author over many pleasant and imagined salami lunches.

Professor Elizabeth Edwards exposed the author to her marvelous work on soil remediation, which ultimately proved to be beyond his interpretive skills.

Professor Honghi Tran is "sans pareil" in his knowledge of the pulp and paper industry.

Professor Donald Kirk, a neighbor, and Professors Brad Saville and Grant Allen acted as indefatigable sounding boards.

Professor Yu-Ling Cheng used her knack for explaining the most complex issues in lucid terms.

Professor Christopher Yip will, surely some fine December day, find himself in Stockholm — at least he ought to.

Professor Donald Mackay proved once again that the Scots can now add infinite patience to their many other qualities.

My thanks go to them all.

Among industrial colleagues, Dr. Jean-Jacques Perraud, Goro Nickel SAS, advised the author, literally, from halfway around the world. His good friend, Dr. Stanley Hatcher, former president of Atomic Energy of Canada and the American Nuclear Association, kept the author abreast of the looming reemergence of nuclear power as a major source of energy. Dr. Kent Knaebel, Adsorption Research Inc., proved once again that the author has no monopoly on wisdom in the field of adsorption. Dr. Kemal Adham, Hatch & Associates, peppered his sage advice with ancient Arab sayings. Their help has been invaluable.

Every department has a collective known as Support Staff. A somewhat dismissive term, perhaps, and yet how appropriate, because it is they who keep the bridges up and the ships afloat. My affectionate salute to them all: Arlene Fillatre, Leticia Gutierrez, Gorette Silva, Paul Jowlabar, and Jacquie Briscoe — you certainly kept this boat afloat!

Diran Basmadjian
Toronto, 2007

Author

Diran Basmadjian is a graduate of the Swiss Federal Institute of Technology, Zurich, and received his M.A.Sc. and Ph.D. degrees in chemical engineering from the University of Toronto. He was appointed assistant professor of chemical engineering at the University of Ottawa in 1960, moving to the University of Toronto in 1965, where he subsequently became professor of chemical engineering. He has combined his research interests in the separation sciences, biomedial engineering, and applied mathematics with a keen interest in the craft of teaching. His most current activities included writing, consulting, and performing science experiments for children at a local elementary school. Professor Basmadjian has authored five books and some fifty scientific publications.

Notations

a	specific surface area, m^2/m^3
A	area, m^2
A	raffinate solvent, kg or kg/s
B	extract solvent, kg or kg/s
Bi	Biot number, dimensionless
BOD	biological oxygen demand, kg/m^3
C	concentration, mol/m^3 or kg/m^3
$\underline{C}$	number of components
$\overline{C}$	average concentration, mol/m^2 or kg/m^3
C_p	heat capacity at constant pressure, J/kg K or J/mol K
d	diameter, m
D	diffusivity, m^2/s
D	distillate, mol/s
D'	cumulative distillate, mol
D_e	effective diffusivity, m^2/s
erf	error function
erfc	complementary error function
E	effectiveness factor, dimensionless
E	extract, kg or kg/s
E	extraction ratio, dimensionless
E	stage efficiency, dimensionless
E_a	activation energy, J/mol
E_h	enhancement or enrichment factor, dimensionless
f	fraction distilled or solidified
F	Faraday number, C/mol
F	degrees of freedom
F	feed, kg or mol, kg/s or mol/s
F	force, N
Fo	Fourier number, dimensionless
g	gravitational constant, m/s^2
G	gas or vapor flow rate, kg/s or mol/s
Gr	Grashof number, dimensionless
G_s	superficial carrier flow rate, kg/m^2s

h	heat transfer coefficient, J/m²s K
h	height, m
H	Henry's constant, Pa m^3 mol^{-1}
H	enthalpy, J/kg, or J/mol
Ha	Hatta number, dimensionless
HETP(S)	height equivalent to a theoretical plate or stage, m
HTU	height of a transfer unit, m
i	electrical current, A
J_w	water flux, m^3/m^2s
k	thermal conductivity, J/m s K
k_0	zero order rate constant, kg/m^3s
k_C, k_G, k_L,	
k_x, k_y, k_Y	mass transfer coefficient, various units
k_e	elimination rate constant, s^1
k_r	reaction rate constant, s^1
K	partition coefficient, various units
K	permeability, m/s, or m^2
K_m	Michaelis-Menton constant, kg/m^3
K_o	overall mass transfer coefficient, various units
	characteristic length, m
L	length, m
L	liquid flow rate, kg/s, or mol/s
L	liquid mass, kg
L_s	superficial solvent flow rate, kg/m^2s
m	distribution coefficient, various units
m	mass, kg
M	mass of emissions, kg, kg/s, or kg/m^2s
M	molar mass, dalton
N	molar flow rate, mol/s
N	number of stages or plates
N_p	number of particles or plates
N_T	number of mass transfer units
NTU	number of transfer units
p	pressure, Pa
P	number of phases
P^o	vapor pressure, Pa
P_T	total pressure, Pa

P_1	permeability, various units
P_w	water permeability, mol/m^{2}s Pa
p_{BM}	log-mean pressure difference, Pa
Pe	Peclet number, dimensionless
q	heat flow, J/s
q	thermal quality of feed, dimensionless
Q	volumetric flow rate, m^3/s
r	radial variable, m
r	recovery, dimensionless
R	gas constant, J/mol K
R	radius, m
R	raffinate, kg or kg/s
R	reflux ratio, dimensionless
R	residue factor, dimensionless
R_1	resistance to mass transfer, s/m
R	electrical resistance, Ω
RO	reverse osmosis
S	amount of solid, kg, or kg/s
S	shape factor, m
S	substrate concentration, kg/m^3
S	solubility, cm^3 STP/cm^3 Pa
Sc	Schmidt number, dimensionless
Sh	Sherwood number, dimensionless
$\mathrm{Sh_w}$	wall Sherwood number, dimensionless
St	Stanton number, dimensionless
t	time, s
T	dimensionless time (adsorption)
T	temperature, K or °C
u	dependent variable
u	velocity, m/s
U	overall heat transfer coefficient, J/m^{2}s K
v	velocity, m/s
v_H	specific volume, m^3/kg dry air
v_t	terminal velocity, m/s
V	voltage, V
V	volume, m^3 or m^3/mole
W	bottoms, mol or mol/s

W	weight, kg
x	liquid weight or mole fraction, dimensionless
x	raffinate weight fraction, dimensionless
x	solid phase weight fraction (leaching), dimensionless
X	adsorptive capacity, kg solute/kg solid
X	liquid-phase mass ratio, dimensionless
y	extract weight fraction, dimensionless
y	vapor mole fraction, dimensionless
Y	humidity, kg water/kg dry air
Y	gas-phase mass ratio, dimensionless
z	distance, m
z_{FH}	heat transfer film thickness, m
z_{FM}	mass transfer film thickness, m
Z	dimensionless distance (adsorption)
Z	flow rate ratio (dialysis)

Greek Symbols

α	relative volatility, dimensionless
α	selectivity or separation factor, dimensionless
α	thermal diffusivity, m^2/s
γ	activity coefficient, dimensionless
$\dot{\gamma}$	shear rate, s^1
δ	film or boundary layer thickness, m
ε	porosity or void fraction, dimensionless
λ	mean free path, m
μ	viscosity, Pa s
ν	kinematic viscosity, m^2/s
π	osmotic pressure, Pa
ρ	density, kg/m^3
σ	liquid film thickness, m
σ^{st}	length of stomatal pore, m
τ	shear stress, Pa
τ	tortuosity, dimensionless
ϕ	pressure ratio, dimensionless
ϖ	angular velocity, s^1

Subscripts

as	adiabatic saturation
b	bed, bulk
c	cold
$\underline{C}$	cross section, condenser
db	dry bulb
D	distillate, dialysate
e	effective
f, F	feed
g, G	gas
h	hot
i	initial
i	inside
i	impeller
L	liquid
m	mean
o	outside
ow	octanol-water
p	particle, pellet
p	permeate
p	pore
v	vessel
w	bottoms
w	water

Superscripts

*	equilibrium
o	initial
o	pure component
′	cumulative

Table of Contents

1 Some Basic Notions: Rates of Mass Transfer **1**
1.1 Gradient-Driven and Forced Transport 2
1.1.1 The Rate Laws 2
1.1.2 The Transport Diffusivities 5
1.1.3 The Gradient 7
1.1.4 Simple Integrations of Fick's Law 14
1.2 Transport Driven by a Potential Difference (Constant Gradient): The Film Concept and the Mass Transfer Coefficient 21
1.2.1 Units of the Potential and of the Mass Transfer Coefficient 24
1.2.2 Equimolar Diffusion and Diffusion through a Stagnant Film: The Log-Mean Concentration Difference 26
1.2.2.1 Equimolar Counterdiffusion 27
1.2.2.2 Diffusion through a Stagnant Film 27
1.3 The Two-Film Theory 33
1.3.1 Overall Driving Forces and Mass Transfer Coefficients 36
1.3.1.1 Comments 38
Practice Problems 42

2 Modeling Mass Transport: The Mass Balances **51**
2.1 The Compartment or Stirred Tank and the One-Dimensional Pipe 51
2.2 The Classification of Mass Balances 62
2.2.1 The Role of Balance Space 62
2.2.2 The Role of Time 63
2.2.2.1 Unsteady Integral Balances 63
2.2.2.2 Cumulative (Integral) Balances 63
2.2.2.3 Unsteady Differential Balances 64
2.2.3 Dependent and Independent Variables 64
2.3 Information Obtained from Model Solutions 76
2.4 Setting Up Partial Differential Equations 78
2.5 The General Conservation Equations 90
Practice Problems 99

3 Diffusion through Gases, Liquids, and Solids **107**
3.1 Diffusion Coefficients 107
3.1.1 Diffusion in Gases 107
3.1.2 Diffusion in Liquids 111
3.1.3 Diffusion in Solids 118
3.1.3.1 Diffusion of Gases through Polymers and Metals 118
3.1.3.2 Diffusion of Gases through Porous Solids 126
3.1.3.3 Diffusion of Solids in Solids 134
Practice Problems 137

4 More about Diffusion: Transient Diffusion and Diffusion with Reaction **143**
4.1 Transient Diffusion 143
4.1.1 Source Problems 145
4.1.2 Nonsource Problems 157
4.1.2.1 Diffusion into a Semi-Infinite Medium 157
4.1.2.2 Diffusion in Finite Geometries: The Plane Sheet, the Cylinder, and the Sphere 161
4.1.2.3 Diffusion in Finite Geometries: The "Short-Time" and "Long-Time" Solutions 166
4.2 Diffusion and Reaction 170
4.2.1 Reaction and Diffusion in a Catalyst Particle 171
4.2.2 Gas–Solid Reactions Accompanied by Diffusion: Moving-Boundary Problems 171
4.2.3 Gas–Liquid Systems: Reaction and Diffusion in the Liquid Film 172
Practice Problems 186

5 More about Mass Transfer Coefficients **195**
5.1 Dimensionless Groups 196
5.2 Mass Transfer Coefficients in Laminar Flow: Extraction from the PDE Model 200
5.2.1 Mass Transfer Coefficients in Laminar Tubular Flow 201
5.2.2 Mass Transfer Coefficients in Laminar Flow around Simple Geometries 203
5.3 Mass Transfer in Turbulent Flow: Dimensional Analysis and the Buckingham π Theorem 206
5.3.1 Dimensional Analysis 206
5.3.2 The Buckingham π Theorem 207
5.4 Mass Transfer Coefficients for Tower Packings 216
5.5 Mass Transfer Coefficients in Agitated Vessels 222
5.6 Mass Transfer Coefficients in the Environment: Uptake and Clearance of Toxic Substances in Animals — The Bioconcentration Factor 226
Practice Problems 231

6 Phase Equilibria **239**
6.1 Single-Component Systems: Vapor Pressure 240
6.2 Multicomponent Systems: Distribution of a Single Component 246
6.2.1 Gas–Liquid Equilibria 246
6.2.2 Liquid and Solid Solubilities 251
6.2.3 Fluid–Solid Equilibria: The Langmuir Isotherm 253
6.2.4 Liquid–Liquid Equilibria: The Triangular Phase Diagram 264
6.2.5 Equilibria Involving a Supercritical Fluid 270
6.2.6 Equilibria in Biology and the Environment: Partitioning of a Solute between Compartments 274

6.3 Multicomponent Equilibria: Distribution of Several Components.....276
6.3.1 The Phase Rule.....276
6.3.2 Binary Vapor–Liquid Equilibria.....277
6.3.2.1 Phase Diagrams.....277
6.3.2.2 Ideal Solutions and Raoult's Law: Deviation from Ideality.....280
6.3.2.3 Activity Coefficients.....282
6.3.3 The Separation Factor α: Azeotropes.....284
Practice Problems.....293

7 Staged Operations: The Equilibrium Stage.....299
7.1 Equilibrium Stages.....301
7.1.1 Single-Stage Processes.....301
7.1.2 Single-Stage Differential Operation.....307
7.2 Staged Cascades.....313
7.2.1 Crosscurrent Cascades.....313
7.2.2 Countercurrent Cascades.....320
7.2.3 Countercurrent Cascades: The Linear Case and the Kremser Equation.....323
7.3 The Equilibrium Stage in the Real World.....330
7.3.1 The Mixer–Settler Configuration.....330
7.3.2 Gas–Liquid Systems: The Tray Tower.....331
7.3.3 Staged Liquid Extraction Again: The Karr Column.....332
7.3.4 Staged Leaching: Oil Extraction from Seeds.....333
7.3.5 Staged Washing of Solids (CCD).....335
7.4 Multistage Distillation.....336
7.4.1 Continuous Fractional Distillation.....337
7.4.2 Mass and Energy Balances: Equimolar Overflow and Vaporization.....339
7.4.3 The McCabe–Thiele Diagram.....341
7.4.4 Minimum Reflux Ratio and Number of Plates.....346
7.4.4.1 Comments.....348
7.4.5 Column and Tray Parameters.....356
7.4.6 Limiting Flow Rates: Column Diameter.....358
7.4.6.1 Gas or Vapor Flow Rates.....359
7.4.6.2 Liquid Velocities.....360
7.4.6.3 Lower Limits.....360
7.4.6.4 Comments.....360
7.4.7 Batch Fractional Distillation: Model Equations and Some Simple Algebraic Calculations.....360
7.4.7.1 Distillation at Constant x_D, Variable R.....362
7.4.7.2 Distillation at Constant R, Variable x_D.....364
7.4.7.3 Multicomponent Batch Distillation (Forget McCabe–Thiele, Part 2).....366
7.5 Percolation Processes.....367

7.6 Stage Efficiencies....370
7.6.1 Distillation and Absorption....370
7.6.2 Extraction....372
7.6.3 Adsorption and Leaching....372
7.6.4 Percolation Processes....373
Practice Problems....376

8 Continuous-Contact Operations....385
8.1 Packed-Column Operation....386
8.1.1 The Countercurrent Gas Scrubber Revisited....387
8.1.1.1 Comments....390
8.1.2 The Countercurrent Gas Scrubber Again: Analysis of the Linear Case....391
8.1.2.1 Comments....394
8.1.3 Packed Column Characteristics....395
8.1.3.1 Main Features....395
8.1.3.2 Relation between HTU and HETP....396
8.1.3.3 Operational Parameters....396
8.1.3.4 Comparison of Packed and Tray Columns....398
8.1.4 Liquid–Liquid Extraction in a Packed Column....403
8.2 Membrane Processes....411
8.2.1 Membrane Structure, Configuration, and Applications....413
8.2.2 Process Considerations and Calculations....418
Practice Problems....433

9 Simultaneous Heat and Mass Transfer....439
9.1 The Air–Water System: Humidification and Dehumidification, Evaporative Cooling....440
9.1.1 The Wet-Bulb Temperature....440
9.1.2 The Adiabatic Saturation Temperature and the Psychrometric Ratio....441
9.1.3 Model for Countercurrent Air–Water Contact: The Water Cooling Tower....448
9.1.3.1 Water Balance Over Gas Phase (kg H_2O/mls)....448
9.1.3.2 Water Balance Over Water Phase....449
9.1.3.3 Gas-Phase Energy Balance (kJ/m^2s)....450
9.1.3.4 Liquid-Phase Energy Balance (kJ/m^2s)....450
9.2 Drying Operations....455
9.3 Heat Effects in a Catalyst Pellet: The Nonisothermal Effectiveness Factor....462
9.3.1 Comments....465
Practice Problems....467

Selected References....469
Appendix A1: The D-Operator Method....475
Appendix A2: Hyperbolic Functions and ODEs....477
Index....479

1

Some Basic Notions: Rates of Mass Transfer

We begin our deliberations by introducing the basic rate laws that govern the transport of mass. In choosing this topic as our starting point, we follow the pattern established in previous treatments of the subject, but depart from it in some important ways. We start, as do other texts, with an introduction to Fick's law of diffusion, but here it is treated as a component of a broader class of processes, which is termed *gradient-driven transport*. This category includes the laws governing transport by molecular motion, Fourier's law of conduction, and Newton's viscosity law, as well as Poiseuille's law for viscous flow through a cylindrical pipe and D'Arcy's law for viscous flow through a porous medium, both of which involve the bulk movement of fluids. In other words, we use as common ground the *form of the rate law*, rather than the underlying physics of the system. This treatment is a departure from the usual pedagogical norm and is designed to reinforce the notion that transport of different types can be drawn together and viewed as driven by a potential gradient (concentration, temperature, velocity, pressure) that diminishes in the direction of flow.

The second departure is the early introduction of the reader to the notion of a linear driving force, or potential *difference*, as the agent responsible for transport. One encounters here, for the first time, the notion of a transport coefficient that is the proportionality constant of the rate law. Its inverse can be viewed as the resistance to transport, and in this it resembles Ohm's law that states that current transport i is proportional to the voltage difference ΔV and varies inversely with the Ohmian resistance R.

Associated with the transport coefficients is the concept of an effective film thickness, which lumps the resistance to transport into a fictitious thin film adjacent to a boundary or interface. Transport takes place through this film driven by the linear driving force across it and impeded by a resistance that is the inverse of the transport coefficient. Note that in these discussions, a conscious effort is made to draw analogies between the transport of mass and heat and to occasionally invoke the analogous case of transport of electricity.

The chapter is, as are all chapters, supplemented with worked examples, which prepare the ground for the practice problems given at the end of the chapter.

TABLE 1.1

Rate Laws Based on Gradients

Name	Process	Flux	Gradient
1. Fick's law	Diffusion	$N / A = -D\frac{dC}{dx}$	Concentration
2. Fourier's law	Conduction	$q / A = -k\frac{dT}{dx}$	Temperature
3. Alternative formulation		$q / A = -\alpha\frac{d(\rho CpT)}{dx}$	Energy concentration
4. Newton's viscosity law	Molecular momentum transport	$F_x / A = \tau_{qx} = -\mu\frac{dv_x}{dy}$	Velocity
5. Alternative formulation		$F_x / A = \tau_{yx} = \nu\frac{d(\rho v_x)}{dy}$	Momentum concentration
6. Poiseuille's law	Viscous flow in a circular pipe	$Q / A = v_x = -\frac{d^2}{32\mu}\frac{dp}{dx}$	Pressure
7. D'Arcy's law	Viscous flow in a porous medium	$Q / A = v_x = -\frac{K}{\mu}\frac{dp}{dx}$	Pressure

TABLE 1.2

Rate Laws Based on Linear Driving Forces

Process	Flux or Flow	Driving Force	Resistance
1. Electrical current flow (Ohm's law)	$i = \Delta V/R$	ΔV	R
2. Convective mass transfer	$N/A = k_C\,\Delta C$	ΔC	$1/k_C$
3. Convective heat transfer	$q/A = h\Delta T$	ΔT	$1/h$
4. Flow of water due to osmotic pressure	$N_A/A = P_w\Delta\pi$	$\Delta\pi$	$1/P_w$

1.1 Gradient-Driven and Forced Transport

1.1.1 The Rate Laws

The physical laws that govern the transport of mass, energy, and momentum, as well as that of electricity, are based on the notion that the spontaneous flow of these entities is induced by a driving potential. This driving force can be expressed in two ways. In the most general case, it is taken to be the *gradient* or *derivative of that potential* in the direction of flow. A list of some rate laws based on such gradients appears in Table 1.1. In the second, more specialized case, the gradient is taken to be *constant*. The driving force then becomes simply the *difference in potential* over the distance covered. This is taken up Section 1.2, and a tabulation of some rate laws based on such potential differences is given in Table 1.2. Ohm's law belongs in this category.

Let us examine how these concepts can be applied in practice by taking up a familiar example of a gradient-driven process, that of the conduction of heat.

The general reader knows that heat flows from a high temperature T, which is the driving potential here, to a lower temperature at some other location. The greater the difference in temperature per unit distance, x, the larger the transport of heat (i.e., we have a proportionality):

$$q \propto -\frac{\Delta T}{\Delta x} \tag{1.1}$$

The minus sign is introduced to convert $\Delta T/\Delta x$, which is a negative quantity, to a positive value of heat flow q. In the limit $\Delta x \to 0$, the difference quotient converts to the derivative dT/dx. Noting further that heat flow will be proportional to the cross-sectional area normal to the direction of flow and introducing the proportionality constant k, known as the thermal conductivity, we obtain

$$\text{Heat flow} \quad q(\text{J/s}) = -kA\frac{dT}{dx} \tag{1.2a}$$

or, equivalently,

$$\text{Heat flux} \quad q/A(\text{J/sm}^2) = -k\frac{dT}{dx} \tag{1.2b}$$

These two expressions, shown graphically in Figure 1.1b, are known as Fourier's law of heat conduction. It can be expressed in yet another alternative form, which is obtained by multiplying and dividing the right side by the product of density ρ (kg/m^3) and specific heat C_p (J/kgK). We then obtain (Item 3 of Table 1.1)

$$q/A = -\alpha\frac{d(\rho C_p T)}{dx} \tag{1.3}$$

where $\alpha = k/\rho C_p$ is termed the thermal diffusivity. We note that the term ρCpT in the derivative has the units of joule per cubic meter (J/m^3) and can thus be viewed as an energy concentration.

The reason for introducing this alternative formulation is to establish a link to the transport of mass (Item 1 of Table 1.1). Here the driving potential is expressed in terms of the molar concentration gradient dC/dx, and the proportionality constant D is known as the (mass) diffusivity of the species, paralleling the thermal diffusivity α in Equation 1.3. Spontaneous transport

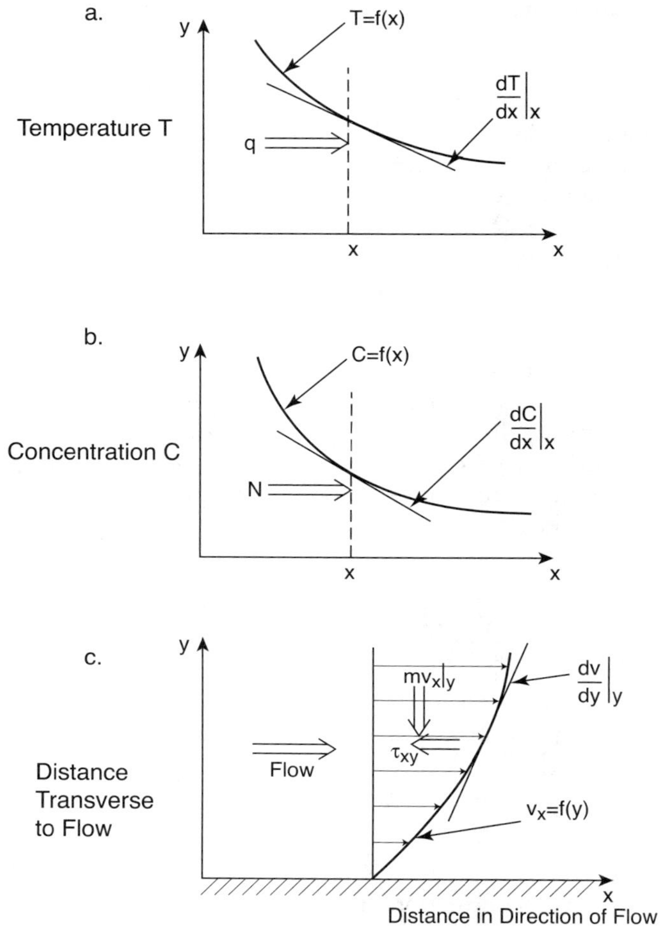

FIGURE 1.1
Diffusive transport: (a) heat; (b) mass; (c) momentum.

takes place from a point of high concentration to a location of lower concentration. Noting, as before, that the molar flow will be proportional to the cross-sectional area A normal to the flow, we obtain

$$\text{Molar flow } N \text{ (moles/s)} = -DA\frac{dC}{dx} \tag{1.4a}$$

and, equivalently,

$$\text{Molar flux } N/A \text{ (moles/m}^2\text{s)} = -D\frac{dC}{dx} \tag{1.4b}$$

These two relations, depicted in Figure 1.1a, are known as Fick's law of diffusion.

There is a third mode of diffusive transport, that of momentum, that can likewise be induced by the molecular motion of the species. Momentum is the product of the mass of the molecular species and its velocity in a particular direction, for example, v_x. As in the case of the flow of mass and heat, the diffusive transport is driven by a gradient, here the velocity gradient dv_x/dy transverse to the direction of flow (Figure 1.1c). It takes place from a location of high velocity to one of lower velocity, paralleling the transport of mass and heat. As the molecules enter a region of lower velocity, they relinquish part of their momentum to the slower particles in that region and are consequently slowed. There is, in effect, a braking force acting on them which is expressed in terms of a shear stress $F_x/A = \tau_{yx}$ pointing in a direction opposite to that of the flow. The first subscript on the shear stress denotes the direction in which it varies, while the second subscript refers to the direction of the equivalent momentum mv_x. The relation between the induced shear stress and the velocity gradient is due to Newton and is termed Newton's viscosity law. It is, like Fick's law and Fourier's law, a linear negative relation and is given by

$$F_x / A = \tau_{yx} = -\mu \frac{dv_x}{dy} \tag{1.5a}$$

Equation 1.5a can be expressed in the equivalent form:

$$\tau_{yx} = -\frac{\mu}{\rho} \frac{d(\rho v_x)}{dy} = -\nu \frac{d(\rho v_x)}{dy} \tag{1.5b}$$

where ν is termed the kinematic viscosity or "momentum diffusivity" in units of square meter per second (m^2/s), and the product of density ρ and velocity v_x can be regarded as a momentum *concentration* in units of kilogram meter per second per cubic meter — ($kg\ m/s)/m^3$. This version of Newton's viscosity law brings it in line with the concentration-driven expressions for diffusive heat and mass transport. Two additional rate processes that are driven by gradients are shown in Table 1.1. The first is Poiseuille's law which applies to viscous flow in a circular pipe, and the second is a similar expression, D'Arcy's law, which describes viscous flow in a porous medium. Both processes are driven by pressure gradients.

1.1.2 The Transport Diffusivities

The analogy among the three modes of transport is further reinforced by noting that all three transport coefficients ν, α, and D have identical units of square meter per second (m^2/s), and all three are referred to as "diffusivities." We have

MOMENTUM DIFFUSIVITY (m^2/s)

$$\nu = \mu/\rho \tag{1.6a}$$

THERMAL DIFFUSIVITY (m^2/s)

$$\alpha = k/\rho Cp \tag{1.6b}$$

MASS DIFFUSIVITY (m^2/s)

$$D \tag{1.6c}$$

Because all three quantities are conveyed by the same molecules that, one assumes, move at the same speed, it is tempting to conclude that the diffusivities of momentum, heat, and mass will be identical, or at least similar, in magnitude. This is in fact the case for transport in low-density gases, but the assumption breaks down in liquids and even more so in solids. Mass diffusivity in particular begins to diverge sharply from its partners and marches on to its own drummer (very slowly). The following illustration examines this important aspect in more detail.

Illustration 1.1: The Transport Diffusivities: A First Look at the Molecular Level

Listed in Table 1.3 are experimental diffusivities for momentum, heat, and mass (taken to be oxygen) in three representation media — air, water, and glycerin.

The first feature of note is the near-identity of values for transport in *air*, and this can be shown to apply to low-density gases in general. The harmony found in gases changes dramatically when we turn to liquids. In water, mass diffusivity has distanced itself from its partners by two to three orders of magnitude, and when we turn to glycerin, with a viscosity one thousand times that of water, mass diffusion has slowed to a crawl, some five to nine orders of magnitude behind its partners.

Why the differences? A partial answer can be found by examining the events at a molecular level. These are sketched in Figure 1.2a and Figure 1.2b.

In low-density gases, the molecules spend their time almost exclusively in transit between collisions. That transit time (~ 10^{-9} s) and the speed at

TABLE 1.3

Transport Diffusivities at 25° (m^2/s)

	ν	α	D (O_2)
Air (1 atm)	1.6×10^{-5}	2.2×10^{-5}	2.0×10^{-5}
Water	8.9×10^{-7}	1.5×10^{-7}	2.4×10^{-9}
Glycerin	1.5×10^{-3}	1.0×10^{-7}	$\sim 10^{-12}$

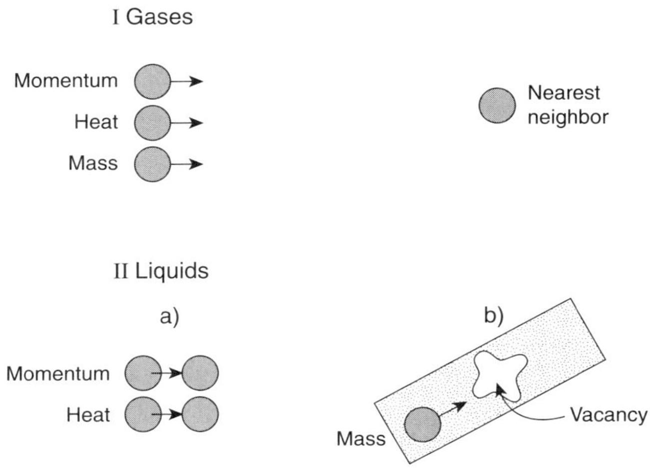

FIGURE 1.2
Molecular Transport Mechanism I, Companions in Flight II (a) transfer to nearest neighbors, (b) transfer to a vacancy.

which they move is the same irrespective of whether the molecules are carrying momentum or heat or are merely conveying themselves. The brief instant of the collision, during which the actual transfer takes place, is insignificant compared to the time of flight, which is the rate-determining step. It follows that all three quantities—momentum, heat, and mass—are conveyed at the same speed. They are "companions in flight" (Figure 1.2a). The distance covered between collisions, the so-called mean face path, will be found in Table 1.4, which also lists various dimensions of relevance to mass transfer processes.

In liquids, the transfer mechanism for momentum and heat is still identical. It is from one molecule to its nearest neighbors which are now very close (Figure 1.2b). In mass diffusion, a different mechanism applies. A vacancy has to open up in the vicinity of the diffusing particle, and it then has to overcome a "viscous drag" (in the classical sense) to reach that opening (Figure 1.2b). This explains the wide divergence of mass diffusivity from its partners. Thus while certain similarities among transport processes may exist at the macroscopic level (identical form of rate laws, etc.), more often than not they break down at the molecular level. We will return to this topic in Chapter 3.

1.1.3 The Gradient

The second component in the rate laws of Table 1.1, the driving gradient, also deserves some closer attention.

Although under normal conditions most transport takes place along a line of diminishing potential (i.e., a negative gradient), situations can arise where

TABLE 1.4

The Microscale of Things

Item	Dimension
Helium atom	0.2 nm (2 Angstrom)
Water molecule	0.3
Hydrated sodium	0.5
Glucose molecule	0.9
Zeolite adsorbent micropores	0.3–1.3
Activated carbon micropores	1–5
Reverse osmosis membrane pore	0.3–3
Proteins: Hemoglobin diameter to fibrinogen length	6–20
Ultrafine aerosols	<100 nm
Nanoparticles for drug delivery	10–100 nm
Micropores of activated carbon, zeolites, catalysts	10–100
Ultrafiltration membrane pores	1–100
Mean free path of gas molecules (1 atm)	100
Coarse aerosols: "2.5 particulate matter (PM)"	0.1–2.5 μm
Red blood cell, diameter	8.5
Blood capillaries	10
Lung air sacs (alveoli)	100
100 mesh particle size, espresso coffee	150
Adsorbent and ion exchange particles, Britta filter	1–10 mm
Aorta, diameter	10

the gradient either vanishes or even becomes positive. These cases are important for a variety of reasons and are taken up in some detail in the following two illustrations.

Illustration 1.2: Transport in Systems with Vanishing Gradients

It frequently happens in transport processes that the driving gradient vanishes at some position in the system, without inhibiting the flow of mass, heat, or momentum. There are two special situations that give rise to such behavior.

First, the potential exhibits a maximum or a minimum at a point or axis of symmetry. These locations can be the centerline of a slab, the axis of a cylinder, or the center of a sphere. Figure 1.3a and Figure 1.3b consider two such cases. Figure 1.3a represents a spherical catalyst pellet in which a reactant of external concentration C_0 diffuses into the sphere and undergoes a reaction. Its concentration diminishes and attains a minimum at the center. Figure 1.3b considers laminar flow in a cylindrical pipe. Here the variable in question is the axial velocity v_x, which rises from a value of zero at the wall to a maximum at the centerline before dropping back to zero at the other end of the diameter. Here, again, symmetry considerations dictate that this maximum must be located at the centerline of the conduit.

The second case of a vanishing derivative arises when flow or flux ceases. Because the proportionality constants in the rate laws cannot vanish, zero

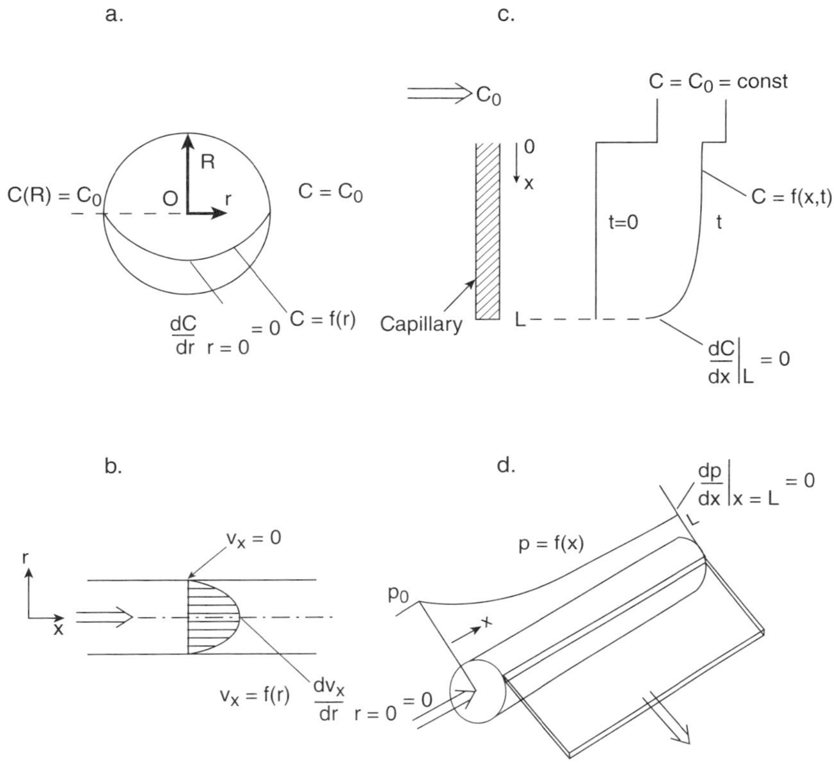

FIGURE 1.3
Systems with vanishing gradients: (a) catalyst pellet; (b) viscous flow in a pipe; (c) diffusion into capillary; and (d) polymer extruder.

flow must perforce imply that the gradient becomes zero. This situation arises when flow or diffusional flux is brought to a halt by a physical barrier. Figure 1.3c and Figure 1.3d depict two such cases. Figure 1.3c shows a capillary filled with a solvent and suddenly exposed to a solution containing a dissolved solute of concentration C_0. This configuration has been used in the past to determine diffusivities. As the solute diffuses into the capillary, a concentration profile develops within it, which changes with time until the concentration in the capillary equals that of the external medium. As these profiles grow, they maintain at all times a zero gradient at the sealed end of the capillary. This must be so because N, the diffusional flow in Equation 1.4, can vanish only if the gradient dC/dx becomes zero. Depicted in Figure 1.3d is a polymer extruder in which molten polymer enters one end of a pipe and exits as a thin sheet through a lateral slit. Here the barrier is the sealed end of the pipe, which prevents an axial outflow of the polymer melt and forces it instead into the lateral channel. The only way for flow to cease, $Q/A = 0$, is for the pressure gradient dp/dx to vanish at this point. The resulting axial pressure profile is shown in Figure 1.3d.

Comments

We singled out vanishing gradients for two reasons: First, they help greatly in visualizing the shape of the profiles and the transport process (see Practice Problems 1.3 and 1.4). Second, because they are confined to a specific location, they can serve as *boundary conditions* in the solution of the model equations. Thus the catalyst pellet shown in Figure 1.3a has two such conditions, one at the center, where the flux vanishes, and a second at the surface, where the reactant concentration attains a constant value. The pellet is encountered again in Chapter 4 (Figure 4.10) where the underlying model is found to be a second-order differential equation. Such equations require the evaluation of two integration constants and must, therefore, be provided with two boundary conditions.

Illustration 1.3: Forced Transport against Positive or Zero Gradients: Reverse Osmosis; Active Transport; The Gas Centrifuge

The notion that mass transport can take place spontaneously in the direction of *increasing* concentration ($C_2 > C_1$) is both counterintuitive and, more fundamentally, contravenes thermodynamic principles. (Recall that in a spontaneous process, free energy decreases — that is, $RT \ln C_2/C_1$ has to be less than zero.) Clearly, what is needed to achieve this is the intervention of an *external* force or an equivalent energy input to "pump" the molecules "up the hill." This happens not only in a natural context ("active transport" in living organisms) but has also led to the development of important industrial separation processes. Figure 1.4a through Figure 1.4c provide three such examples.

Figure 1.4a illustrates the principal feature of *reverse osmosis*, a process that became possible with the development of high-performance selective membranes (good chemistry often precedes good engineering). Here water is forced by an applied external pressure from a solution of low water "concentration" (high salt content) through a membrane to yield a product of high water concentration (low salt content). The size of the membranes pores is such that they exclude, or nearly exclude, the ions while allowing free passage of water. We will return to the subject of reverse osmosis in Chapter 8, dealing with membrane processes.

Active transport occurs in nearly all cells of the body, but most notably in the kidney, liver, and the intestine. The species transported are primarily ions, Na^+ in particular, and also include amino acids and sugars. The sketch in Figure 1.4b shows the basic mechanism of active Na^+ transport in the kidney, which is crucial to its proper function and the production of urine. The process shown, a very crude approximation of the actual intricacies, involves a protein carrier C picking up Na^+ from the low concentration (urine) side and then delivering and releasing it on the high concentration (peritubular) side. Both the release of Na^+ and the return trip of the carrier

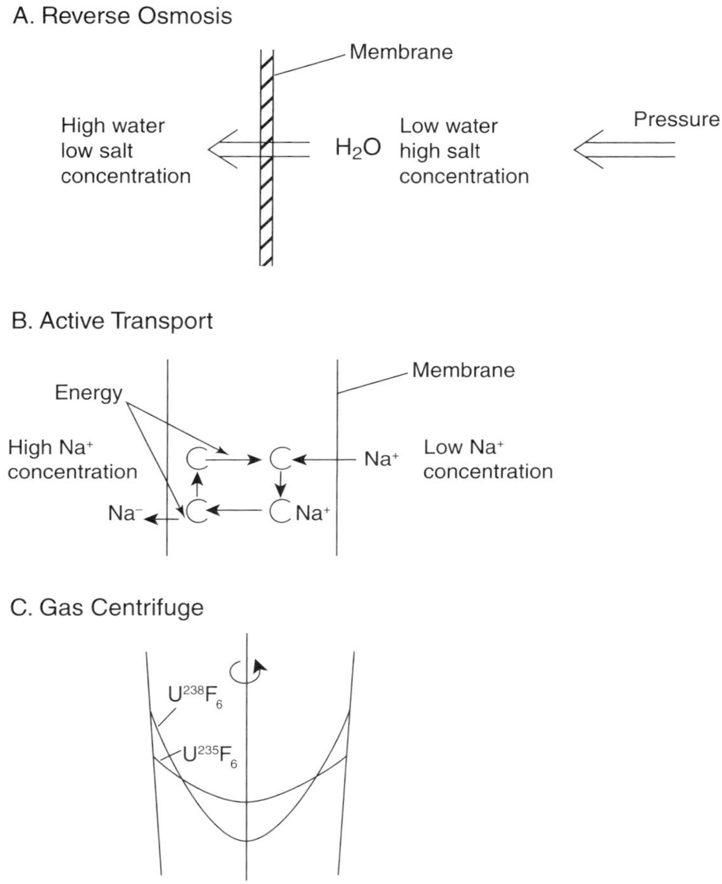

FIGURE 1.4
Examples of forced transport.

require enzymatic (i.e., catalytic) action and the input of metabolic energy. The reader is referred to Chapter 8, Illustration 8.8, for a more detailed discussion of mass transport in the kidney.

In Figure 1.4c, external intervention is provided by a *centrifugal force* that — starting from a zero gradient — establishes a rising concentration profile that reaches its maximum at the periphery of the centrifuge. Its most prominent current use is in the enrichment of uranium isotopes. The lighter fissionable components of interest, here in the form of gaseous hexafluoride, $U^{235}F_6$, are enriched at the center while $U^{238}F_6$ predominates at the periphery (i.e., has a higher partial pressure).

The calculation of the degree of enrichment or "separation factor" obtained in a gas centrifuge is particularly intriguing and is quite different from similar calculations seen in later chapters which are based on phase equilibria. It proceeds in a somewhat roundabout but highly ingenious way by first considering the pressure distribution of a gas in the gravitational field of the earth. That distribution is obtained by integration of the hydrostatic formula of fluid mechanics:

$$dp = -\rho g dz = -\frac{pMg}{RT}dz \tag{1.7a}$$

$$p(z) = p(0)\exp\left(-\frac{Mgz}{RT}\right) \tag{1.7b}$$

where Mgz is the potential energy per mole of the gas at a height z above the surface of the earth. For two gases with molar masses $M_1 > M_2$, this becomes

$$\frac{p_1(z)}{p_2(z)} = \frac{p_1(0)}{p_2(0)}\exp\frac{(M_2 - M_1)gz}{RT} \tag{1.7c}$$

(This equation is used in Practice Problem 1.7 to calculate greenhouse gas concentrations in the upper atmosphere.)

If we now replace *without benefit of a formal proof* the potential energy by the kinetic energy of the rotating gas $M(\omega r)^2$, Equation 1.7c transforms into the expression

$$\frac{p_1(R)}{p_2(R)} = \frac{p_1(0)}{p_2(0)}\exp\frac{(M_1 - M_2)(\omega R)^2}{2RT} \tag{1.7d}$$

where ω = angular velocity (s^{-1}) = 2π ($RPM/60$).

In terms of mole fractions y_1, this becomes

$$\alpha = \frac{(y_{2345} / y_{238})_0}{(y_{235} / y_{238})_R} = \exp\frac{(M_1 - M_2)(\omega R)^2}{2RT} \tag{1.7e}$$

It is in this semi-intuitive fashion that physicists were first able to deduce the enrichment obtained in a gas centrifuge. To the knowledge of the author, this derivation has not been superceded by a more rigorous approach.

α can be viewed as a "separation factor," indicative of the degree of enrichment one can achieve in a single centrifuge. $\alpha = 1$ signifies no enrichment, $\alpha = 1.01$ would indicate a difficult separation requiring extensive staging, but anything above $\alpha = 1.1$ is considered respectable. Centrifuge radius and rotational speed enormously affect the separation factor that varies *exponentially and with the square* of these quantities.

A.

B.

FIGURE 1.5
The gas centrifuge: (a) foreground — sixth-generation composite centrifuge; background — prototype all-metal centrifuge (1978). (b) A uranium enrichment cascade. (Courtesy Uranium Enrichment Company (Urenco).)

The use of gas centrifuges was briefly considered during the development of the atomic bomb in World War II but rejected because of various mechanical difficulties. Their acceptance and use for peaceful purposes beginning in the 1960s came about only after the development of gas-filled bearings and high-strength casings. Most details remain classified, but Figure 1.5 provides an indication of centrifuge size.

Let us assume a centrifuge radius of $R = 0.5$ m, and set RPM at 10,000. We obtain the following from Equation 1.7e:

$$\alpha = \exp \frac{3 \times 10^{-3}(\text{kg / mole})(2\pi \times 10^{4} \times 0.5 / 60)^{2}}{2 \times 8.31 \times 298} \tag{1.7f}$$

$$\alpha = 1.18 \tag{1.7g}$$

This is a respectably high separation factor, but, as shown in Practice Problem 1.6, it leads to a single centrifuge enrichment of only 10%. Because nuclear power plants require U^{235} to be enriched from its natural abundance level of 0.7% to about 3.5%, considerable staging would still be required. Such staging is achieved in so-called countercurrent "cascades," an example of which is shown in Figure 1.5b. Much more about cascades and their crucial role in separation processes will appear in Chapter 7.

1.1.4 Simple Integrations of Fick's Law

The mainstay of our discussions so far, Fick's law of diffusion, can be incorporated in mass transfer models of varying degrees of complexity, up to and including the level of partial differential equations. It can also be used, at a simple level, as a one-equation model for diffusion in simple geometries, leading to results of practical importance. The following two illustrations will serve as examples.

Illustration 1.4: Underground Storage of Helium: Diffusion through a Spherical Surface

Helium is present in air at a concentration of about 1 ppm, which is far too small for the economical recovery of this gas. It also occurs in natural gas (methane CH_4), where its concentration is considerably higher, of the order of 0.1 to 5%, making economical extraction possible. Because helium is a nonrenewable resource, regulations were put in place starting in the early 1960s which required all shipped natural gas to be treated for helium recovery. With supply by far outweighing the demand, ways had to be found to store the excess helium. One suggested solution was to pump the gas into abandoned and sealed salt mines where it remained stored at high pressure.

The problem here will be to estimate the losses that occur by diffusion through the surrounding salt and rock, assuming a solid-phase diffusivity D_s of helium of 10^{-8} m²/s (i.e., more than three orders of magnitude less than the free-space diffusivity in air). The helium is assumed to be at a pressure of 10 MPa (~ 100 atm) and a temperature of 30°C. The cavity is taken to be spherical and of radius 100 m (see Figure 1.6a). Applying Fick's law, Equation 1.4a, and converting to pressure, we obtain

$$N = -D_s 4\pi r^2 \frac{dC}{dr} = -\frac{D_s 4\pi r^2}{RT}\frac{dp}{dr} \tag{1.8a}$$

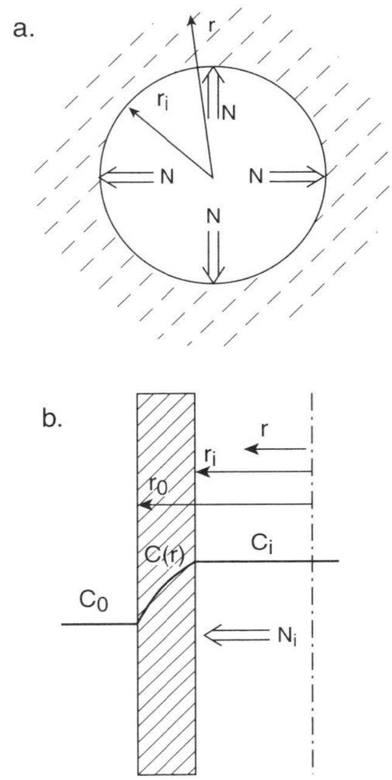

FIGURE 1.6
Diffusional flow from (a) a spherical cavity and (b) a hollow cylinder.

Separating variables and integrating yields

$$-\frac{4\pi D_s}{RT}\int_p^0 dp = N\int_{r_1}^{\infty}\frac{dr}{r^2} = N\frac{1}{r_1} \tag{1.8b}$$

and, consequently,

$$N = \frac{4\pi D_s r_1 p}{RT} = \frac{4\pi 10^{-8}\times 10^2\times 10^7}{8.314\times 303} \tag{1.8c}$$

$$N = 0.05\ \text{mol/s} \tag{1.8d}$$

Comments

This is an example of some practical importance, which nevertheless yields to a simple application and integration of Fick's law. Two features deserve some mention. The first is the formulation of the upper integration limit in Equation 1.8b. We use the argument that "far away" from the spherical cavity (i.e., as $r \to \infty$), the concentration and partial pressure of helium tend to zero (i.e., we assume that the cavity to be embedded is an infinite region). The second point that needs to be examined is the assumption of a constant cavity pressure. We compute for this purpose the yearly loss and show that even over this lengthy period, the change in cavity pressure will be negligibly small. Thus,

$$\text{Yearly loss} = 0.05\ (\text{mole/s}) \times 3600 \times 24 \times 365 = 1.58 \times 10^6\ \text{moles/year}$$

that is, about 100 kg per year.

By comparison,

$$\text{Cavity contents: } n = \frac{pV}{RT} = \frac{10^7\ \frac{4}{3}\pi 10^6}{8.31 \times 303} = 1.17 \times 10^{10}\ \text{mol}$$

and, therefore,

$$\%\ \text{loss/year} = 1.58 \times 10^6/(1.7 \times 10^{10})100 = 1.3 \times 10^{-2}\%$$

We will examine this problem again in Illustration 2.9. The recently introduced process of "sequestering" CO_2 in exhausted oil wells yields to a similar analysis.

Illustration 1.5: Diffusion through a Composite Cylindrical Wall: Principle of Additivity of Resistances

Although the basic features of this process are the same as those of the spherical cavity (Fickian diffusion through a variable area), we use it to cast a somewhat wider net by considering a *composite* cylinder made up of *different* materials with *different diffusivities*. This leads us to the concept of *resistances in series*, and the principle that flows from it, the *additivity of resistances*. Both the concept and the principle are so pervasive in mass (and heat) transfer that they deserve an early introduction.

THE SIMPLE CYLINDRICAL WALL

The starting point here is again Fick's law of diffusion, which is applied to a cylindrical surface of radius r and length L (Figure 1.6b). We obtain

$$N = -D2\pi rL\frac{dC}{dr} \tag{1.9a}$$

where N = constant because we assume steady operation.

Separating variables and formally integrating between the limits of internal and external concentrations C_i and C_o we obtain

$$\int_{C_i}^{C_o} dC = -\frac{N}{D2\pi L}\int_{r_i}^{r_o}\frac{dR}{r} \quad (1.9b)$$

and after evaluation of the integrals and rearrangement,

$$N = D2\pi L\frac{(C_i - C_o)}{\ln r_o / r_i} \quad (1.9c)$$

where i and o denote the inner and outer conditions. This relation expresses diffusion rate N in terms of a driving force $C_i - C_o$ and the geometry of the system.

By multiplying numerator and denominator by $(r_o - r_i)$, Equation 1.8c can be cast into the frequently used alternative form:

$$N = DA_m\frac{(C_i - C_o)}{r_o - r_i} = \frac{(C_i - C_o)}{R} = \frac{\text{Concentration Change}}{\text{Resistance}} \quad (1.9d)$$

where A_m is the so-called logarithmic mean of the inner and outer areas, given by

$$A_m = \frac{A_o - A_i}{\ln A_o / A_i} \quad (1.9e)$$

and R is a resistance defined by

$$R = \frac{r_o - r_i}{DA_m} \quad (1.9f)$$

Note that the mass transfer rate is now expressed in terms of a *constant* gradient $\Delta C/\Delta r$.

THE COMPOSITE CYLINDRICAL WALL

Suppose now that the wall is made up of two *different* materials with resistances R_1 and R_2. Diffusion is from a higher inner concentration C_i to the lower outside level C_o. We can then write

$$N(\text{moles / s}) = \frac{C_i - C_j}{R_1} = \frac{C_j - C_0}{R_2} \quad (1.10a)$$

where C_j = concentration of the material interface. A bit of inspired algebra will then yield

$$1+\frac{C_i - C_j}{C_j - C_o} = 1+\frac{R_1}{R_2} \tag{1.10b}$$

$$\frac{C_i - C_o}{C_j - C_o} = \frac{R_1 + R_2}{R_2} \tag{1.10c}$$

and, hence,

$$N(\text{moles / s}) = \underset{\text{Outer wall}}{\frac{C_j - C_o}{R_2}} = \underset{\text{Both walls}}{\frac{C_i - C_o}{R_1 + R_2}} \tag{1.10d}$$

This result can be extended to an arbitrary set of resistances and is expressed in the following general form:

$$N(\text{moles / s}) = \frac{\Delta C}{\Sigma R} = \frac{\text{Overall Concentration Change}}{\text{Sum of Resistances}} \tag{1.10e}$$

This both states and proves the Principle of Additivity of Resistances.

THE COMPOSITE PLANAR WALL

For a planar wall, the log mean area of the cylindrical wall becomes unity. One can then write, for a single material,

$$N / A(\text{moles / m}^2\text{s}) = \frac{C_i - C_o}{\dfrac{x_o - x_i}{D}} = \frac{\Delta C}{R} \tag{1.10f}$$

and in general for n materials in series,

$$N / A = \frac{C_i - C_o}{\sum_j \dfrac{\Delta x_j}{D_j}} = \frac{\text{Overall Concentration Change}}{\text{Sum of Resistances}} \tag{1.10g}$$

The ratio of diffusivity to wall thickness, $D/\Delta x$, has units of meter per second (m/s) and consequently provides a sense of the *speed* at which the process takes place. It is termed the *permeability* of the material to a particular species, or the *mass transfer coefficient* k_C of the system. k_C will surface again in the next section when we have our first encounter with film theory.

Comments

In precise work, one prefers to identify and quantify each resistance to obtain an exact value of the diffusion rate. This is often not possible, and it then becomes necessary to use one of the following important simplifications:

1. One lumps the resistances into a single *empirical overall resistance*. This is the empiricism used in Illustration 1.6. It will be seen time and again throughout the text whenever mass transfer between two phases is expressed in terms of a single *overall mass transfer coefficient* K_{OC}.
2. Alternatively, one identifies a *single dominant* resistance that determines the rate of mass transfer. This approach is also used throughout the text, particularly in Chapter 5 and Chapter 8 (membrane processes).

These simple principles are vital to the successful application of mass transfer theory, and much else beyond.

Illustration 1.6: Multiple Resistances in Biology: The Lung–Blood Interface

Multiple resistances to mass transfer are all pervasive in living organisms and organs. At the cellular level, the membrane wall is composed of several different layers (lipid, protein, polysaccharide), each with its own resistance. The complexity escalates when transport takes place between adjacent compartments. For example, the transfer of oxygen from the alveoli (tiny sacs in the lung) to neighboring blood capillaries, and ultimately to the hemoglobin in the red cells, involves at least *six major resistances* composed of membranes and fluids (Figure 1.7). To provide a sense of scale, we note that the lung contains some 250 million alveoli with an astonishingly high surface area of ~70 m^2. This factor makes for high mass transfer rates. The other factor is the short distances (Δx) involved. The blood capillaries and cell are of the order of a few micrometers (μm) in width; the membranes range from 0.1 to 1 μm in thickness. The body is thus well equipped to provide high rates of oxygen transfer.

In passing from the alveoli to the hemoglobin, the oxygen partial pressure drops by 10 mmHg or 1.3×10^3 Pa. It is also known that oxygen is supplied to the blood at a rate of 200 mL/min or 1.5×10^{-4} moles/s. We can use these data to calculate an overall resistance to oxygen transfer, R_{O_2} using Equation 1.10a. The results are as follows:

$$N / A = \frac{\Delta C}{R_{O_2}} = \frac{\Delta p / RT}{R_{O_2}} \tag{1.11a}$$

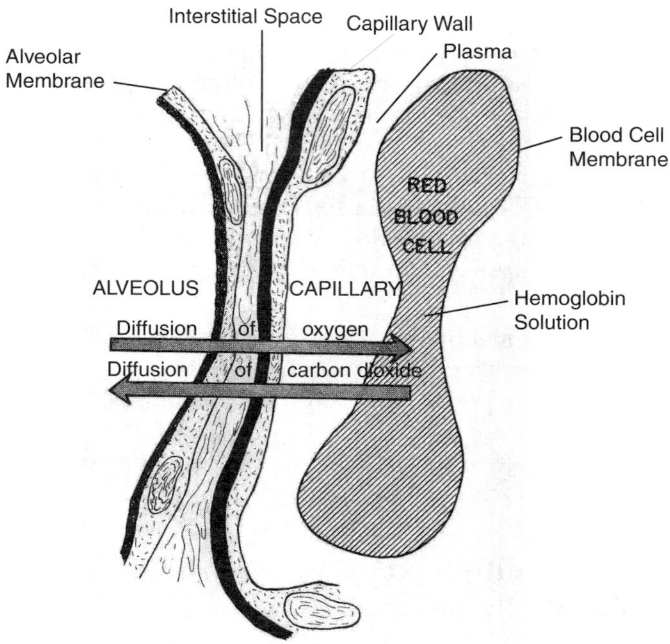

FIGURE 1.7
Transfer of oxygen and carbon dioxide between alveoli of the lung and blood capillaries. There are a total of six major mass transfer resistances in series.

$$R_{O_2} = \frac{\Delta p / RT}{N / A} = \frac{1.3 \times 10^3 / 8.31 \times 300}{1.5 \times 10^{-4} / 70} \tag{1.11b}$$

and

$$R_{O_2} = 2.36 \times 10^4 \text{ s/m} \tag{1.11c}$$

Its inverse, the overall mass transfer coefficient, is then

$$K_{OC} = 1 / R_{O_2} = 4.2 \times 10^{-5} \text{ m/s} \tag{1.11d}$$

How does this compare to oxygen transfer in other contexts? We will show in Illustration 1.8 that a reasonable upper limit for mass transfer to a liquid in highly turbulent flow over a flat surface (say, a fast shallow river) is of the order 10^{-5} m/s. This value would apply to oxygen transfer to flowing water assuming negligible resistance in the air.

It comes as somewhat of a surprise that the much slower diffusive process in the lung yields a similar value, in fact exceeds it by a factor of 4. We attribute this to the factors mentioned before (high A, low Δx) which make for efficient mass transfer in the lung.

1.2 Transport Driven by a Potential Difference (Constant Gradient): The Film Concept and the Mass Transfer Coefficient

At the outset of this chapter, transport driven by a *constant* gradient, or "potential difference," was identified as an important subcase of Fick's law (variable gradient). It was shown in Illustration 1.5 that integration of Fick's law for various geometries likewise yielded a constant gradient, $\Delta C/\Delta r$ or $\Delta C/\Delta x$. Even Ohm's law can be cast into a constant gradient form by noting that electrical resistance varies directly with length L and inversely with the cross-sectional area of the conductor, A_C. One can then write

$$\text{Current } i = \frac{A_C}{R_S}\frac{\Delta V}{L}$$

where R_S is the so-called specific resistivity (Ωm).

Items 2 and 3 of Table 1.2 concern what we term *convective mass* and *heat transfer*. Let us illustrate these terms by making use of Figure 1.8a. Depicted in this figure is turbulent flow of either a gas or liquid past a liquid or solid boundary shown crosshatched on the left. That boundary can be the confining wall of a duct or the interface separating two phases. Mass transfer is assumed to occur from a concentration C_{A2} of the boundary to a lower concentration C_{A1} in the bulk of the flowing fluid. This can come about if the boundary consists of a soluble substance or if a volatile liquid evaporates into a flowing gas stream.

These two operations, as well as the reverse processes of condensation and crystallization, are shown in Figures 1.9a–d. In all four cases shown, the concentrations and partial pressures in the fluid phase are in equilibrium with the neighboring condensed phase. This condition is denoted by an asterisk. Thus, p^* is the equilibrium vapor pressure of the liquid, and C^* is the equilibrium solubility of the solid.

Mass transfer takes place initially through a laminar sublayer, or boundary layer, which is located immediately adjacent to the interface. Transfer through this region, also known as an "unstirred layer" in biological applications, is relatively slow and constitutes the preponderant portion of the resistance to mass transport. This layer is followed by a transition zone where the flow gradually changes to the turbulent conditions prevailing in the bulk of the fluid. In the main body of the fluid, we see macroscopic packets of fluid or eddies moving rapidly from one position to another, including the

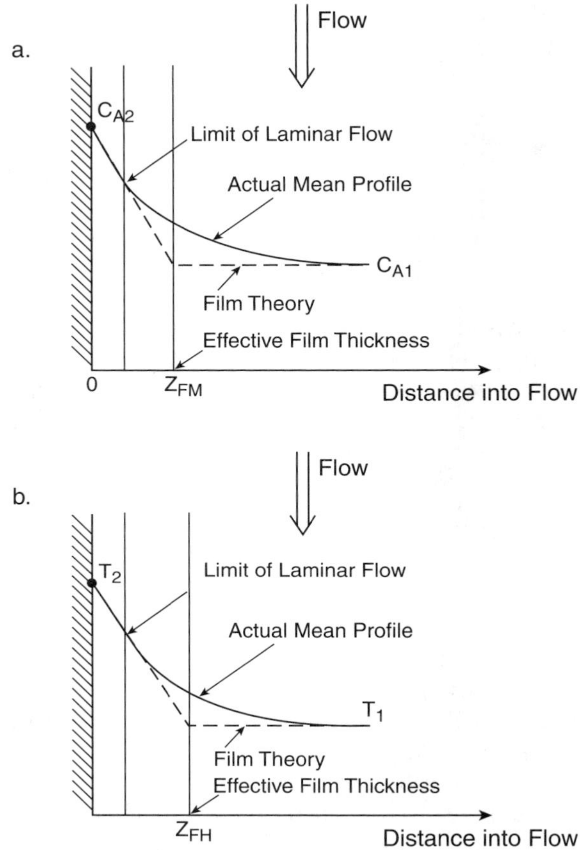

FIGURE 1.8
The effective film in the transport of (a) mass and (b) heat.

directions toward and away from the boundary. Mass transfer in both the transition zone and the fully turbulent region is relatively rapid and contributes much less to the overall transport resistance than the laminar sublayer. In addition, with an increase in fluid velocity, there is an attendant increase in the degree of turbulence, and the eddies are able to penetrate more deeply into the transition and boundary layers. The latter consequently diminish in thickness, and the transport rate experiences a corresponding increase in magnitude. Thus, high flow rates mean a greater degree of turbulence and, hence, more rapid mass transfer.

Concentrations in the turbulent regime typically fluctuate around a mean value shown in Figure 1.8a and Figure 1.8b. These fluctuations cannot be easily quantified, and they do not lend themselves readily for the formulation of a rate law. This can be overcome by postulating the existence of an equivalent linear concentration profile that extends from the boundary into the bulk fluid. This postulate is enshrined in the concept known as *film theory,*

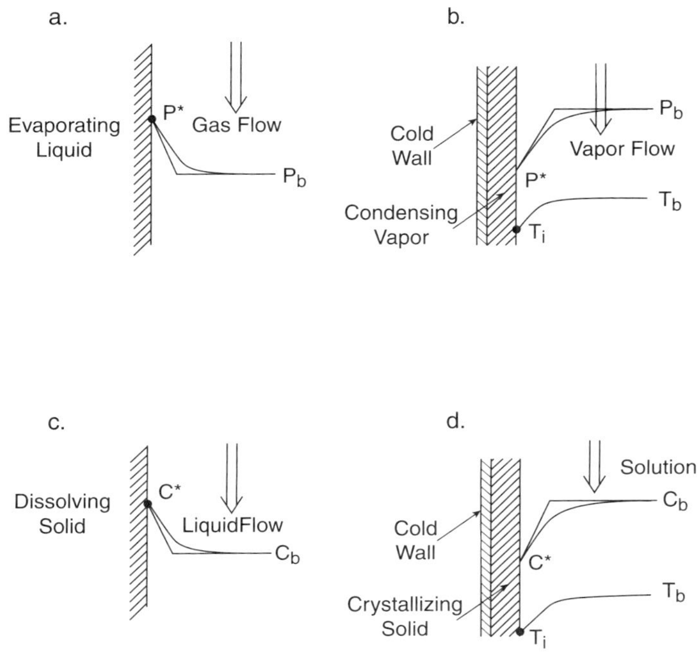

FIGURE 1.9
Four types of single-film mass transfer: (a) evaporation; (b) condensation; (c) dissolution; and (d) crystallization.

and the dimension of the film in question is termed the *effective film thickness*, denoted as z_{FM} in Figure 1.8a.

Let us see how this concept can be quantified into a rate law. We start with Fick's law, and applying it to the constant gradient of film theory, we obtain

$$N_A / A = -D\frac{dC_A}{dz} = D\frac{(C_{A2} - C_{A1})}{z_{FM}} \tag{1.12a}$$

The ratio of diffusivity to film thickness D/z_{FM} is coalesced into a single term called the mass transfer coefficient k_C, and we obtain

$$N_A / A = \frac{D}{z_{FM}}(C_{A2} - C_{A1}) = k_C(\Delta C_A) \tag{1.12b}$$

as shown in Table 1.2.

A similar film theory can be postulated for the case of heat transfer, shown in Figure 1.8b. The conditions here parallel those shown for mass transfer with temperature replacing concentration as the driving potential. Starting with Fourier's law, Equation 1.2b, we then obtain

$$q / A = -k\frac{dT}{dz} = k\frac{(T_2 - T_1)}{z_{FH}} \tag{1.12c}$$

Coalescing the ratio k/z_{FH} into a single term h then leads to

$$q / A = \frac{k}{z_{FH}}(T_2 - T_1) = h\Delta T \tag{1.12d}$$

where h is now termed the heat transfer coefficient.

The effective film thicknesses for the two cases, z_{FM} and z_{FH}, are not generally equal but depend in a complex functional form on the physical properties, the geometry, and the velocity of flow of the system. That functional form will be explored in greater detail in Chapter 5. In addition, the transport rate depends *linearly* on the potential difference, a feature that is often referred to as a *linear driving force*. All three items have this characteristic in common.

A special type of driving force arises in Item 4 of Table 1.2. The process here is the selective transport of water through a semipermeable membrane from a dilute solution (high water concentration) to a more concentrated solution (low water concentration), introduced in Illustration 1.3. The driving force is in this case the difference of the so-called osmotic pressure π, which makes its appearance in transport through cell membranes as well as in industrial membrane processes. We will take a closer look at osmotic-pressure-driven processes in Chapter 8.

1.2.1 Units of the Potential and of the Mass Transfer Coefficient

In deriving the mass transfer rate law, Equation 1.12b, we started with Fick's law, which uses molar concentration C in units of mole per cubic meter (mol/m^3) as the driving potential. This quantity was retained to describe the driving force in the final expression (Equation 1.12b). It is a convenient quantity to use in many gas–liquid operations and carries the advantage of imparting units of meter per second (m/s) to the mass transfer coefficient. Thus, k_C can be viewed as the velocity with which the rate process proceeds. It frequently happens, however, that molar concentrations are inconvenient to use in the description of certain mass transfer operations. In distillation, for example, the preferred concentration unit is the mole fraction, because the associated vapor–liquid equilibrium is commonly expressed in liquid and vapor mole fractions (x, y). In the evaporation of liquids, the vapor pressure is the potential of choice, and it then becomes convenient to use a pressure difference as the driving force. Yet another operation that calls for a change in concentration units is gas absorption, in which the preferred concentration is the mass ratio of the diffusing species. In each of these cases, the change in concentration units carries with it a change in the units of the mass transfer coefficient. The pertinent rate laws,

TABLE 1.5

Rate Laws and Transfer Coefficients for Diffusion through a Stagnant Film

Flux (mol/m²s)	Driving Potential	Mass Transfer Coefficient
Gases		
$N_A/A = k_G \Delta p_A$	p_A (Pa)	k_G (mol/m²s Pa)
$N_A/A = k_y \Delta y_A$	y_A (mole fraction)	k_y (mol/m²s mole fraction)
$N_A/A = k_C \Delta C_A$	C_A (mol/m³)	k_C (m/s)
$W_A/A = k_Y \Delta Y_A$ (kg/m²s)	Y_A (kg A/kg B)	k_Y (kg/m²s ΔY_A)
Liquids		
$N_A/A = k_L \Delta C_A$	C_A (mol/m³)	k_L (m/s)
$N_A/A = k_x \Delta x_A$	x_A (mole fraction)	k_x (mol/m²s mole fraction)
Conversion Factors		
Gases	$k_G = k_y/P_T = k_C/RT = k_Y/Mp_{BM}$	
Liquids	$k_L C = k_x$	

driving forces, and mass transfer coefficients, together with their units, are summarized in Table 1.5.

Also listed in Table 1.5 are conversion factors for the transformation of mass transfer coefficients from one set of units to another. These are frequently required to convert literature values of k given in a particular set of units, to one needed in a different application. This type of conversion is discussed in Illustration 1.7. Also of note in Table 1.5 is the appearance of the term p_{BM}, the so-called logarithmic mean, or log-mean driving force, defined by

$$p_{BM} = \frac{p_{B2} - p_{B1}}{\ln \dfrac{p_{B2}}{p_{B1}}} \tag{1.13}$$

where the subscript B denotes the second component in a binary system; the first was the component A being transferred. Derivation of this quantity, and its appearance in the conversion factor, is addressed in Section 1.4.

Illustration 1.7: Conversion of Mass Transfer Coefficients

In a particular application related to air flowing over a water surface, the following data were reported at $T = 317$ K:

$p_{A1} = 2487$ Pa

$p_{B1} = 101{,}300 - 2487 = 98{,}813$ Pa

$p_{B2} = P_T = 101{,}300$ Pa

$k_{CA} = 0.0284$ m/s

Water evaporates into the dry airstream at a total pressure $P_T = 101.3$ kPa and is denoted by the subscript *A*; *B* refers to the air component; and k_C is the mass transfer coefficient, here of water. We wish to calculate the corresponding value for k_Y in units of kilogram H_2O per square meter second ΔY (kg H_2O/m^{2}s ΔY). The conversion formula given in Table 1.3 is

$$k_Y = k_C \frac{M_B}{RT} p_{BM} \tag{1.14}$$

where M_B is the molar mass of air = 29×10^{-3} kg/mole.

We obtain

$$k_Y = 0.0284 \frac{29 \times 10^{-3}}{8.31 \times 317} \frac{98,813 - 101,300}{\ln \dfrac{98,313}{101,300}}$$

and, therefore,

$$k_Y = 0.0312 \, \text{kg H}_2\text{O} / \text{m}^2\text{s}\,\Delta Y$$

1.2.2 Equimolar Diffusion and Diffusion through a Stagnant Film: The Log–Mean Concentration Difference

So far our treatment has been confined to mass transfer due to diffusion only. We considered diffusion in a stationary or unmixed medium, which led to the use of Fick's Equation 1.4. When a stirred or turbulent medium was involved, we invoked film theory and the linear driving force concept to describe transport in such situations. This led to the formulation of Equation 1.12b.

Mass transport can, however, also come about as a result of the bulk motion or flow of a fluid. To take this factor into account, we postulate the total flux of a component *A* to be the sum of a diffusive flux term and a bulk flow term. Thus, for a gaseous mixture,

$$\underset{\text{Flux of } A}{N_A / A} = \underset{\text{Diffusional Flux}}{-CD_{AB}\frac{dy_A}{dz}} + \underset{\text{Bulk Flow}}{y_A(N_A / A + N_B / A)} \tag{1.15a}$$

Here we replaced the C_A, which appears in Fick's law, with the equivalent term Cy_A, where C = total molar concentration, assumed to be constant.

For ideal gases, we have $y_A = p_A/P_T$ and $C = P_T/RT$, where P_T = total pressure, so that Equation 1.15a becomes

$$N_A / A = -\frac{D_{AB}}{RT}\frac{dp_A}{dz} + \frac{p_A}{P_T}(N_A / A + N_B / A) \quad (1.15b)$$

This is the form we wish to develop and simplify.

Two special cases of Equation 1.15a are to be noted: *equimolar counter-diffusion and diffusion through a stagnant film.*

1.2.2.1 Equimolar Counterdiffusion

In this case, we have

$$N_A = -N_B \quad (1.16)$$

and Equation 1.15b reduces to Fick's law. This situation arises in the inter-diffusion of pure fluids of equal molar volume or in binary distillation processes of substances with identical molar heats of vaporization. Straightforward integration of Fick's law then leads to, for a gaseous system,

$$N_A' / A = \frac{D_{AB}}{RT(z_2 - z_1)}(p_{A_1} - p_{A_2}) = \frac{D_{AB}}{RTz_{FM}}(p_{A_1} - p_{A_2}) \quad (1.17a)$$

or in short,

$$N_A'/A = k_G' \Delta p_A \quad (1.17b)$$

where we use the prime symbol to denote equimolar counterdiffusion. Similarly, for a liquid system, using mole fraction as the driving potential,

$$N_A' / A = \frac{CD_{AB}}{(z_2 - z_1)}(x_{A_1} - x_{A_2}) = \frac{CD_{AB}}{z_{FM}}(x_{A_1} - x_{A_2}) \quad (1.17c)$$

or in short,

$$N_A'/A = k_x' \Delta x_A \quad (1.17d)$$

Because $\Delta p/z_M$ and $\Delta x/z_{FM}$ are constant, the concentration profiles in both cases are linear. This is shown for a liquid system in Figure 1.10a. A summary of the pertinent rate laws, transfer coefficients, and conversion factors that apply to equimolal counterdiffusion appears in Table 1.6.

1.2.2.2 Diffusion through a Stagnant Film

Here the flux of the species B is zero, and we have

$$N_B/A = 0 \quad (1.18a)$$

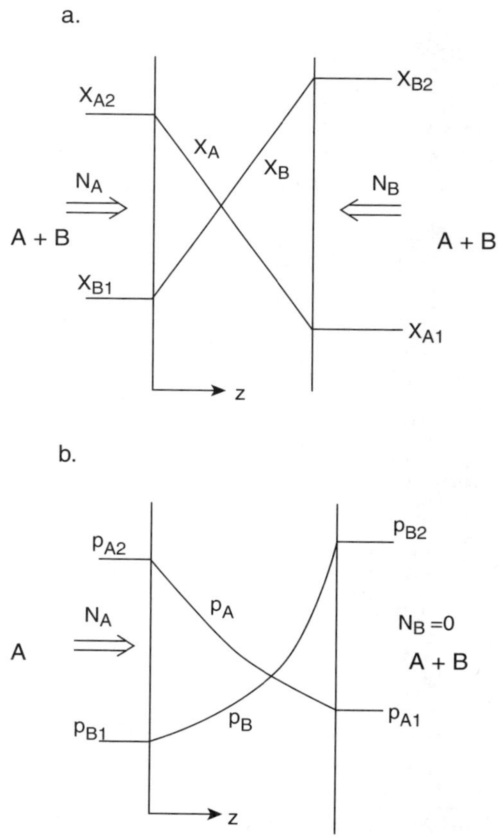

FIGURE 1.10
Two modes of transport: (a) equimolar counterdiffusion and (b) diffusion through a stagnant film.

so that Equation 1.15b, after solving for N_A/A, is reduced to the expression

$$N_A / A = -\frac{D_{AB}P_T}{RT(P_T - p_A)}\frac{dp_A}{dz} \tag{1.18b}$$

which can be integrated to yield

$$N_A / A = -\frac{D_{AB}P_T}{(z_2 - z_1)RT}\ln\frac{(P_T - p_{A1})}{(P_T - p_{A2})} = \frac{D_{AB}P_T}{(z_2 - z_1)RT}\ln\frac{p_{B1}}{p_{B2}} \tag{1.18c}$$

Because of the logarithmic terms, the profiles for component A and component B are nonlinear. This is depicted in Figure 1.6b.

We now introduce a clever device to reduce the nonlinear terms in Equation 1.18c to the product of a linear driving force in the diffusing

TABLE 1.6

Rate Laws and Transfer Coefficients for Equimolar Diffusion

Flux (mol/m²s)	Driving Potential	Mass Transfer Coefficient
Gases		
$N_A'/A = k_G'\Delta p_A$	p_A (Pa)	k_G' (mol/m²s Pa)
$N_A'/A = k_y'\Delta y_A$	y_A (mole fraction)	k_y' (mol/m²s mole fraction)
$N_A'/A = k_C'\Delta C_A$	C_A (mol/m³)	k_C' (m/s)
Liquids		
$N_A'/A = k_L'\Delta C_A$	C_A (mol/m³)	k_C' (m/s)
$N_A'/A = k_x'\Delta x_A$	x_A (mole fraction)	k_x' (mol/m²s mole fraction)
Conversion Factors		
Gases	$k_G' = \frac{k_y y}{P_T} = \frac{k_C'}{RT}$	
Liquids	$k_L' C = k_y'$	
Conversion from Equimolar to Stagnant Film Coefficients		
Gases		
$k_G' = k_G \frac{p_{BM}}{P_T}$	$k_y' = k_y \frac{p_{BM}}{P_T} = k_y y_{BM}$	$k_C' = k_C \frac{p_{BM}}{P_T} = k_C \frac{C_{BM}}{C}$
Liquids		
$k_C' = k_C \frac{C_{BM}}{P} = k_C x_{BM}$	$k_x' = k_x x_{BM}$	

species A, $p_{A1} - p_{A2}$, and a constant mass transfer coefficient k_G. This is done by writing the following:

$$1 = \frac{p_{B2} - p_{B1}}{p_{B2} - p_{B1}} = \frac{P_T - p_{A2} - P_T + p_{A1}}{p_{B2} - p_{B1}} = \frac{p_{A1} - p_{A2}}{p_{B2} - p_{B1}} \tag{1.18d}$$

Using the definition of the log-mean pressure difference p_{BM} given by Equation 1.13 and setting $z_2 - z_1 = z_{FM}$ as before, we obtain

$$N_A / A = \frac{D_{AB} P_T}{z_{FM} RT p_{BM}} (p_{A2} - p_{A1}) \tag{1.18e}$$

or in short,

$$N_A/A = k_G \Delta p_A \tag{1.18f}$$

We have thus reduced a complex nonlinear situation to one that fits the linear driving force and film concepts and agrees with the tabulations of Table 1.3.

Illustration 1.8: Estimation of Mass Transfer Coefficients and Film Thickness

It is of some interest to the practicing engineer to have a sense of the order of magnitude both of the mass transfer coefficient and its associated film thickness. This would appear to be an impossible task, given the wide range of flow conditions, geometrical configurations, and physical properties encountered in practice. Surprisingly, one can arrive at some reasonable estimates of upper and lower bounds in spite of this diversity. This is due to three factors: First, it is common engineering practice to associate the upper limit of normal turbulent flow with velocities of the order 1 m/s in the case of liquids and 10 m/s for gases. This applies to industrial systems (pipe and duct flow) as well as within an environmental context (wind, river flow) and holds even in extreme cases. Hurricane-force winds, for example, as high as to 100 km/h, are still within the order of magnitude cited. Second, the diffusivities for a wide range of substances are, as we will see in Chapter 3, surprisingly constant. They cluster, in the case of gases, around a value of 10^{-5} m^2/s and for liquids around 10^{-9} m^2/s. Third, if we confine ourselves to flow over a plane as a representative configuration, it will be found that mass transfer coefficients vary inversely with the two-thirds power of the ratio ($\mu/\rho D$) (see Table 5.5). That ratio, termed the *Schmidt number*, Sc, is again surprisingly constant. It is of the order 1 for gases, and some three orders of magnitude higher for transport of modest-sized solutes in liquids of normal viscosity.

Drawing on the correlation given in Table 5.5:

$$\frac{k_C}{v} = 0.036\ \mathrm{Re}^{-0.2}\mathrm{Sc}^{-0.67} \qquad (\mathrm{Re} > 10^6)$$

and noting the extremely weak dependence on Reynolds number, Re, the following order-of-magnitude estimates for turbulent flow mass transfer can be obtained:

For gases: $k_C \sim 1$ cm/s
For liquids: $k_C \sim 10^{-3}$ cm/s

The k_C values given here represent the *order of magnitude* of the outer limits of what can be accomplished (i.e., the maximum rate of mass transfer obtainable or the minimum time required to achieve the transfer of a given mass to or from a flat surface).

TABLE 1.7

Upper and Lower Bounds on Mass Transfer (Order of Magnitude)

		k_C (MTC)	z_{FM} (Film Thickness)
Upper bound (turbulent flow)			
	Gas film	1 cm/s	1 mm
	Liquid film	10^{-3}	10^{-1} mm
Upper bound (diffusion)			
After 1 msec	Gas film	10 cm/s	10^{-1} mm
	Liquid film	10^{-1}	10^{-3}
After 10 sec	Gas film	10^{-1}	10
	Liquid film	10^{-3}	10^{-1}
Lower bound (diffusion)			
After 3 h	Gas film	10^{-3} cm/s	1 m
	Liquid film	10^{-5}	10^{-2}
After 10 days	Gas film	10^{-4}	10
	Liquid film	10^{-6}	10^{-1}
After 1000 days	Gas film	10^{-5}	100
	Liquid film	10^{-7}	1

Mass transfer by (transient) molecular diffusion resides at the other end of the spectrum. It yields, at least in the long term, the *lowest* possible mass transfer rate and sets an upper limit on time requirements. The solvent spill considered in Practice Problem 4.6, for example, requires several *days* to complete evaporation by diffusion into stagnant air. The same data applied to turbulent air flow at the same temperature yield an estimate of several *minutes*, lower by three orders of magnitude.

To make a valid comparison between these two cases, one needs to express transient diffusion in terms of an equivalent mass transfer coefficient k_C. That is provided by the Higbie equation that we derive in Chapter 4 and reproduce below:

$$(k_C)_{\text{diffusion}} = (D / \pi t)^{1/2} \tag{4.11g}$$

The expression predicts high mass transfer coefficients during the first *fractions of a second* when the boundary layer is still extremely thin. Thereafter they decline rapidly, falling to 1/100th to 1/1000th of the turbulent flow values after 3 h of elapsed time.

The results of this analysis are summarized in Table 1.7. They are to be regarded as *order of magnitude estimates* and apply to mass transfer between a plane surface and a neighboring unbounded medium. Table 1.8 provides a comparison of corresponding heat and mass transfer parameters.

TABLE 1.8

Mass and Heat Transfer Parameters for Two-Phase Transport

	Mass Transfer[a]	Heat Transfer
Driving force	Δy, Δx	ΔT
Single-film coefficient	k_x, k_y	h_h, h_C
Overall coefficient	K_{ox}, K_{oy}	U
Single-film resistance	$1/k_x$, $1/k_y$	$1/h_h$, $1/h_C$
Overall resistance	$1/K_{oy} = 1/k_y + m/k_x$	$1/U = 1/h_h + 1/h_C + L/k$
Single-film rate of transfer	$N_A/A = k_y\ (y_A - y^*_A)$	$q/A = h_h\ (T_h - T_{w2})$
	$= k_x\ (x_{Ai} - x_A)$	$= h_C\ (T_{w1} - T_C)$
Overall rate of transfer	$N_A/A = K_{oy}\ (y_A - y_A^*)$	$q/A = U\ (T_h - T_C)$
	$= K_{ox}\ (x_A^* - x_A)$	
Equilibrium relation	$y^* = mx$	—

[a] Items listed are based on mole fraction concentration units. For conversion to other units, see Table 1.4.

Illustration 1.9: The Environment: Analysis of an Ocean Spill

Suppose that on the ocean under windy conditions a spill occurred of a solvent with a molar mass of 100 Da, solubility in water of 1 g/L = 10 mol/m^3, and a vapor pressure of 10^3 Pa. These values are representative of an aromatic solvent such as toluene. The spill density is 100 mol/m^2 (approximately 1 cm thick layer).

As a first crude approximation, air may be viewed as flowing over a flat surface. No such claim can be made for the ocean water underlying the spill. We do assert, however, that the solvent is in equilibrium with the water at the interface, and that this is followed by a thin stagnant film beneath which a considerable degree of mixing takes place. Our aim is to establish the proportions of solvent transfer to the atmosphere and ocean, respectively, using the guidelines of Table 1.7 and to estimate a minimum time for the dissipation of the spill.

We have, for transfer to the water,

$$(N/A)_w = k_C C^* = 10^{-5}\ \text{m/s} \times 10\ \text{mol/m}^3 \tag{1.19a}$$

$$(N/A)_w = 10^{-4}\ \text{mol/m}^2\text{s} \tag{1.19b}$$

and for transfer to the atmosphere,

$$(N/A)_a = k_G p^* = k_C p^*/RT \tag{1.19c}$$

$$(N/A)_a = 10^{-2}\ \text{m/s} \times 10^3\ \text{Pa}/8.31 \times 298 \tag{1.19d}$$

$$(N/A)_a = 4 \times 10^{-3}\ \text{mol/m}^2\text{s} \tag{1.19e}$$

Transfer to the water is thus only 2.5% that of the transfer to the atmosphere, even if one assumes highly turbulent conditions in the ocean phase. Using the spill density of 100 mol/m², we obtain for the time of complete dissipation,

$$t = \frac{100\ \text{mol}/\text{m}^2}{4 \times 10^{-3}\ \text{mol}/\text{m}^2\text{s}} \tag{1.19f}$$

$$t = 2.5 \times 10^5\ \text{s} = 69\ \text{hours} \tag{1.19g}$$

It is of interest to compare these results with the empirical correlations used in environmental work. For transfer from the ocean surface to the atmosphere, k_C is expressed in the form

$$k_C\ (\text{m/h}) = 3.6 + 5\ \nu^{1.2} \tag{1.19h}$$

where ν is the wind velocity in meter per second (m/s). Setting $\nu = 10$ m/s, the value used in arriving at the "upper bound" of Table 1.7, one obtains $k_C = 2.2$ cm/s, in agreement with our order of magnitude estimates of 1 cm/s. Similar correlations cited for transfer to the ocean phase confirm the small magnitude of that component arrived at here.

Comments

There are two hopeful signs in these results. One is the relatively short dispersion time (the environmental correlation, Equation 1.19h, predicts even shorter times). The second is the relatively small amount of the spill, which passes into the water phase. The aquatic life is thus less at risk than if it had to bear the full impact of the spill.

Crude oil does not have nearly the volatility of toluene and is much less soluble in water. It would take months to evaporate and must be dispersed or collected by other means instead (detergents, skimmers, etc.). The Exxon Valdez spill of 1989 required 3 years to clean up at a cost of $2.5 billion.

1.3 The Two-Film Theory

Our considerations so far have been limited to transport through a single phase — that is, in postulating the film theory, it was assumed that a single film resistance was operative. This was the case for a pure liquid evaporating into a gas stream, or when a solid dissolved into a solvent stream. We now consider the extension of this process to simultaneous transport in two adjacent phases. This leads to the formation of two film resistances and brings us to the so-called *two-film theory,* which is taken up below.

Consider two phases, I and II, in turbulent flow and in contact with each other, as shown in Figure 1.11a. Transport takes place in the first place, from a high concentration y_A through the effective film associated with Phase II to the interface. Here the Phase II concentration y^*_{Ai} is assumed to be in equilibrium with the Phase I interfacial concentration x_{Ai}, so that

$$y^*_{Ai} = mx_{Ai} \tag{1.20a}$$

where m is the local slope of the equilibrium curve, and the asterisk serves to denote equilibrium conditions. Transport then continues from the interface, through the second film, to the bulk of Phase I of concentration x_A. One can write the following for the entire process:

$$N_A/A = k_y\,(y_A - y_{Ai}) = k_x\,(x_{Ai} - x_A) \tag{1.20b}$$

These expressions, while valid under the constraints of two-film theory, are nevertheless ill-suited for practical use, as neither of the interfacial concentrations x_{Ai} or y_{Ai} is generally known. We avoid this difficulty by postulating an equivalent rate law, given by

$$N_A/A = K_{yo}\,(y_A - y_A^*) \tag{1.21a}$$

$$N_A/A = K_{xo}\,(x_A^* - x_A) \tag{1.21b}$$

where K_{oy} and K_{ox} are termed *overall mass transfer coefficients*, and the asterisked quantities are the concentrations in equilibrium with the bulk concentration of the neighboring phase. These are generally known and are displayed in Figure 1.12. It is shown in the following section that the equations are valid provided the overall coefficients are related to the film coefficients as follows:

$$\frac{1}{K_{oy}} = \frac{1}{k_y} + \frac{m}{k_x} \tag{1.22a}$$

$$\frac{1}{K_{ox}} = \frac{1}{mk_y} + \frac{1}{k_x} \tag{1.22b}$$

The reciprocal terms appearing in these expressions may be regarded as *resistances* to the mass transfer process. Thus, in Equation 1.22a, $1/k_y$ is the resistance due to the gas film, m/k_x represents the liquid film resistance, and the sum of the two yields the overall resistance $1/K_{oy}$. This is often referred to as the law of *additivity of resistances*, which was previously encountered in Illustration 1.5. Similar arguments apply to Equation 1.22b.

Two limiting cases are of note: When m is small compared to k_x, one obtains the approximate relation

$$\frac{1}{K_{oy}} \doteq \frac{1}{k_y} \qquad m << 1, k_L >> 1 \tag{1.22c}$$

This implies that the gas is highly soluble and that most of the resistance is likely to reside in the gas phase. We speak of the process as being *gas-film controlled*. Conversely, if m is large (i.e., the gas is sparingly soluble), we obtain

$$\frac{1}{K_{ox}} \doteq \frac{1}{k_x} \qquad m << 1 \tag{1.22d}$$

and we refer to the process as being *liquid-film controlled*. Depending on which resistance predominates, we choose either Equation 1.22a or Equation 1.22b to express the transfer process.

Overall coefficients are widely used to describe transport between two flowing phases in contact with each other. They are usually determined experimentally and reported as *lumped averages* over the span of equilibrium constants m encountered in the operation. Alternatively and much less frequently, empirical correlations of the film coefficients, if available, can be used to compute K from Equation 1.22b. Such correlations are discussed in Chapter 5.

The two-film concept is also applied to heat transfer operations, as shown in Figure 1.11b. The process is similar to that of mass transport but differs from it in two important aspects. First, the two fluids (hot and cold) are usually, but not always, separated by a solid partition. This is in contrast to mass transfer operations where direct contact of the phases is the norm. Second, no phase-equilibrium relation needs to be invoked at the interface. Instead, convergence of the two temperature profiles on either side of an interface leads to the same temperature at this point. No jump-discontinuities in temperature occur at any location along an interface. We can, consequently, express the rate of heat transfer in the following alternate ways:

$$q / A = h_h(T_h - T_{w_2}) = \frac{k}{L}(T_{w_2} - T_{w_1}) = h_C(T_{w_1 - T_c}) \tag{1.23a}$$

Transfer from hot fluid	Transfer through wall	Transfer to cold fluid

where the subscripts h and c represent hot and cold fluids, respectively; k is the thermal conductivity; and L is the thickness of the partition.

Individual resistances can be added to obtain an overall heat transfer coefficient U, given by

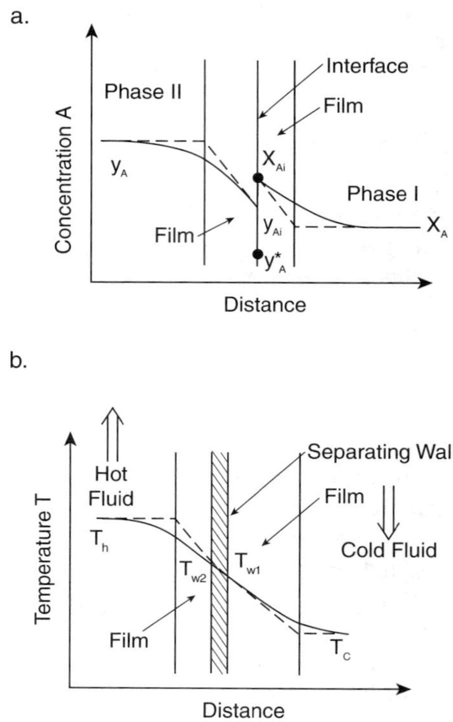

FIGURE 1.11
The two-film concept: (a) mass transfer and (b) heat transfer.

$$\frac{1}{U} = \frac{1}{h_h} + \frac{1}{h_c} + \frac{L}{k} \tag{1.23b}$$

which is associated with an overall heat transfer driving force — that is, we have

$$q/A = U(T_h - T_c) \tag{1.23c}$$

The corresponding mass transfer rate expression is given by Equation 1.25. For convenience, the various parameters that appear in the foregoing equations are summarized in Table 1.7.

1.3.1 Overall Driving Forces and Mass Transfer Coefficients

We embark here on the proof of the validity of Equation 1.21 and Equation 1.22 — that of the rate law based on an overall driving force, and the law of additivity of resistances. Figure 1.12 represents a plot of the gas-phase mol fraction of the diffusing species, y_A, against its liquid-phase counterpart, x_A.

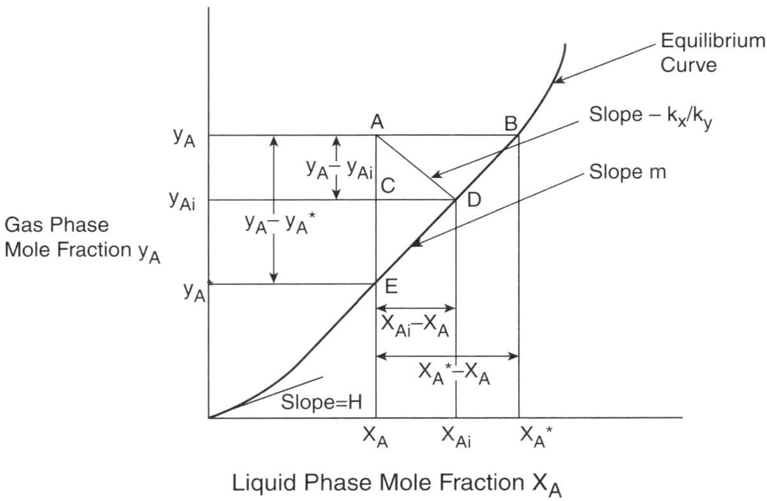

FIGURE 1.12
The driving forces.

It contains an equilibrium curve, which is generally nonlinear but is assumed to contain a short segment BE which is considered linear. Also indicated on the diagram are the various film and overall driving forces. For the gas phase, AC represents the film driving force $y_A - y_{Ai}$, and the distance AE equals the overall gas-phase driving force $y_A - y_A^*$.

We start by noting that the slope of the line AD is given by the ratio AC/CD; hence, by virtue of Equation 1.20b,

$$\frac{AC}{CD} = \frac{(y_A - y_{Ai})}{-(x_{Ai} - x_A)} = -\frac{k_x}{k_y} \tag{1.24a}$$

Similarly, we have for the local slope of the equilibrium curve

$$\frac{CE}{CD} = m \tag{1.24b}$$

The overall driving force y – y* is given by the sum of the two segments AC and CE:

$$y_A - y_A{}^* = AC + CE = \frac{k_x}{k_y}(x_{Ai} - x_A) + m(x_{Ai} - x_A) \tag{1.24c}$$

Because we postulated an overall rate law of the form $N_A/A = K_{oy}(y_A - y_A^*)$, it follows from Equation 1.20 and Equation 1.24c that

$$N_A / A = K_{oy}(y_A - y_A^*) = K_{oy}\left(\frac{k_x}{k_L} + m\right)(x_{Ai} - x_A) = k_x(x_{Ai} - x_A) \quad (1.25a)$$

Consequently,

$$K_{oy}\left(\frac{k_x}{k_y} + m\right) = k_x \quad (1.25b)$$

and, therefore,

$$\frac{1}{K_{oy}} = \frac{1}{k_y} + \frac{m}{k_x} \quad (1.25c)$$

This proves the validity of Equation 1.21a and Equation 1.21b. Similar arguments can be used to verify the validity of the liquid-phase counterparts.

Comments

Recall that this entire development arose from the need to replace the interfacial concentrations in Equation 1.20b, which are generally unknown, by some other known or measurable quantity. Inspection of the diagram in Figure 1.12 shows that there is only one such quantity for the gas phase, mainly y_A^*, the mole fraction in equilibrium with the bulk liquid concentration x_A. Hence, it was natural to replace the film driving force by an overall driving force $y_A - y_A^*$.

This was accomplished by making clever use of the diagram and Equation 1.20b to relate the segments AC and CE to k_x, k_y, and m, and to establish that the sum of the two equals the overall driving force $y_A - y_A^*$. Introduction of the rate law $N_A/A = K_{oy}(y - y^*)$ then culminated in the derivation of the law of additivity of resistances, Equation 1.22a and Equation 1.22b. Thus, it was possible to resolve, by a series of simple moves, a seemingly intractable problem.

The two-film (or two-resistance) theory was first proposed in a landmark paper by W.K. Lewis and W.G. Whitman over 80 years ago (*Ind. Eng. Chem. 16*, 1215 [1924]). In spite of sporadic criticism, particularly of the assumption of interfacial equilibrium, it has survived remarkably well and continues to serve as one of the mainstays in the design of industrial separation processes.

The following two illustrations reinforce the notion of its importance.

Illustration 1.10: Percent Resistance in Two-Film Mass Transfer

Consider a turbulent gas stream containing ammonia, in contact with water, also in turbulent flow. NH_3, a fairly soluble gas, transfers to the interface and dissolves in the water. This type of operation is carried out on a large scale in "packed towers" to remove or recover the NH_3 from process streams (see Chapter 8).

At sufficiently low concentrations of NH_3 (<5 mol%), the phase equilibrium is linear, with $m = y/x = 1.4$ at $P_T = 100$ kPa and $T = 25°C$. The task will be to calculate the percent resistance residing in the gas film. We have, first,

$$\frac{1}{K_{oy}} = \frac{1}{k_y} + \frac{m}{k_x} \tag{1.22a}$$

Overall Gas film Liquid film

and using the conversions of Table 1.4,

$$\frac{1}{K_{oy}} = \frac{1}{k_C P_T / RT} + \frac{m}{k_L C} \tag{1.26a}$$

where both k_L and k_C are in units of m/s.

If we now introduce the "upper bounds" of Table 1.7, there results

$$\frac{1}{K_{oy}} = \frac{1}{10^{-2}(\text{m/s})10^5 / 8.31 \times 298} + \frac{1.4}{10^{-5}(\text{m/s})55000(\text{mole/m}^3)} \tag{1.26b}$$

$$\frac{1}{K_{oy}} = 2.48 + 2.55 \tag{1.26c}$$

so that

$$\%\text{ of resistance in gas film} = \frac{2.48}{2.48 + 2.55} \times 100 = 49\% \tag{1.26d}$$

In industrial gas absorbers, the gas film resistance for ammonia is usually higher, because gas velocities are much lower, of the order 1 m/s. That translates into a tenfold *decrease* in k_C, and we obtain

$$\frac{1}{K_{oy}} = 24.8 + 2.55 \tag{1.26e}$$

Consequently,

$$\%\text{ resistance in gas film} = \frac{24.8}{24.8 + 2.55} \times 100 = 91\% \tag{1.26f}$$

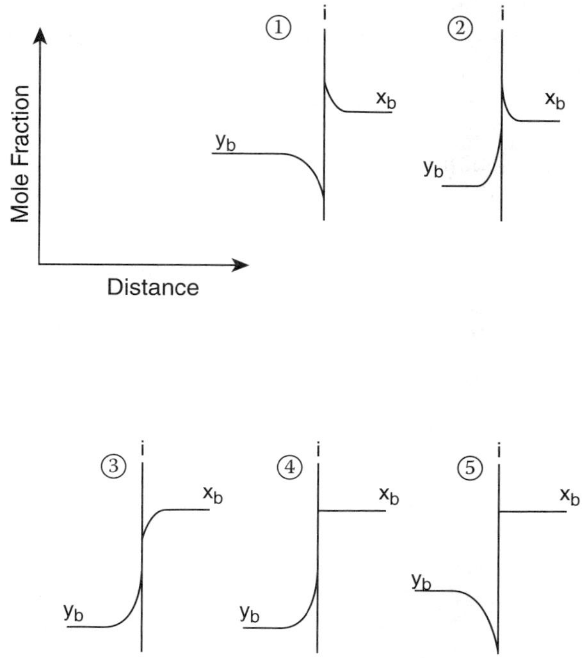

FIGURE 1.13
Hypothetical concentration profiles near a gas–liquid interface.

We pause at this point to take stock of the principles established so far and to test our grasp of those principles with the following example.

Illustration 1.11: Qualitative Analysis of Concentration Profiles and Mass Transfer

The reader is here confronted with a series of hypothetical concentration profiles near a gas–liquid interface, which are sketched in Figure 1.13. The task to be addressed is twofold: We wish to establish whether in each of these instances mass transfer does in fact take place and, if so, in which direction it will proceed. The main principle we have to apply is that mass transfer can occur only along *a negative concentration gradient*. With this fact firmly in mind, one can proceed as follows:

Case 1: Here the concentrations in both the gas and liquid phase diminish in the positive direction, causing solute to transfer from the gas to the liquid phase. The fact that the interfacial liquid mole fraction is higher than the gas concentration is no impediment. It is merely an indication of high gas solubility, a perfectly normal and acceptable phenomenon. Because Figure 1.13 indicates that the gas solubilities are quite large ($x_i \gg y_i$), we assume gas-phase resistance to be the controlling factor so that Equation 1.22c applies:

$$\frac{1}{K_{oy}} \cong \frac{1}{k_y} \tag{1.22e}$$

Case 2: The gas-phase concentration here increases in the positive direction so that no transfer of solute from gas to liquid can take place. No transfer can occur in the opposite sense, because the liquid concentration rises in the negative direction. Such profiles arise only in cases when solute is generated by chemical reaction at the gas–liquid interface. The product solute then diffuses from the interface into the bulk fluids using the profiles shown. The case is a relatively rare one because the products of a chemical reaction — typically an acid–base neutralization — tend to be nonvolatile and do not transfer back into the gas phase.

Case 3: This case involves decreasing concentrations in both phases, but the decrease is in the negative direction. Solute will therefore desorb from the liquid into the gas phase. Because the solubility here appears to be at an intermediate level ($x_i \approx y_i$), one suspects that both resistances are operative. We have

$$\frac{1}{K_{oy}} = \frac{1}{k_y} + \frac{m}{k_x} \tag{1.22a}$$

or, alternatively,

$$\frac{1}{K_{ox}} = \frac{1}{mk_y} + \frac{1}{k_x} \tag{1.22b}$$

Case 4: The flat liquid-phase profile indicates that the liquid phase is well stirred and shows no mass transfer resistance. Because the gas-phase concentration diminishes in the negative direction, the transfer will be from liquid to gas. There is also a total absence of liquid-phase resistance (flat x-profile), and the same equation applies as in Case 1 but in a precise sense marked by the absence of an approximation sign:

$$\frac{1}{K_{oy}} = \frac{1}{k_y} \tag{1.22c}$$

Case 5: What was said for Case 4 applies here as well, but the transfer this time is from the gas to the liquid. This follows from the fact that gas-phase concentration decreases in the positive direction. The same equation applies as in Case 4 — that is,

$$\frac{1}{K_{oy}} = \frac{1}{k_y} \tag{1.22c}$$

The question is now asked whether the results would still be the same if the gas-phase profiles had in each case been located *above* the liquid-phase counterparts. The answer is yes; transfer would still take place as indicated before. The only change would be in the equilibrium solubility of the gas, which would now be lower than before.

Comments

The answers being sought here require a firm understanding of the principles involved. There were several pitfalls to be avoided on the way. We may be led to think, for example, that transfer from gas to liquid is not possible because of the higher level of x_B. That level, as we had seen, merely indicates a high equilibrium solubility of the gas but does not preclude movement from gas to liquid. A second pitfall lies in the tendency to focus on transfer in the positive direction only. Evidently, movement in the opposite direction is equally possible and should be kept in mind as an alternative.

Practice Problems

1.1 Gradient-Driven Processes

a. Do the flux relations listed in Table 1.1 apply to time-varying processes?

b. Under what conditions is the concentration gradient in Fick's law a constant, and when does it become a variable?

c. Under what conditions does the diffusivity become a variable?

d. The gradients that appear in Fick's law and Fourier's law, dC/dx and dT/dx, are normally negative, because the potentials of C and T diminish in the direction of increasing values of x. Do these gradients ever become positive for Fickian diffusion?

e. Under steady flow conditions in a cylindrical pipe, flow rate Q and velocity v_x in Poiseuille's law are constants and, hence, so is the gradient dp/dx which becomes $\Delta p/\Delta x$. This is the form commonly

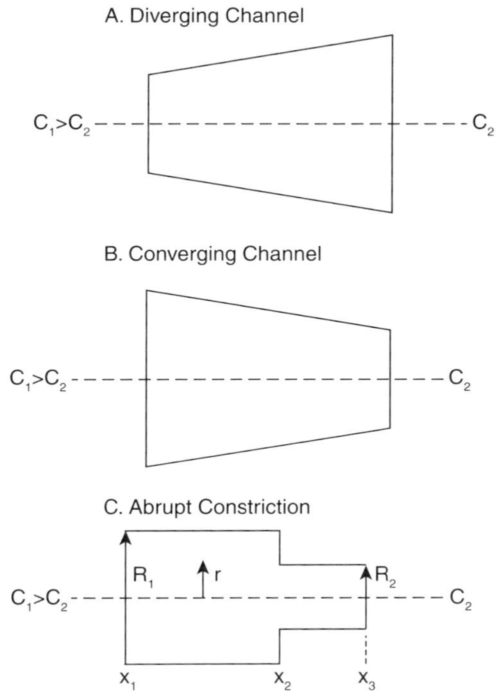

FIGURE 1.14
Diffusion in variable area geometries.

encountered in pipeline calculations. Can you envisage conditions that would lead to a *variable* gradient?

1.2 Oinology

Give a reason for the following practices well known to wine connoisseurs:

a. Prior to consumption, wine is poured into decanters with wide, flared bottoms. The contents of a standard wine bottle cover the bottom to a height of only about 3 cm.
b. The recently introduced plastic corks are used for the cheaper wines that will be drunk relatively quickly (say, within a year).

1.3 Diffusion in Converging and Diverging Channels

Consider diffusion in the geometries shown in Figure 1.14A and Figure 1.14B. Give a qualitative description of the concentration gradients and profiles that result.

1.4 Diffusion in an Abrupt Constriction

Analyze the diffusion process taking place in a channel with a *jump change* in radius. (Figure 1.14C)

a. Establish an upper limit to the diffusional rate *N/A*.
b. What is the value of the gradient at the vertical constriction wall?
c. Show that diffusion takes place in both the radial and axial directions.
d. Give a qualitative description of the radical and axial concentration profiles.

1.5 The Hypsometric Formula: Greenhouse Gases in the Upper Atmosphere

Equation 1.7b, known as the barometric or hypsometric formula, gives a good description of the pressure distribution in the troposphere (i.e., up to a height of 11 km), provided the local temperature is used in the formula. That temperature is given by the empirical relation $T(K) = 288.15 - 6.5 \times 10^{-3}$ (K/m)h.

At a height of 3 km, the temperature has an average value of 270 K.

a. Calculate the atmospheric pressure at this height, assuming that it is 100 kPa at ground level.
b. If the carbon dioxide concentration at the ground is 380 ppm, what will it be 3 km above the earth's surface? Comment on the result for Part b.

Answer:

a. 68 kPa
b. 313 ppm

1.6 U^{235} Enrichment by Centrifuge

In a typical uranium enrichment cascade operating at steady state, the feed coming into a centrifuge is exactly balanced by equal amounts of gas withdrawn from the periphery and the center. The peripheral stream is sent "backward" to a parallel array of centrifuges for further *stripping*, while the central portion moves forward into a similar succession of units for further *enrichment*.

Assuming a 50–50 split of the product streams and a separation factor $\alpha = 1.18$, calculate the mole fraction y of U^{235} in the central product. Feed is assumed to enter at the natural abundance level of $y_F = 7 \times 10^{-3}$. (*Hint*: Make a component and total mass balance.)

Answer:

7.6×10^{-3}

1.7 Diffusional Concentration Profiles in a Spherical Geometry

Derive the concentration profile $C = f(r)$ in a spherical shell, which arises when a solute with uniform internal concentration C_i diffuses through the shell to an external medium held at a fixed concentration C_o.

Answer:

$$C(r) = C_i - (C_i - C_o)\frac{(1 - r_i / r_o)}{(1 - r_i / r)}$$

What is unusual about this result?

1.8 Resistances in Series: The Transdermal Patch

Transdermal patches are devices attached to the skin which are designed to deliver a medication to the body at a controlled rate and over an extended period of time. The medication, which is maintained at a constant internal concentration, passes through a containing membrane, the adhesive, and the skin into the neighboring blood capillaries (Figure 1.15). A well-known example of such devices is the Nicoderm patch.

In order to maintain a constant rate and not be hostage to variations in skin permeability, one often aims to locate the controlling resistance *within* the membrane itself.

Suppose that a particular medication has a membrane diffusivity of 10^{-7} cm^2/s and its permeability through the skin varies over the range 10^{-5} – 2×10^{-4} cm/s. What should be the membrane thickness so that it contains 80% of the overall resistance?

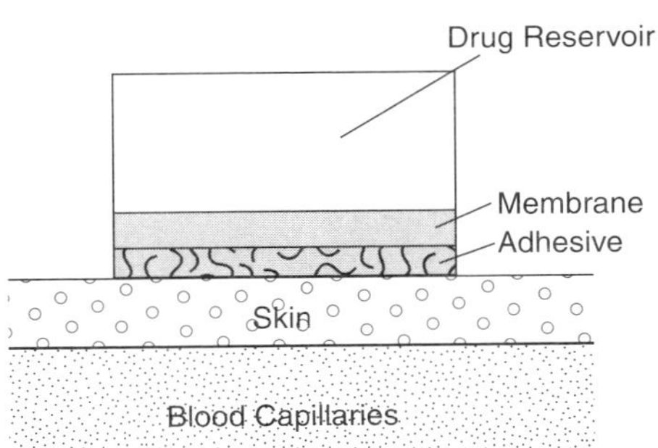

FIGURE 1.15
The transdermal patch.

Answer:

0.04 cm

1.9 Diffusion through a Stagnant Film

a. In Figure 1.9b, the stagnant component *B* exhibits a considerable concentration gradient. Why, then, is there no Fickian diffusion along it?

b. Show that the bulk-flow component xANA/A of the diffusing species is given by

$$x_A N_A / A = D_{AB} \frac{dC_B}{dz}$$

(*Hint*: Use the answer to Part a.)

c. Show that for dilute gas mixtures (i.e., low concentrations of the diffusing species *A*) the log-mean pressure difference p_{BM} tends to the total pressure P_T.

1.10 Conversion Factors for Mass Transfer Coefficients

Prove the following relation:

$$k_G' = k_G \frac{p_{BM}}{p_T}$$

given in Table 1.5 and show that for dilute gases, $k_G' \doteq k_G$.

1.11 More about Driving Forces and Transport Coefficients

a. Would you expect the effective film thickness to increase or diminish with an increase in velocity of the flowing fluid?

b. Liquid-phase diffusivities are some four orders of magnitude smaller than the corresponding gas-phase diffusivities. Would you therefore expect all gas–liquid operations to be liquid-film controlled?

c. Gases with high solubility have a low slope *m* of the equilibrium curve $p_A^* = f(C_A)$. Does this imply that the liquid-phase driving force is small? (Hint: Consult Figure 1.11.)

d. Consider evaporation from a falling water droplet. Would you expect the local mass transfer coefficient to vary with angular position φ? If so, where would it be highest, and where would it be the lowest?

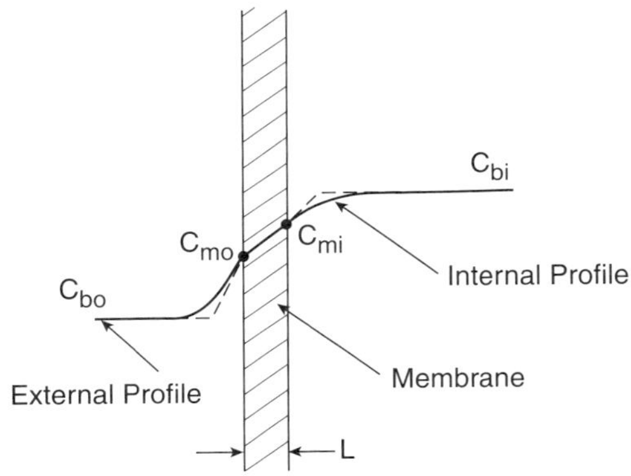

FIGURE 1.16
Mass transport through a membrane (Practice Problem 1.12).

1.12 Mass Transfer through a Membrane

In the most general case of mass transfer through a permeable membrane, transport proceeds from an internal bulk concentration C_{bi}, through an internal film resistance, to an internal membrane concentration C_{mi}, then passes through the membrane of thickness L to an external membrane concentration C_{mo} and from there through an external film resistance to its final destination, C_{bo} (Figure 1.16). Note that concentrations are continuous (i.e., no interfacial equilibrium needs to be invoked). Using the principle of additivity of resistances, derive the following relation:

$$\frac{1}{K_{OC}} = \frac{1}{k_i} + \frac{1}{k_o} + \frac{L}{D_m}$$

where K_{OC} = overall mass transfer coefficient, and k_i and k_o are the internal and external film coefficients.

1.13 Mass Transfer between Ocean Waters and the Atmosphere

The following have been determined for the transport of carbon monoxide (CO) between ocean waters and air:

Overall Mass Transfer Coefficient: K_{OL} = 20 cm/h

Equilibrium Constant: H = 62,000 atm/mol fraction

Global Ocean Surface Area: A = 3.6×10^{18} cm^2

Mean Concentration of CO in Air: 0.13 ppm by volume

Mean Concentration of CO in Water: 6×10^{-8} cc STP/cc H_2O

a. In what direction is the transfer of carbon monoxide?
b. What is the transfer rate in g/yr? (Hint: Transform the carbon monoxide concentrations to units of mol/cm^3, and derive the equilibrium concentration C^*.)

Answer:

4.5×10^{13} g/yr

1.14 Neutron Diffusion and Critical Mass

A nuclear chain reaction of a fissionable material can in principle occur whenever the number of neutrons produced by fission per unit volume of material exceeds the number of neutrons absorbed unproductively by the fuel matrix. Suppose that a particular material has been found to satisfy this criterion. Why then is it necessary to exceed a certain critical mass of the material before the chain reaction can proceed unhindered? (*Hint*: Neutron transport obeys Fick's law of diffusion — that is, its rate is proportional to the neutron density gradient per unit cross-sectional area.)

1.15 The Blood Coagulation Trigger

Blood coagulation, which takes place at the site of an injury or in response to exposure to a foreign surface, is triggered by a series of enzymatic reactions that culminate in the production of the enzyme thrombin. Thrombin is responsible for the formation of fibrin, which together with the platelets present in blood is a key ingredient of a *blood clot*. Most of these events take place at the contact site.

Assume that in response to an event requiring blood coagulation, thrombin is produced in accordance with an overall first-order rate $k_r C$. Its concentration in the flowing blood can be taken as constant. Show that the likelihood of coagulation increases dramatically as k_r approaches the value of the mass transfer coefficient from the site to the flowing blood.

Note: To prevent coagulation, grafts are suitably modified chemically or coated to render them "inert."

1.16 A Fermi Problem: Estimation of the Mass Transfer Coefficient for a Hand Dryer

The familiar hand dryer acts by blowing a high-velocity warm airstream over the hands. The task is to estimate the evaporation rate N/A, and from it the associated mass transfer coefficient k_C.

Guessed values:

Thickness of water layer: $L = 0.1$ mm
Drying time: $t = 100$ s
Temperature (hand): 35°C
Vapor pressure: 35 mmHg

Note: The type of problem addressed here is often referred to as a Fermi problem. Nobel laureate Enrico Fermi used to regale his students by showing them how to estimate the number of piano tuners in Chicago. He was also responsible for estimating the yield of the first atomic bomb immediately after the explosion and long before the pertinent instrument readings had been analyzed. He did this by dropping pieces of paper into the path of the oncoming shock wave and pacing off the distance of entrainment. His estimate came remarkably close to the actual value of 10,000 tons of TNT.

2

Modeling Mass Transport: The Mass Balances

Problems involving mass transport, and the solution and analysis of such problems, almost always require the formulation of a mathematical model of the process. The term *model*, as used here, refers to the equation, or set of equations, that describes the physical system or process under consideration. Such models were already encountered, at a modest level, in the solution of Illustration 1.4 and Illustration 1.5. In both these examples, a single expression, Fick's law of diffusion, was applied and integrated to model the process. No other basic equations were required, and we were able to proceed to a solution of the problem without invoking any additional principles.

This simple procedure, requiring only the application of a single and established expression, is the exception rather than the rule in problems involving mass transport. In the vast majority of cases, we have to draw on additional tools to complete the mathematical formulation of the process. The tools required comprise various forms of the law of conservation of mass, supplemented by what we term *auxiliary relations*. These latter relations are largely empirical in nature and include the equations of transport seen in the previous chapter, as well as expressions describing chemical reaction rates and phase equilibria.

Frequently, one has to make use of more than one conservation law. Thus, if mass transport is accompanied by heat effects, we may have to invoke both the law of conservation of mass and the law of conservation of energy. We will see some examples of this dual case in Chapter 9. The law of conservation of momentum, on the other hand, is much more sparingly used in mass transport problems and can, for the purposes of this book, be set aside.

2.1 The Compartment or Stirred Tank and the One-Dimensional Pipe

The question that has to be addressed now is how these laws and auxiliary relations should be cast into mathematical expressions (i.e., into mathematical models). We consider, for this purpose, the two physical entities shown in Figure 2.1a, which are widely used to model transport of mass and energy,

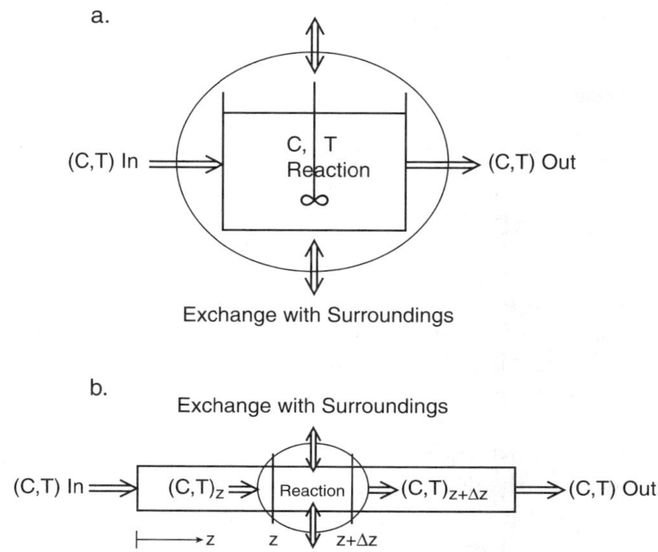

FIGURE 2.1
Two basic physical models: (a) the stirred tank or compartment and (b) the one-dimensional pipe.

and to a lesser degree, momentum. Figure 2.1a represents a well-stirred tank, also often referred to as a compartment. In the most general case, mass or energy flow by bulk movement into and out of the tank is generated or consumed by chemical reactions or is exchanged with the surroundings. Within the tank or compartment, concentrations, temperature, and the physical properties are generally uniform — they vary at most with time, and not at all with distance. To apply the law of conservation of mass to the system, we take an inventory of the mass of a particular species, or make what is termed a *mass* or *material balance*. To do this, we argue that for mass to be conserved, the difference between input and output must equal the change undergone by the tank contents. We include under *input* the mass generated by a chemical reaction within the tank, or that received from the surroundings, and under *output* the mass consumed by reaction or that transferred to the surroundings. This leads to the following general scheme:

$$\text{Rate of mass in} \quad - \quad \text{Rate of mass out} = \begin{array}{c}\text{Rate of change}\\ \text{of mass in}\\ \text{the tank}\end{array} = \frac{d}{dt}(\text{mass}) \qquad (2.1a)$$

$$\begin{bmatrix}\text{by bulk flow,}\\ \text{by reaction,}\\ \text{by transfer}\\ \text{from the surroundings}\end{bmatrix} \quad \begin{bmatrix}\text{by bulk flow,}\\ \text{by reaction,}\\ \text{by transfer}\\ \text{to the surroundings}\end{bmatrix}$$

A similar procedure can be applied to the device we call a one-dimensional (1-D) pipe, shown in Figure 2.1b. In this configuration, the properties of the system are time invariant, but they vary with distance in the direction of flow. The system is said to be at steady state and distributed in one spatial coordinate. This is the exact reverse of the conditions that prevailed in a compartment. The physical phenomena that take place in the two cases, however, are similar. Mass is again transported by bulk flow, enters and leaves the device by exchange with the surroundings, and is generated or consumed by chemical reaction. The only difference here is that mass can also enter and leave by diffusion, which was not the case in the compartmental model.

To obtain an expression for the steady-state distribution of the system variables, the mass balance must now be taken over an incremental element extending from z to $z + \Delta z$. This is necessary in order to bring the distance variable into the model. The increment Δz need not worry us here because we will ultimately allow it to go to zero, thereby transforming the original difference equation into the more familiar form of a differential equation. This leads to the following representation of the mass balance:

$$\underset{\left[\begin{array}{c}\text{by bulk flow and diffusion,}\\ \text{by reaction,}\\ \text{by transfer}\\ \text{from the surroundings}\end{array}\right]}{\begin{array}{c}\text{Rate of mass in at } z\\ \text{and over } \Delta z\end{array}} - \underset{\left[\begin{array}{c}\text{by bulk flow and diffusion,}\\ \text{by reaction,}\\ \text{by transfer}\\ \text{to the surroundings}\end{array}\right]}{\begin{array}{c}\text{Rate of mass out at } z + \Delta z\\ \text{and over } \Delta z\end{array}} = 0 \qquad (2.1b)$$

Expression 2.1a and Expression 2.1b are our principal starting tools for modeling mass transport. We make extensive use of them in the sections and chapters that follow. To convey an idea of how these tools are applied in practice, we start with a simple compartmental model drawn from biology and follow this up with two cases, one compartmental, the other distributed, drawn from the field of environmental engineering.

Illustration 2.1: A Simple Biological Compartment: The Dye Dilution Method

The case considered here involves, in its most general form, the injection of a square tracer pulse into a flowing fluid entering a well-mixed compartment where it is diluted. It then exits the compartment as an attenuated concentration pulse that is monitored by repeated sampling (see Figure 2.2). The data collected can be used to estimate flow rates and compartment volumes or, at a more sophisticated level, to gauge the degree of mixing ("residence time distribution").

The first tracers used in physiological investigations were inert dyes that were injected "upstream" from a compartment (e.g., the heart) and

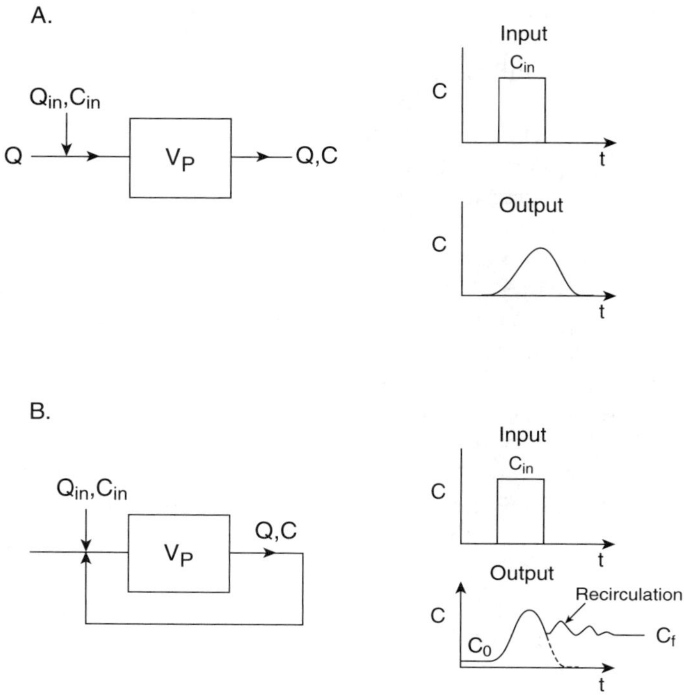

FIGURE 2.2
Plasma volume, *V*, and cardiac output, *Q*, by dye dilution: (a) without recirculation and (b) with recirculation.

monitored "downstream" by repeated sampling. They yielded, among other results, the first estimates of *plasma volume* and *cardiac output*. The data typically consisted of a strong first peak, representing the first pass through the compartment, followed by several secondary peaks due to recirculation, which ultimately subside to a steady-state concentration C_t. At this point, the dye is completely mixed within the recirculating plasma. The plasma volume V_p then follows from a simple dye mass balance, as follows:

Mass of dye injected = Mass of dye in plasma at steady state

$$m_d = V_p\, C_t \tag{2.2a}$$

Thus, the plasma volume is given simply by the total mass of dye injected, divided by its final steady-state concentration. Because blood volume "turns over" once every minute, a short time interval is sufficient to obtain reliable values of C_t.

In determining cardiac output, use is made only of the first and principal peak, which is extrapolated to the time axis (Figure 2.1b). In that form the peak represents the hypothetical case of *no recirculation* (i.e., the dye makes a single passage through the heart chamber, exits, and vanishes). A simple unsteady dye balance over the heart chamber will then yield its output (i.e., the blood flow rate). We have

$$\text{Rate of dye in} - \text{Rate of dye out} = \begin{array}{c}\text{Rate of change}\\ \text{of contents}\end{array}$$

$$QC_u - QC_d = V\frac{dC_d}{dt} \tag{2.2c}$$

where the subscripts u and d refer to upstream and downstream conditions and Q is the volumetric flow rate. Integrating by separation of variables, we obtain

$$VC_d = Q\int_0^t (C_d - C_u)dt \tag{2.2d}$$

and in the limit $t \to \infty$ (i.e., when C_d has become zero),

$$0 = Q\int_0^\infty (C_d - C_u)dt \tag{2.2e}$$

The upstream integral in Equation 2.2d simply represents the total amount of dye injected — that is,

$$Q\int_0^\infty C_u dt = m_d \tag{2.2f}$$

Substituting this expression into Equation 2.2e, we obtain the final form:

Cardiac output

$$Q = \frac{m_d}{\int_0^\infty C_d dt} \tag{2.2g}$$

Q is the total mass of injected dye divided by the area under the first extrapolated peak.

It speaks to the genius of these early investigators that they were able, with the crude tools then available, to determine two of the most important physiological parameters in a single experiment.

Illustration 2.2: Evaporation of a Solute to the Atmosphere

Consider a body of water such as a lake, which receives an inflow of Q m³/s of water and discharges it at the same rate (Figure 2.2a). The volume of water in the lake is constant. Dissolved in the intake is a pollutant, such as a pesticide, at a concentration level of C_f (kg/m³). The lake is initially devoid of any contaminant.

During its passage through the basin, which is assumed to be uniform in concentration, the pollutant is partially transferred to the atmosphere by evaporation and leaves at the concentration C (kg/m³), which prevails in the basin. The task here is to determine the time course of the pollutant concentration.

Drawing on the scheme represented by Equation 2.1a and assuming the basin contents to be well mixed, we write

Rate of pollutant in	–	Rate of pollutant out	=	Rate of change of pollutant content	
QC_f	–	$[QC + N]$	=	$\frac{d}{dt}VC$	(2.3a)

If one assumes mass transfer to be controlled by the water phase, we can write

$$N = K_{OL}A(C - C^*) \tag{2.3b}$$

where A = surface area (m²), K_{OL} = overall mass transfer coefficient (m/s), and C^* is the pollutant concentration in the water which is assumed to be in equilibrium with the concentration in the atmosphere (kg/m³). Equation 2.3a now becomes

$$(QC_f + K_{OL}AC^*) - (QC + K_{OL}AC) = V\frac{dC}{dt} \tag{2.3c}$$

which, after separating variables and formally integrating the result, yields

$$\int_0^C \frac{VdC}{(QC_f + K_{OL}AC^*) - (Q + K_{OL}A)C} = \int_0^t dt \tag{2.3d}$$

and, consequently,

$$\frac{V}{(Q + K_{OL}A)}\ln\frac{QC_f + K_{OL}AC^*}{(QC_f + K_{OL}AC^*) - (Q + K_{OL}A)C} = t \tag{2.3e}$$

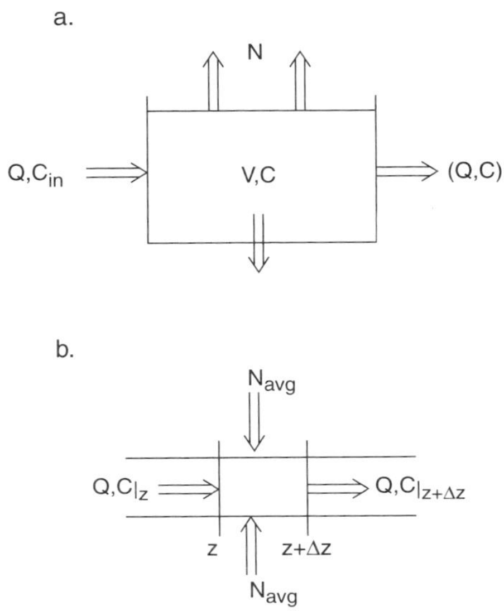

FIGURE 2.3
Two environmental models: (a) evaporation from a basin and (b) reaeration of a river.

This expression can be written in the equivalent exponential form:

$$C = \frac{QC_f + K_{OL}AC^*}{Q + K_{OL}A}\left[1 - \exp\left(-\frac{Q + K_{OL}A}{V}t\right)\right] \tag{2.3f}$$

which gives us the desired final result, the time dependence of the pollutant concentration in the water basin.

Comments

Before setting aside the solution of a model, it is always useful to verify its validity and to examine it for unusual features. This is done by first setting time $t = 0$. This yields $C = 0$ as required. Next we allow t to go to infinity. This reduces the exponential term to zero, and we obtain

$$C_{ss} = \frac{QC_f + K_{OL}AC^*}{Q + K_{OL}A} \tag{2.3g}$$

Equation 2.3g represents the ultimate steady-state value of the pollutant concentration C_{ss}, which is attained after a long period of time. Note that this result is also obtained by setting the time derivative in Equation 2.3c equal to zero. The entire solution curve is shown in Figure 2.3a and demonstrates the asymptotic approach to steady-state conditions as time goes to infinity.

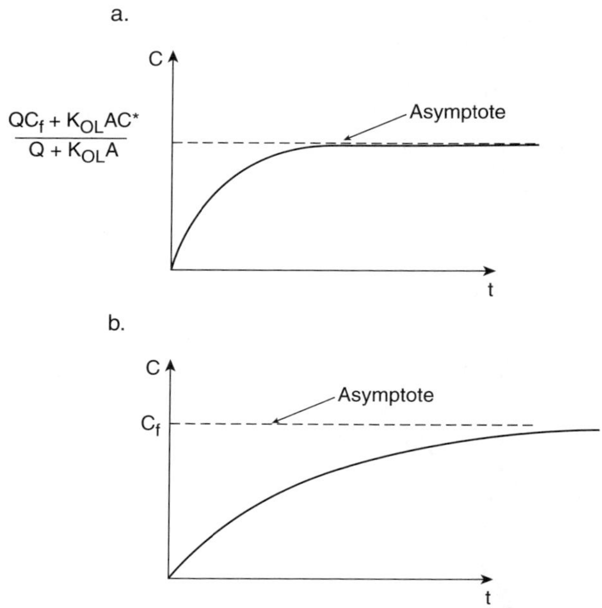

FIGURE 2.4
Time course of a pollutant concentration in a water basin: (a) with evaporative loss and (b) without evaporative loss.

A third way of verifying the validity of the solution is to eliminate the evaporative term N in Equation 2.3a or to set $K_{OL} = 0$. Equation 2.3f and Equation 2.3g then become

$$C / C_f = 1 - \exp\left(-\frac{Q}{V}t\right) \tag{2.4a}$$

and for the steady state, $t \to \infty$,

$$C_{ss} = C_f \tag{2.4b}$$

Equation 2.4a and Equation 2.4b represent the response of pollutant concentration in the water basin to a jump increase in the inflow from $C = 0$ to $C = C_f$ (Figure 2.2b). The difference in the two responses lies mainly in the time of approach to steady state. This can be shown by computing the time required to attain 95% of the respective steady-state values. For Figure 2.4a with evaporative loss, we obtain by division of Equation 2.3f and Equation 2.3g,

$$\frac{C}{C_{ss}} = 0.95 = 1 - \exp\left(-\frac{Q + K_{OL}A}{V}t_{0.95}\right) \tag{2.5a}$$

and, consequently,

$$t_{0.95} = -\frac{V}{Q + K_{OL}A}\ln 0.05 = 3.0\frac{V}{Q + K_{OL}A} \quad (2.5b)$$

For Figure 2.3b (no evaporative loss), we obtain the following by setting $K_{OL} = 0$:

$$t_{0.95} = 3.0\ V/Q \quad (2.5c)$$

The result shows that the approach to steady state is faster with evaporation loss than when there is none. This was not immediately anticipated on intuitive grounds and demonstrates the power of models to reveal the unexpected, or at least to resolve uncertainties.

Equation 2.5b and Equation 2.5c reveal another interesting feature of system behavior. To show this, we recast the equations in the following form:

FIGURE 2.3A

$$\frac{Q\,t_{0.95}}{V} = \frac{3.0}{1 + K_{OL}A\,/\,Q} \quad (2.5d)$$

FIGURE 2.3B

$$\frac{Q\,t_{0.95}}{V} = 3.0 \quad (2.5e)$$

Because $Qt_{0.95}$ equals the total volume that has passed through, it is apparent that the ratio $Qt_{0.95}/V$ represents the number of volume changes, or turnover, undergone by the basin in the course of attaining 95% of the steady-state pollutant level. For Figure 2.3b, three volume changes are required, whereas for Figure 2.3a, the turnover number is less than three. Depending on the size of the basin, this may take from days to weeks or even years.

How realistic is the model we just presented in an environmental or "real-world" context? What are its limitations? We start by noting that the model assumes uniform distribution of the pollutant within the basin. No account is taken of dead water zones or of portions of the flow that may proceed rapidly from inlet to outlet, in effect short-circuiting the mixing process. The best one can expect from the model is that it will give us the course in time of the *average* concentration in the basin, ignoring local high and lows. This is useful information to have, as it provides us with a semiquantitative time frame for the contamination process.

A second limitation resides in the fact that we confined pollutant losses to those that occur by evaporation to the atmosphere. This ignores the role played by bottom sediments as well as solids suspended in the water of the basin in removing solute by *adsorption*, and of possible *biodegradation* of the solute by bacterial action. By ignoring these processes, we have in effect set an upper limit to the pollutant concentration in the water. In other words, things will not be as bad as our model predicts, at least as far as the aqueous phase is concerned.

We turn in the next illustration to an examination of the reverse process, that of uptake of a substance by a body of water. The substance here is a benign one, namely, oxygen, which is taken up by the water from the atmosphere. This process is termed *reaeration*, and it is highly desirable, as it helps maintain aquatic life and aids in the biodegradation of objectionable substances. Also, the body of water is assumed to be in steady flow, such as a river. The system is no longer considered to be well mixed except in the vertical direction, and we can expect a steady increase in oxygen concentration in the direction of flow. This calls for the use of our 1-D pipe model (i.e., one in which the mass balance is performed over an increment Δz in the direction of flow — see Figure 2.1b). There are no variations with time, and distance z now becomes the independent variable (Figure 2.2b).

Illustration 2.3: Reaeration of a River

Consider a river that has, at a point we refer to as $z = 0$, an oxygen content C_o that is below the saturation solubility of oxygen. This means that oxygen concentration will steadily increase in the direction of flow due to uptake from the atmosphere and will ultimately, as $z \to \infty$, attain the equilibrium solubility of oxygen, C^*. Application of Equation 2.2 yields, in the first instance,

$$\begin{matrix} \text{Rate of } O_2 \text{ in at } z \\ \text{and over } \Delta z \end{matrix} - \begin{matrix} \text{Rate of } O_2 \text{ in at } z + \Delta z \\ \text{and over } \Delta z \end{matrix} = 0$$

$$(QC_z + N_{avg}) \quad - \quad (QC_{z+\Delta z}) \quad = 0 \tag{2.6a}$$

If we assume the liquid-phase resistance to be controlling, N_{avg} will be of the form

$$N_{avg} = K_{OL}A\,(C^* \quad C)_{avg} \tag{2.6b}$$

where C^* is the aqueous oxygen concentration that would be in equilibrium with the atmosphere.

A second question concerns the definition of the interfacial area A. It is relatively rare to find that this quantity coincides with the area of a flat surface. More commonly, the river has surface ripples and waves of an unknown interfacial area. The value becomes particularly uncertain in the case of rapids in which air is entrained into the water phase in the form of bubbles or foam, again, of an unknown interfacial area. It has become common practice in these cases to lump this unknown factor into the mass transfer coefficient in the form of an average specific area a with units of square meter (m^2) interfacial area per cubic meter (m^3) river volume. We write

$$N = K_{OL}a(C^* \quad C)_{avg}\, x\Delta V \tag{2.6c}$$

or

$$N = K_{OL}a(C^* \quad C)_{avg}\, A_C\Delta z \tag{2.6d}$$

where A_C is the cross-sectional area of flow of the river, and $K_{OL}a$ is the so-called *volumetric mass transfer coefficient* with units of reciprocal time (s^1). The mass balance (Equation 2.6a) then becomes

$$QC_z - QC_{z+\Delta z} - K_{OL}a(C^* \quad C)_{avg}A_C\Delta z = 0 \tag{2.6e}$$

Dividing by $A_C\Delta z$, and letting Δz go to zero, the difference quotient $\Delta C/\Delta z$ is converted into a derivative, and $(C^* \quad C)_{avg}$ becomes a point quantity:

$$Q/A_C \frac{dC}{dz} + K_{OL}a(C^* - C) = 0 \tag{2.6f}$$

Because Q/A_C equals the river flow velocity v, we can also write

$$v\frac{dC}{dz} + K_{OL}a(C^* - C) = 0 \tag{2.6g}$$

Integrating by separation of variables yields

$$\int_{Co}^{C} \frac{dC}{C^* - C} = \frac{K_{OL}a}{v}\int_0^z dz \tag{2.6h}$$

and, hence,

$$\ln\frac{C^* - C_o}{C^* - C} = \frac{K_{OL}a}{v}z \tag{2.6i}$$

Solving for C leads to the final exponential oxygen distribution:

$$C(z) = C^* - (C^* - C_o)\exp\left(-\frac{K_{OL}a}{v}z\right) \tag{2.6j}$$

Comments

A number of underlying assumptions need to be noted. First, it was assumed that the river cross section, and hence its flow velocity, was constant. In general, there will be some variation in these factors, which can be easily incorporated into the model provided their functional dependence on distance is known or can be estimated. One merely has to recast Equation 2.6h into the form

$$\int_{Co}^{C} \frac{dC}{C^* - C} = K_{OL}a \int_{0}^{1} v(z)dz \tag{2.6k}$$

Second, we neglected any consumption of oxygen in the river (i.e., the use of oxygen by aquatic life and its consumption in the biodegradation of dissolved substances). These two processes will act to delay the reaeration process so that the result (Equation 2.6j) is to be regarded as a lower threshold value. A complete model, which takes account of both processes, is presented in Practice Problem 2.4.

2.2 The Classification of Mass Balances

It will have been noted that a number of different types of mass balances exist, and their use is dictated by the type of process under consideration. Thus, in Illustration 2.2, the mass balance was taken over a finite entity, the water basin, and the balance space in Illustration 2.3 was an incremental quantity Δz, which ultimately shrank to a point in space. A further distinction is the appearance of time in Illustration 2.2 and an absence of spatial coordinates. The reverse was the case in Illustration 2.3. Clearly, both time and spatial geometry determine the type of mass balance that has to be made.

2.2.1 The Role of Balance Space

The space over which a mass balance is taken generally falls into two categories:

1. The space is finite in size. This occurs when the balance is taken over a finite entity such as a tank or compartment, a finite length of pipe, a column, or a sphere. We speak of the balance as being an "integral" or "macroscopic" balance. Balances involving compartments are invariably of this type and lead to either algebraic equations (AEs) or ordinary differential equations (ODEs).
2. The balance is taken over an incremental space element, Δx, Δr, or ΔV. The mass balance equation is then divided by these quantities

and the increments are allowed to go to zero. This reduces the difference quotients to derivatives, and the mass balance now applies to an infinitesimal point in space. We speak in this case of a "difference" or "differential" balance, or alternatively of a "microscopic" or "shell" balance. Such balances arise whenever a variable such as concentration undergoes changes in space. They occur in all systems that fall in the category of the device we termed a 1-D pipe (Figure 2.1b). When the system does not vary with time (i.e., is at steady state), we obtain an ODE. When variations with time occur, the result is a partial differential equation (PDE), because we are now dealing with two independent variables, t and z. Finally, if we discard the simple 1-D pipe for a multidimensional model, the result is again a PDE.

2.2.2 The Role of Time

When a process is time dependent, one speaks of "unsteady," "unsteady-state," or "dynamic" systems and balances. If, on the other hand, there are no variations with time, the process is said to be at steady state. There are several categories of unsteady balances depending on the balance space and time framework we use. They are summarized below.

2.2.2.1 Unsteady Integral Balances

Here the balance is taken over a well-stirred tank or compartment such as the water basin we considered in Illustration 2.1. To describe the process, we use the scheme Rate in Rate out = d/dt contents, which was given by Equation 2.1. It follows from this expression that all unsteady integral balances lead to first-order ODEs. When the time derivative in Equation 2.1 is zero, the system reverts to a steady state, and the result is an AE.

2.2.2.2 Cumulative (Integral) Balances

This is a special type of balance involving time as a variable, which is rarely, if ever, singled out for discussion in textbooks. It consists of considering a finite time interval $(0, t)$ and making an inventory using the mass of a species present originally at time $t = 0$, the mass consumed, or lost to the surroundings over the time interval $(0, t)$, and that left over at the end. We put this formally as follows:

$$\text{Mass initially present } (t = 0) = \text{Mass consumed or lost over } (0,t) + \text{Mass left over at } t \tag{2.7}$$

Cumulative balances invariably lead to AEs. Paradoxically, and in spite of the occurrence of time in the statement (Equation 2.7), they may arise in both

TABLE 2.1

Categories of Balances and the Resulting Equations

Type of Balance	Equation
A. Integral balances	
1. Steady-state balance	AE
2. Unsteady balance	ODE
3. Cumulative balance	AE
B. Differential balances	
1. Steady-state 1-D balance	ODE
2. Unsteady 1-D balance	PDE
3. Steady-state multidimensional balance	PDE
4. Unsteady multidimensional balance	PDE

steady- and unsteady-state processes. We will demonstrate this in a number of illustrations and practice problems.

2.2.2.3 *Unsteady Differential Balances*

It was previously noted that when the 1-D pipe model is applied to an unsteady process, the mass balance will lead to a PDE. We can generalize this into the following statements: Whenever a variable is distributed in both time and distance — that is, varies with t and x (y, z) — the resulting mass balance will be a PDE.

We can summarize these results as follows: A mass balance can be time-independent or time-dependent, and it can be applied over a finite entity or a differential increment of either time or distance. It can further depend on a single space variable, several such variables, or none. Depending on which combination of factors applies to a particular problem, this will result in either an AE, an ODE, or a PDE. These features are summarized in Table 2.1.

2.2.3 Dependent and Independent Variables

Before concluding our discussion, remember the distinction that has to be made between dependent and independent variables. This is necessary in the context of differential equations, which for ODEs generally lead to solutions of the following form:

$$u = f(x) \tag{2.8a}$$

or

$$u = f(t) \tag{2.8b}$$

and in the case of PDEs and rectangular coordinates,

$$u = f(x, y, z) \tag{2.8c}$$

$$u = f(x, t)$$

and

$$u = f(x, y, z, t)$$

Here u is the dependent variable, and x, y, z, and t are the independent variables.

The dependent variable u — also termed a *state variable* — is, for mass transfer operations, usually represented by the concentration of a system, or its total mass. Concentration can be expressed in a variety of ways, the most common being kilogram per cubic meter (kg/m^3) or mole per cubic meter (mol/m^3), or in terms of mole and mass fractions and ratios. Remember that the number of dependent variables equals the number of unknowns. For a system to be fully specified, *the number of equations must therefore equal the number of unknowns*, in other words, the number of dependent variables. The model is then said to be complete.

The independent variables are usually represented by time t and distance x, y, z or, in the case of radial coordinates, by the radial distance variable r. Occasionally, distance may depend on time and then becomes the dependent variable. This is the case, for example, with spherical particles that undergo a change in size due to reaction, dissolution, or deposition of material. The attendant change in mass is then expressed by the derivative

$$\text{Rate of change of mass} = \frac{dm}{dt} = \rho_p \frac{d}{dt} 4/3\pi r^3 = \rho_p 4\pi r^2 \frac{dr}{dt} \tag{2.9}$$

where radial distance r is now the dependent variable.

Another departure from the normal definition of variables occurs when two first-order differential equations are combined by division into a single ODE.

Consider, for example, the system

$$\frac{du}{dt} = f(u,v) \tag{2.10a}$$

$$\frac{dv}{dt} = g(u,v) \tag{2.10b}$$

which may be the result of two unsteady mass balances.

Division of the two equations leads to the result

$$\frac{du}{dv} = \frac{f(u,v)}{g(u,v)} \tag{2.10c}$$

which, given suitable forms of f and g, can be integrated by separation of variables. In the process of dividing the two equations (Equation 2.10a and Equation 2.10b), we eliminated the independent time variable t and replaced it by the former dependent variable v. Such transformations are frequently used to obtain *partial* solutions of systems of simultaneous ODEs (see Illustration 2.9).

Illustration 2.4: The Countercurrent Gas Scrubber: Genesis of Steady Integral and Differential Mass Balances

Gas scrubbers are widely used devices designed to remove impurities or recover valuable substances from gases by contacting them with a suitable solvent, such as water. A gas scrubber typically consists of a cylindrical shell filled with plastic or ceramic particles designed to enhance the contact area between the two phases. Solvent enters the column at the top and trickles down through the packing where it contacts the gas phase, which enters the scrubber at the bottom and flows upward countercurrent to the solvent stream. The purified gas stream leaves the column at the top while used solvent containing the impurity exits at the bottom.

In the present illustration we use the gas scrubber as a vehicle to demonstrate the genesis of steady-state integral and differential balances. Four such balances are shown in Figure 2.5a through Figure 2.5d, with the balance space indicated by an envelope drawn around it. Solvent enters the envelope with a flow rate L_s (kg solvent/m^{2}s) and a solute concentration X_2 (kg solute/kg solvent) and leaves with the same solvent flow rate but an increased concentration X_1. For the gas stream, the corresponding quantities are G_s (kg carrier gas/m^{2}s) and Y_2, Y_1 in units of kilogram solute per kilogram carrier gas. The term *carrier* denotes the gaseous component, which is not absorbed by the solvent. Typically that component is air or some other inert gas, such as hydrogen or nitrogen.

We start by considering the two integral balances shown in Figure 2.5a and Figure 2.5b.

The balance space in these two cases is finite, consisting either of a part of the column (Figure 2.5a) or the entire column (Figure 2.5b). A solute balance for the two cases then leads to the following expression:

FIGURE 2.3A

$$\text{Rate of solute in} - \text{Rate of solute out} = 0$$

$$(L_sX_2 + G_sY) - (L_sX + G_sY_2) = 0 \tag{2.11a}$$

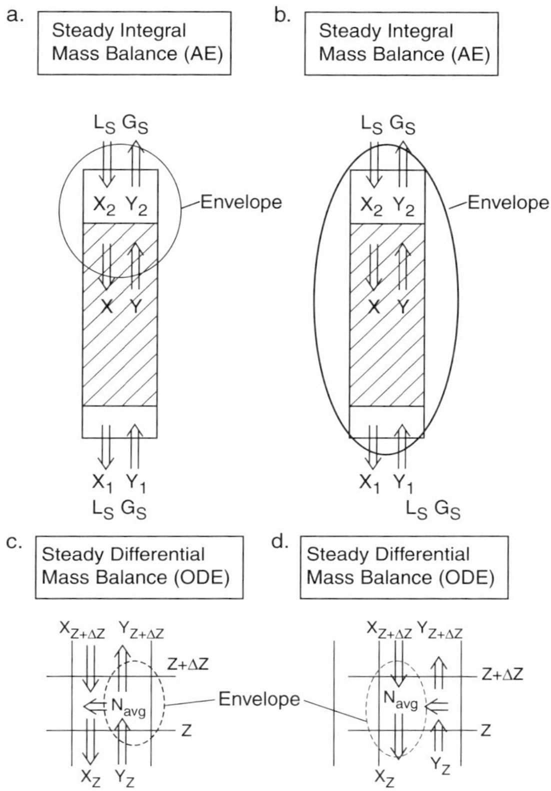

FIGURE 2.5
The packed-gas absorber: (a, b) types of mass balances leading to algebraic equations and (c, d) types of mass balances leading to ordinary differential equations.

FIGURE 2.3B

$$\text{Rate of solute in} - \text{Rate of solute out} = 0$$

$$(L_s X_2 + G_s Y_1) - (L_s X_1 + G_s Y_2) = 0 \tag{2.11b}$$

These two expressions are processed further by casting them in the following difference form:

$$\frac{Y - Y_2}{X - X_2} = \frac{L_s}{G_s} \tag{2.11c}$$

and

$$\frac{Y_1 - Y_2}{X_1 - X_2} = \frac{L_s}{G_s} \tag{2.11d}$$

We will return to these expressions, and those that follow, for a more detailed examination in Chapter 8, Illustration 8.1. It will be shown there that Equation 2.11c and Equation 2.11d represent the so-called *operating lines* used in the graphical representation of scrubber performance. For our present purposes, we limit ourselves to the observation that these balances resulted in AEs, as predicted and stipulated in Table 2.1. We note further that neither Equation 2.11c nor Equation 2.11d contains the distance variable z. They can therefore tell us nothing about the concentration variations as a function of column height and cannot help us establish the size of column required to effect a reduction of solute content from Y_1 to Y_2.

We must, for those purposes, turn to differential balances that upon integration will yield the desired functional dependence of solute concentration as column height — that is, $Y = f(z)$. Two such balances are sketched in Figure 2.5c and Figure 2.5d. To set up the corresponding equations, we write the following for the case of the gas-phase differential balance (Figure 2.5c):

$$\text{Rate of solute in} - \text{Rate of solute out} = 0$$

$$(G_s Y_z) - (G_x Y_{z+\Delta z} + N_{avg}) = 0 \tag{2.12a}$$

For the liquid-phase balance (Figure 2.4d),

$$(L_s X_{z+\Delta z} + N_{avg}) - (L_s X_z) = 0 \tag{2.12b}$$

We must next formulate an expression for the mass transfer rate N_{avg}, and here we encounter the same difficulty we had in Illustration 2.3 dealing with the reaeration of rivers. In both cases, the interfacial area is unknown; therefore, we must resort again to the use of a volumetric mass transfer coefficient $K_o a$ where the unknown interfacial area a (m^2/m^3 column volume) is lumped together with K_o. If the gas phase is assumed to be controlling, we can write

$$N_{avg} = K_{oY} a(Y - Y^*)_{avg} A_C \Delta z) \tag{2.12c}$$

where A_C is the column cross-sectional area, and Y^* is the gas-phase solute content in equilibrium with the liquid-phase concentration X.

Introducing this expression into Equation 2.12a and Equation 2.12b, dividing by $A_C\Delta z$, and letting Δz go to zero, we obtain the twin result:

$$G_s \frac{dY}{dz} + K_{oY} a(Y - Y^*) = 0 \tag{2.12d}$$

and

$$L_s \frac{dX}{dz} + K_{oY} a(Y - Y^*) = 0 \tag{2.12e}$$

where G_s and L_s are the mass velocities of carrier gas and solvent, respectively, with units of kilogram per second square meter (kg/s m^2) column cross sectional area.

Because the two equations, 2.12d and 2.12e, contain three dependent variables, X, Y, and Y^*, we require a third relation to complete the model. This will be given by the equilibrium relation, which can be written in the general form:

$$Y^* = f(X) \tag{2.12f}$$

The model, consisting of the three equations (Equation 2.12d, Equation 2.12e, and Equation 2.12f), can now be said to be complete. We return to it in Illustration 8.1, where the solution to this model is discussed. We also examine the role of the integral balances (Equation 2.11c and Equation 2.11d), which, at the moment at least, are seemingly adrift with no apparent use in modeling the system.

Comments

This example was intended to draw attention to the multitude of mass balances that can be applied even in cases of only modest complexity. In fact, there are three additional balances that can be performed on the scrubber:

DIFFERENTIAL MASS BALANCE OVER BOTH PHASES

$$\text{Rate of solute in} - \text{Rate of solute out} = 0$$

$$(L_s X_{z+\Delta z} + G_s Y_z) - (L_s X_z + G_s Y_{z+\Delta z}) = 0 \tag{2.12g}$$

INTEGRAL MASS BALANCE OVER THE GAS PHASE

$$\text{Rate of solute in} - \text{Rate of solute out} = 0$$

$$(G_s Y_1) - (G_s Y_2 + N_{Tot}) = 0 \tag{2.12h}$$

INTEGRAL MASS BALANCE OVER THE LIQUID PHASE

$$\text{Rate of solute in} - \text{Rate of solute out} = 0$$

$$(L_s X_2 + N_{Tot}) - (L_s X_1) = 0 \tag{2.12i}$$

This profusion of balances will have made it clear that the choice of the proper balances and balance space is often not a straightforward one and calls for good judgment or leads to some trial-and-error work.

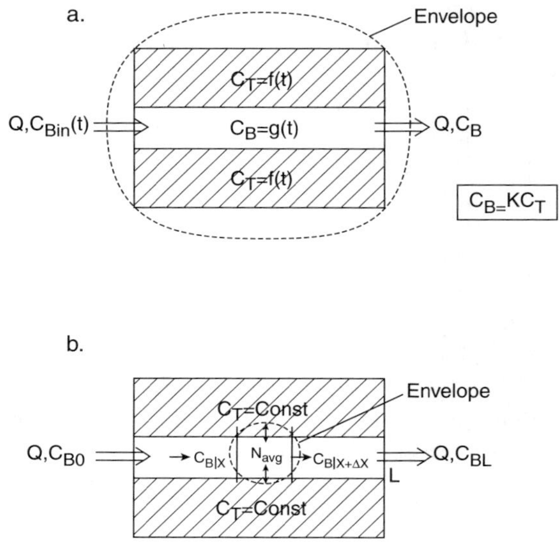

FIGURE 2.6
A well-mixed tissue region surrounding a blood vessel: (a) uniform concentration in blood and (b) concentration in blood distributed and in quasi-steady state.

Illustration 2.5: Two Examples from Biology: The Quasi-Steady-State Assumption

In simple biological models, the human or animal body is assumed to be composed of a number of well-stirred compartments of uniform concentration. These compartments include body fluids (plasma, intercellular fluids) as well as body tissues, such as fat, muscle, and bones, and are described by one or more first-order ODEs in time. The first example presented below is of this type and is depicted in Figure 2.6a.

A second simple model assumes that the concentrations in one or more of the compartments are distributed in one direction but vary only slowly with time so that the time derivative of the concentration is very small and can be neglected. This is referred to as the *quasi-steady-state assumption* and leads to an ODE in distance. The second example presented below belongs to this category and is shown in Figure 2.6b.

1. Two well-mixed compartments in instantaneous equilibrium

The first example to be considered involves a "well-mixed" tissue region surrounding a blood vessel whose contents are likewise taken to be well mixed and uniform in concentration. Blood entering and leaving the heart muscle and conveying to it a drug dissolved in the blood can be considered representative of this type of situation. If passage of the solute into the muscle tissue is rapid enough, the two phases may be taken to be at equilibrium at all times. It is common at the low concentrations involved to assume a linear equilibrium relation of the form

$$C_B = KC_T \tag{2.13a}$$

where K is termed the *partition coefficient*. The physical configuration and pertinent variables for the present case are displayed in Figure 2.6a. We proceed to write an integral mass balance over the entire system, as shown in the figure, and we obtain

$$\text{Rate of solute in} - \text{Rate of solute out} = \begin{matrix}\text{Rate of} \\ \text{change of} \\ \text{solute contents}\end{matrix}$$

$$QC_{B\,in}(t) - QC_B = V_T \frac{dC_T}{dt} + V_B \frac{dC_B}{dt} \tag{2.13b}$$

where $C_{B\,in}(t)$ is, in general, an arbitrary function of time; Q = volumetric flow rate in cubic centimeter per second (cm^3/s); and V_B and V_T are the blood and tissue volumes, respectively (cm^3).

Substitution of Equation 2.13a into Equation 2.13b yields a single expression in either blood or tissue concentrations, which can be integrated in straightforward fashion by separation of variables. We obtain for C_B in the first instance

$$QC_{B\,in} - QC_B = \frac{(V_T + V_B)}{K}\frac{dC_B}{dt} \tag{2.13c}$$

For the case where a drug is suddenly introduced at a constant rate into the blood (i.e., C_{Bin} goes from zero to a constant value C_{Bo} at $t = 0$), integration results in

$$\int_0^{C_B} \frac{dC_B}{C_{Bo} - C_B} = \frac{V_T + V_B}{KQ}\int_0^t dt \tag{2.13d}$$

and, consequently,

$$-\ln\frac{C_{Bo} - C_B}{C_{Bo}} = \frac{V_T + V_B}{KQ}t \tag{2.13e}$$

or, equivalently in exponential form,

$$C_B = C_{Bo}\left[1 - \exp\left(-\frac{QKt}{V_B + V_T}\right)\right] \tag{2.13f}$$

Comments

A first comment concerns the use of the model solutions, Equation 2.13e and Equation 2.13f. We start by noting that a semilog plot of Equation 2.13e, $\ln(C_{Bo} - C_B)$ versus t, using measured values of the drug concentration C, yields a slope with the value $(V_T + V_B)/KQ$. In many cases of interest, the blood volume is much smaller than that of the surrounding tissue, $V_T \gg V_B$, so that the slope becomes V_T/KQ. The flow rate Q is usually known from independent measurements (see Illustration 2.1), while the partition coefficient can often be determined *in vitro* (i.e., in the laboratory using extracted tissue). The value of the slope can then be used to determine the unknown tissue volume V_T. Conversely, if tissue volume is known from independent measurements, the slope will yield a value for the partition coefficient $K\alpha$. Note that the tissue concentration is then given by

$$C_T = C_B / K = (C_{Bo} / K)\left[1 - \exp\left(-\frac{QKt}{V_B + V_T}\right)\right] \tag{2.13g}$$

Once these parameters have been established from appropriate experiments, we can use Equation 2.13g to establish the time required to obtain a desired therapeutic concentration.

2. A well-mixed tissue compartment in contact with flowing blood with a varying concentration

The case in question is sketched in Figure 2.6b and involves the exchange of a substance between a tissue region of uniform concentration and blood flowing in a capillary along which concentration varies. The membrane separating the two has a transport resistance represented by the mass transfer coefficient k_m, which equals diffusivity divided by the thickness of the membrane. Let us consider first the case of substance uptake by the tissue. If that substance enters the capillary at a flow rate Q, we can write for the difference element shown in Figure 2.6b,

$$\text{Rate of solute in} - \text{Rate of solute out} = 0$$

$$QC_B\big|_z - (QC_B\big|_{z+\Delta z} + N_{avg}) = 0 \tag{2.14a}$$

or, using an explicit expression for N_{avg} and setting $C_B^* = KC_T$, as given by Equation 2.13a,

$$QC_B\big|_z - QC_B\big|_{z+\Delta z} - k_m P\Delta x(C_B - KC_T) = 0 \tag{2.14b}$$

where P = perimeter of capillary. Here, use has been made of the quasi-steady-state assumption — that is, the process was taken to be slow enough that over a short finite period of observation (say, a few minutes) neither the tissue nor the blood concentrations vary significantly.

Dividing by Δx and going to the limit $\Delta x \to 0$, we obtain the ODE

$$Q\frac{dC_B}{dz} + k_m P(C_B - \alpha C_T) = 0 \tag{2.14c}$$

Integration by separation of variables then leads to

$$\int_{C_{Bo}}^{C_B} \frac{dC_B}{C_B - KC_T} = -\frac{k_m P}{Q}\int_0^L dz \tag{2.14d}$$

and, consequently,

$$\ln\frac{C_B - KC_T}{C_{Bo} - KC_T} = -\frac{k_m A}{Q} \tag{2.14e}$$

or, alternatively,

$$\text{Uptake:} \quad \frac{C_B - KC_T}{C_{Bo} - KC_T^{0}} = \exp\left(-\frac{k_m A}{Q}\right) \tag{2.14f}$$

where A = interfacial area.

For the reverse process (i.e., when the substance is removed by blood that enters the capillary devoid of it), we have

$$C_B / KC_T = 1 - \exp\left(-\frac{k_m A}{Q}\right) \tag{2.14g}$$

This process is referred to as *clearance* of the tissue.

Comments

Expression 2.14f and Expression 2.14g are used particularly for the determination of the mass transfer coefficient, assuming that values of K and Q were determined independently. This can be done in the usual fashion by making a semilog plot of the concentration fraction in Equation 2.14e against the reciprocal $1/Q$ (i.e., by running a series of experiments at different volumetric flow rates Q). For clearance experiments, C_{Bo} is set equal to zero. The slope of the plot then yields the product $k_m A$. Because detailed anatomical information for the determination of the interfacial area A is rarely available, we must be content to deal with the product $k_m A$. That quantity, however, is still

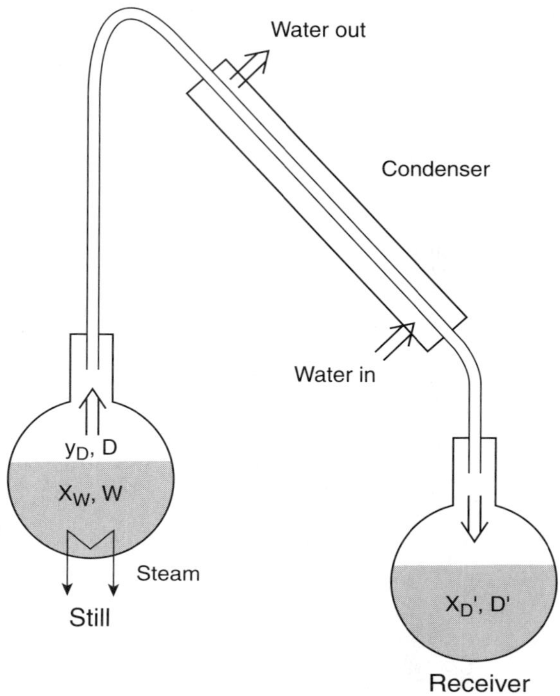

FIGURE 2.7
A batch-distillation apparatus.

highly useful because it provides a measure of the overall permeability of the tissue–capillary interface.

Experiments of the type just described can be used in evaluating the effectiveness of drugs designed to increase the passage of metabolically important materials into tissue. Such drugs are referred to as *vasoactive,* and they act by increasing the area available for transfer or by increasing the permeability of the membrane. This is reflected in an increase in the slope of the plot of Equation 2.14e.

Illustration 2.6: Batch Distillation: An Example of a Cumulative Balance

Batch distillation is practiced with considerable frequency on both laboratory and industrial scales for the purpose of separating and purifying liquid mixtures.

A simple version of the process which comes close to what one uses in laboratory practice is shown in Figure 2.7. The still, shown on the left, is loaded with a liquid mixture of composition x_W^0 and total mass W^0 and subsequently brought to a boil by internal or external heating. The vapor produced at a rate D mol/s is passed into a water-cooled condenser, and the resulting liquid condensate is collected in a receiver shown on the right of Figure 2.7. The composition and total mass in the still at any instant are denoted by x_W and W and that in the receiver by x_D' and D'.

As is usual in systems of some complexity, a number of different balances can be made depending on the choice of balance space. We may choose, for example, to make an unsteady integral balance about the still, or to make a similar balance about the receiver. They can be instantaneous or cumulative in time and can involve total or component mass balances.

The balance considered here is the cumulative balance up to some point in time t. We apply the scheme previously given in Equation 2.7 and write

Mass initially present = Mass left in still + Mass in receiver

$$W^0 = W + D' \tag{2.15a}$$

and for the component mass, assuming a binary system,

$$x_W^0 W^0 = x_W W + x_D' D' \tag{2.15b}$$

where the total mass is expressed in kilogram (kg) or in mole (mol), and the compositions represent mass or mole fractions. The receiver contents D' have been primed to distinguish them from the *rate* of distillation (mol/s), which is commonly given the symbol D.

Cumulative balances such as the two simple expressions (Equation 2.15a and Equation 2.15b) are often overlooked in modeling or are written out without much thought as to their origin. It is important to note that they are quite *independent* of the unsteady integral balances mentioned previously and consequently serve as *additional tools* that can be used to supplement the model equations. Typically, they are used as adjuncts to instantaneous balances in batch distillation. We can, for example, solve Equations 2.15 for the distillate composition and obtain

$$x_D' = \frac{x_W^0 - x_W(1-f)}{f} \tag{2.15c}$$

where $f = 1 - W/W_0$ = fraction distilled. The term f can be related to the final bottoms concentration x_W by the solution of the *independent* unsteady balances, which will be encountered in Chapter 7 (Rayleigh equation). It follows that one only needs to specify x_W (or fraction distilled f) to calculate the distillate composition x_D'. This is a considerable simplification of a complex process which will be used again in Section 7.2.6 to analyze multistage batch distillation.

2.3 Information Obtained from Model Solutions

In several of the illustrations previously presented, the model and its solution were used to extract a particular piece of information about the underlying physical system or process. We now wish to generalize this aspect of modeling and ask the question: What types of information can we expect to find in the solution of a model? The question can be answered in a simple way by stating that *any of the quantities appearing in the solution, including the dependent and independent variables, can be the unknown or the information being sought.*

Time t and distance z are the most common unknowns. We may wish, for example, to determine the time or distance necessary to bring about a prescribed concentration change. Designing a scrubber (i.e., calculating its height z), falls into the latter categories.

Next in importance as unknowns are the parameters, which we can accommodate in the following broad categories:

1. Transport coefficients, such as diffusivities, mass transfer coefficients, and permeabilities
2. Rate constants pertaining to chemical or biological reactions taking place in the system
3. Flow rates, including those in and out of compartments, pipes, and columns; carrier and solvent flow rates
4. Volumes, in particular, those pertaining to a stirred tank or compartment
5. Parameters describing phase equilibria such as partition coefficients and Henry's law constants, relative volatilities, equilibrium solubilities, and vapor pressures
6. Inlet concentrations to compartments, pipes, and columns

There is yet another type of information contained in the model which is obtained by manipulation of the primary results. We call this *derived information*. The manipulations involved typically consist of differentiation or integration of the results. If these come in the form of concentration distributions, we obtain the following derived quantities:

BY DIFFERENTIATION

$$N = -DA\frac{dC}{dz} \quad \text{Diffusional flow} \tag{2.16a}$$

TABLE 2.2

Models, Model Solutions, and the Information Contained in Them

Model	Model Solutions
A. Integral balances	
1. Steady-state balance	$u = f(v, \text{parameters})$
2. Unsteady balance	$u = F(t, \text{parameters})$
3. Cumulative balance	$u = g(v, \text{parameters})$
B. Differential balances	
1. Steady-state 1-D balance	$u = G(z, \text{parameters})$
2. Unsteady-state 1-D balance	$u = g(z, t, \text{parameters})$
3. Steady-state multidimensional balance	$u = H(z, t, \text{parameters})$
4. Unsteady multidimensional balance	$u = k(x, y, z, t, \text{parameters})$
C. Parameters	
1. Transport coefficients	2. Rate constants
3. Flow rates	4. Volumes
5. Phase equilibrium parameters	6. Inlet concentrations
D. Derived information	
1. Diffusional flow	
2. Average concentration	

BY INTEGRATION

$$\bar{C} = \frac{\int C dA}{A} \tag{2.16b}$$

or,

$$\bar{C} = \frac{\int C dV}{V} \tag{2.16c}$$

where $\bar{C}$ = average concentration.

Use is made of these expressions in Illustration 4.9 and Practice Problem 4.11 and again in Section 5.2. They can be regarded as part of the information package provided by the model solution. We summarized both the model solutions and the information contained in them for convenient reference in Table 2.2.

2.4 Setting Up Partial Differential Equations

The mass balances considered up to this point are confined to cases involving a single independent variable, time or distance. Suppose, now, that diffusion takes place from an external medium into a porous cylindrical or spherical particle, which is initially devoid of diffusing spears. Concentrations will then vary both with time and radial distance and in the case of the cylindrical particle, with axial distance as well. This is a system that is distributed in both time and distance and that consequently leads to a PDE (see Table 2.1).

Let us assume that the variations are with respect to one distance variable x and to time t. We previously considered, at the ODE level, distributions in time only, or in distance only. This led to the schemes represented by Equation 2.1 and Equation 2.2. To deal with simultaneous variations in both time and distance, we superpose the two expressions — that is, we write

$$\begin{matrix}\text{Rate of mass in} \\ \text{at } x \text{ and over } \Delta x\end{matrix} - \begin{matrix}\text{Rate of mass out} \\ \text{at } x+\Delta x \text{ and over } \Delta x\end{matrix}$$

$$= \begin{matrix}\text{Rate of change} \\ \text{of mass in } \Delta x\end{matrix} = \left[\frac{\partial}{\partial t}(\text{mass})\right]_{\Delta x} \tag{2.17}$$

On occasion, mass will enter at $x + \Delta x$ and leave at x, in which case the scheme is adjusted accordingly.

This is the formulation that has to be used when the system is distributed over distance x and time t. When variations occur in more than one direction, we merely add appropriate terms to the left side of Equation 2.17, for example, "Rate of mass in at y and over Δy and so on. Note that the time derivative in Equation 2.17 is now a *partial* derivative, because we are dealing with more than one independent variable. The following illustration provides an example of the application of Equation 2.17.

Illustration 2.7: Unsteady Diffusion in One Direction: Fick's Equation

Consider diffusion to be taking place into a rectangular slab, which is infinitely wide in the y and z directions and of finite width L in the x direction. The slab can be a stagnant gas or liquid, or a porous solid, and initially contains the diffusing species at a concentration level $C = C_0$. At time $t = 0$, the two sides of the slab at $x = L/2$ and $x = L/2$ are suddenly exposed to a higher external concentration $C = C_e$. Diffusion into the slab commences, with the diffusing species entering each face and moving simultaneously

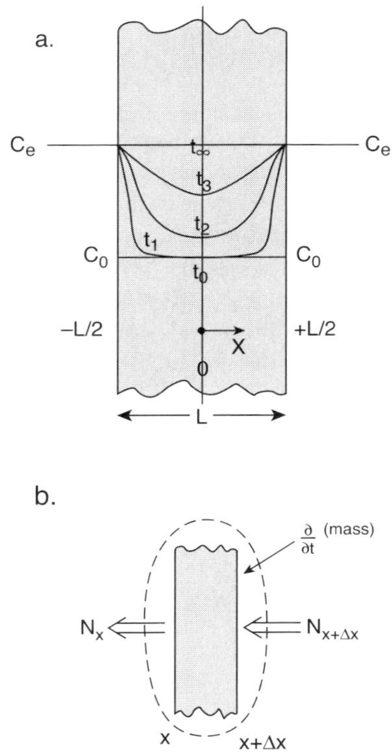

FIGURE 2.8
Diffusion into a slab: (a) development of concentration profiles and (b) difference element for the mass balance.

toward the centerline. The resulting profiles and their development with time are shown in Figure 2.8a. Initially, at $t = t_o$ the profile is flat and uniform at the level $C = C_0$. As time progresses and solute penetrates into the interior, the profile assumes a parabolic shape, which becomes increasingly flatter until at time $t = t_\infty$ it has reached the level of the external concentration $C = C_e$. Diffusion then comes to a halt.

To model this process, we choose a difference element to the right of $x = 0$ (see Figure 2.8b). Solute movement will then be from right to left, entering at the position $x + \Delta x$ and exiting at x. Applying the scheme of Equation 2.18, the following result is obtained:

$$\underset{\text{at } x+\Delta x}{\text{Rate of mass in}} - \underset{\text{at } x}{\text{Rate of mass out}} = \left(\frac{\partial}{\partial t}\text{mass}\right)_{\Delta x} \tag{2.18a}$$

or, introducing Fick's law of diffusion, Equation 1.4a,

$$DA\left.\frac{\partial C}{\partial x}\right|_{x+\Delta x} - DA\left.\frac{\partial C}{\partial x}\right|_{x} = \frac{\partial}{\partial t} A\Delta x C_{\text{avg}} \tag{2.18b}$$

Remember that in the original formulation of Fick's law, concentration diminished in the direction of increasing x (see Figure 1.1a), resulting in a negative gradient. The reverse is true here: Flow is in the direction of diminishing values of x and C, yielding a positive gradient $\partial C/\partial s$. The need for a negative sign in Fick's law is thus removed, and we simply write $N = DA\partial C/\partial x$. This ensures that the diffusional flow will be a positive quantity.

To obtain the final result, Equation 2.18b is divided by $A\Delta x$, and Δx is allowed to go to zero. This yields

$$D\frac{\partial^2 C}{\partial x^2} = \frac{\partial C}{\partial t} \tag{2.18c}$$

This is Fick's equation in one dimension. Its solution yields the time-dependent concentration profiles shown in Figure 2.8a. Illustrations dealing with Fick's law will appear in Chapter 3.

Comments

Although we confined ourselves to Fick's law in one dimension, its extension to three dimensions is straightforward. We merely have to extend Equation 2.18a to three dimensions in the increments Δx, Δy, Δz, which now represent the sides of a cube. The result is

$$D\Delta y\Delta z\left[\left(\frac{\partial C}{\partial x}\right)_{x+\Delta x} - \left(\frac{\partial C}{\partial x}\right)_{x}\right] + D\Delta x\Delta z\left[\left(\frac{\partial C}{\partial y}\right)_{y+\Delta y} - \left(\frac{\partial C}{\partial y}\right)_{y}\right] +$$

$$D\Delta x\Delta y\left[\left(\frac{\partial C}{\partial z}\right)_{z+\Delta z} - \left(\frac{\partial C}{\partial z}\right)_{z}\right] = \frac{\partial}{\partial t}\Delta x\Delta y\Delta z C \tag{2.18d}$$

On dividing by Δx, Δy, Δz and allowing the increments to go to zero, we obtain the three-dimensional version of Fick's law in rectangular coordinates:

$$D\left(\frac{\partial^2 C}{\partial x^2} + \frac{\partial^2 C}{\partial y^2} + \frac{\partial^2 C}{\partial z^2}\right) = \frac{\partial C}{\partial t} \tag{2.18e}$$

For radial, spherical, and cylindrical coordinates, the distance derivatives are somewhat more complicated.

Although we will not, in this text, take up the actual solution of PDEs, it is important to examine one of the tools needed for this purpose — the boundary and initial conditions. Recall that at the ODE level, the number of boundary conditions required equals the *order* of the equation. This concept can be extended to PDEs as follows: Each set of partial derivatives requires a number of conditions equal to its highest order. Thus, the 1-D Fick's equation (Equation 2.18c) requires two boundary conditions for the distance derivative $\partial^2 C/\partial x^2$, and one condition, also called an initial condition, for the time derivative $\partial C/\partial t$. These conditions are obtained from an examination of the physical system. Thus, from Figure 2.8a we have the following conditions:

For the distance derivative:

1. $C = C_e$ at $x = +L/2$ and any time
2. $C = C_e$ at $x = L/2$ and any time

For the time derivative:

3. $C = C_0$ at $t = 0$ and any position

These conditions can be put in the following terse form:

1. $$C(t, L/2) = C_e \quad (2.18f)$$

2. $$C(t, -L/2) = C_e \quad (2.18g)$$

3. $$C(0, x) = C_0 \quad (2.18h)$$

These, and other boundary conditions, will be encountered again in Chapter 4 where they will be used for a number of different processes and geometries. As noted, we do not undertake the actual solutions of the PDEs but rather present them in graphical or tabular form and use them to address a number of practical problems. At this, the reader will wish to breathe a sigh of relief.

Let us next consider a process at steady state, but one in which the dependent variable varies in two directions. This, too, leads to a PDE.

Illustration 2.8: Laminar Flow and Diffusion in a Tube: The Graetz Problem for Mass Transfer

When a solute is released from a soluble tubular wall into a flowing fluid, two cases need to be distinguished:

1. The fluid is in turbulent flow. This implies that the core is well mixed and has a uniform concentration $C_b(x)$, which varies in the direction

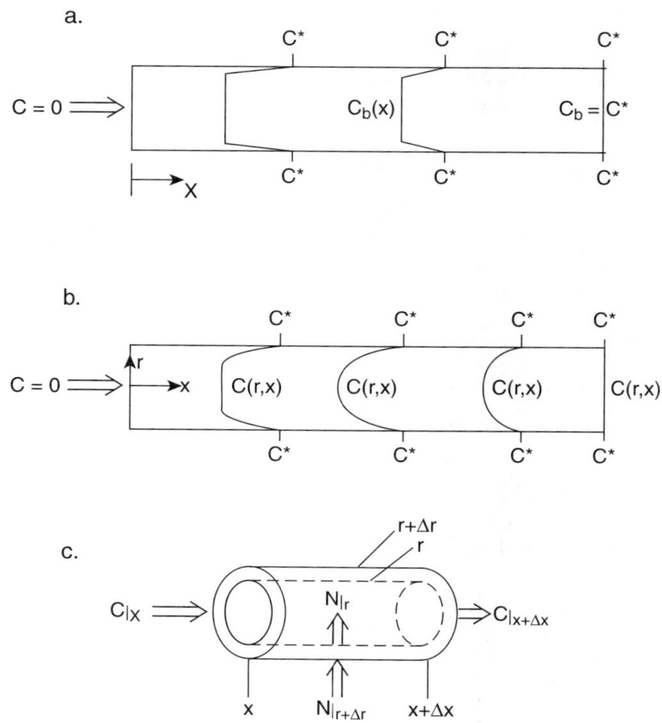

FIGURE 2.9
Solute dissolution into a flowing fluid: (a) turbulent flow; (b) laminar flow; and (c) difference element for laminar flow.

of flow. That bulk concentration is initially zero at the inlet if we assume the feed to be pure solvent, and it gradually increases in level as material dissolves into the flowing fluid. Ultimately, at long distances from the inlet, the fluid becomes fully saturated with solute, and the bulk concentration equals the equilibrium solubility, $C_b = C^*$. Mass transfer then comes to a halt. This situation is depicted in Figure 2.9a and leads to an ODE when a mass balance is applied to an increment Δx in the direction of flow. A situation similar to this case was encountered in Illustration 2.2, where oxygen from the atmosphere entered and dissolved in a river that was in turbulent flow.

2. When the fluid is in laminar flow, the core is no longer well mixed, and we see instead a gradual variation of concentration in the radial direction as solute from the wall enters and dissolves in the flowing fluid. The concentration profile assumes the shape of a parabola with height that gradually diminishes in the direction of flow until the fluid is fully saturated. At this point the concentration has become uniform ($C(r, x) = C^*$), and mass transfer again ceases. This situation is depicted in Figure 2.9b.

Because we are here dealing with concentration variations in two directions, r and x, the mass balance has to be taken over an element with increments Δx and Δr. This leads to the doughnut shape shown in Figure 2.9c, and ultimately yields a PDE.

The volumetric flow rate into and out of the element is given by the product of velocity and area — $[v(r)]_{avg}\, 2\pi r \Delta r$, where the bracketed term represents the local average velocity. The mass of solute entering or leaving the element is then obtained by multiplication by the solute concentration C at the two locations. Radial transport is by diffusion, and here we note that, once again, as in the previous illustration, flow is in the direction of diminishing values of the distance variable and of C. The minus sign in Fick's law is consequently dropped, and we use instead the form $N = DA(\partial C/\partial r)$. With these expressions in place, the mass balance becomes

$$\begin{matrix}\text{Rate of solute in} \\ \text{at } x \text{ and } r+\Delta r\end{matrix} - \begin{matrix}\text{Rate of solute out} \\ \text{at } x+\Delta x \text{ and } r\end{matrix} = 0$$

$$\begin{bmatrix} [v(r)]_{avg} 2\pi r \Delta r C_{1x} \\ +D\Delta x \left(2\pi r \dfrac{\partial C}{\partial r}\right)_{r+\Delta r} \end{bmatrix} - \begin{bmatrix} [v(r)]_{avg} 2\pi r \Delta r C_{1x+\Delta x} \\ +D\Delta x \left(2\pi r \dfrac{\partial C}{\partial r}\right)_{r} \end{bmatrix} = 0 \qquad (2.19a)$$

By dividing by $2\pi r \Delta x \Delta r$ and letting both increments go to zero, the PDE is obtained:

$$v(r)\frac{\partial C}{\partial x} = D\frac{1}{r}\frac{\partial}{\partial r}\left(r\frac{\partial C}{\partial r}\right) \qquad (2.19b)$$

or in expanded form,

$$v(r)\frac{\partial C}{\partial x} = D\left[\frac{\partial^2 C}{\partial r^2} + \frac{1}{r}\frac{\partial C}{\partial r}\right] \qquad (2.19c)$$

where $[v(r)_{avg}]$ has now become a point quantity. The boundary conditions are again three in number, the radial derivatives requiring two such conditions (highest order 2) and the axial derivative one. They are the following:

CONDITION AT THE WALL

$$C(x, R) = C^* \qquad (2.19d)$$

CONDITION AT THE AXIS

$$\left.\frac{\partial C}{\partial r}\right|_{r=0} = 0 \tag{2.19e}$$

CONDITION AT THE INLET

$$C(0, r) = 0 \tag{2.19f}$$

The condition at the axis is of a type encountered before in Illustration 1.2 (see Figure 1.3a and Figure 1.3b) and reflects the fact that the concentration profile must be symmetrical about the central axis and must consequently vanish there. This system of equations (Equation 2.19c through Equation 2.19f) yields, on solution, the radial and axial concentration gradients shown in Figure 2.8b.

Comments

The Graetz problem, also known as the Graetz–Nusselt problem, was originally formulated for the corresponding heat transfer case represented by the PDE:

$$v(r)\frac{\partial T}{\partial x} = \alpha\left[\frac{\partial^2 T}{\partial r^2} + \frac{1}{r}\frac{\partial T}{\partial r}\right] \tag{2.19g}$$

where α = thermal diffusivity.

In the original version, first put forward in the 1880s, two boundary conditions were considered: constant wall temperature $T(x, R) = T_w$ and constant flux q. Some initial results were given by Graetz, but it was not until 1956 that the complete analytical solution became available.

Since its inception, the Graetz problem has been applied to a host of related problems in both heat and mass transfer with a variety of boundary conditions encompassing Newtonian and non-Newtonian flow. In Chapter 5, we show how the solution profiles of the Graetz problem can be cast into equivalent mass transfer coefficients, which can then be used to model the process at the ODE level.

Illustration 2.9: A Metallurgical Problem: Microsegregation in the Casting of Alloys and How to Avoid PDEs

One method of casting alloys is to pour the molten charge into a mold and allow it to cool in contact with the ambient air. The process of solidification that results is complex, involving the transport of heat to the external medium

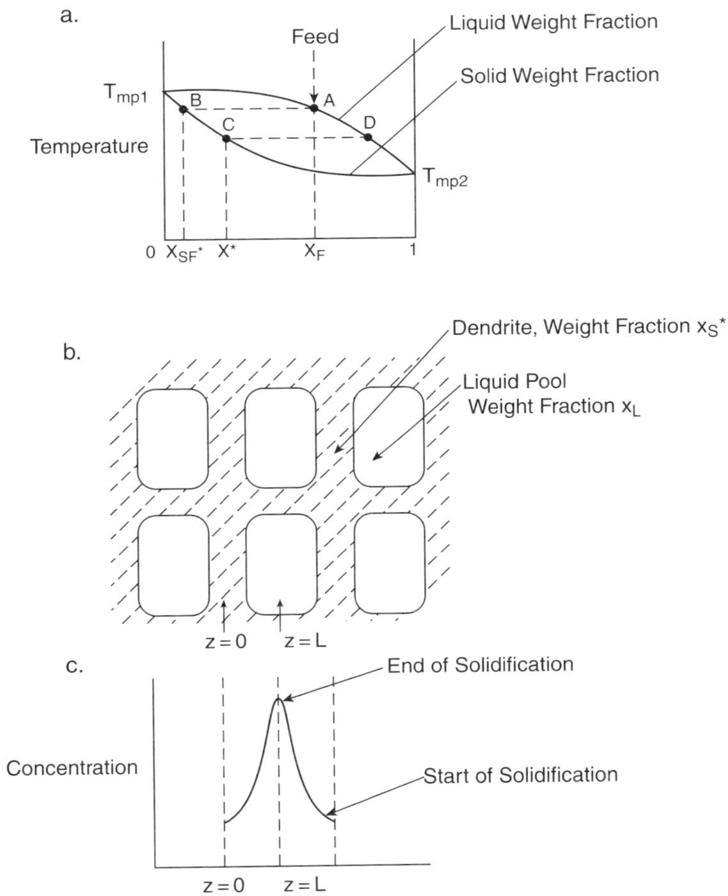

FIGURE 2.10
Microsegregation in the casting of alloy: (a) phase diagram; (b) dendrite formation; and (c) concentration profile.

and a simultaneous transfer of mass from the liquid to the solid phase. The progress of the proceedings is best visualized by means of a plot of temperature versus liquid and solid compositions termed a *phase* or *melting-point diagram*. A simple version of this diagram for a binary (i.e., two-component) system appears in Figure 2.10a, and we will see in Chapter 6 that similar diagrams can be constructed for vapor–liquid systems.

The upper curve, referred to as the *liquidus*, represents a plot of liquid composition versus temperature, while the lower curve, termed the *solidus*, shows the corresponding solid compositions. Horizontal lines drawn through the diagram intersect the two curves at points representing liquid and solid compositions in equilibrium with each other. The extremities of the diagram denote the melting point of the metal components of the alloy, $T_{mp1} > T_{mp2}$.

On cooling, the liquid charge proceeds along the vertical line of constant composition x_F until it reaches the liquidus at point *A*. Here the first solid crystals of composition x_{sF}^* are formed, which are in equilibrium with the feed composition x_F. On further cooling and solidification, a gradual shift of these compositions along the liquidus and solidus curves takes place, resulting in ever-increasing concentrations of component 2 in the liquid and solid phases. These two phases are at all times in equilibrium with each other, with the two compositions located at the end points x_S^* and x_L of a horizontal line drawn through the diagram. These two concentrations gradually approach each other and with further cooling ultimately converge to a single point representing the pure metal component 2 with the lower melting point T_{mp2}. Solidification is then complete.

The physical structure of the charge during this solidification process is shown in Figure 2.10b. Typically, the solid phase is initially confined to narrow, fingerlike protrusions termed *dendrites*. These regions are surrounded by small pools of liquid, which gradually diminish in size as solidification progresses. Simultaneously, the concentration of component 2 in the thickening dendrite increases until it peaks as pure metal 2. The resulting concentration profile is a single dendrite shown in Figure 2.10c. These dendrites form a repetitive pattern of microsegregation throughout the cast and are an undesirable feature because they lead to nonuniform properties of the material. To remove these nonhomogeneities, the cast is subjected to a thermal treatment termed *homogenization* in which the concentrations are smoothed by a slow process of solid-phase interdiffusion. This process is an unsteady one and is described by Fick's equation (Equation 2.19c). Its solution requires, as an initial condition, the concentration profile shown in Figure 2.10c. It is this distribution that initiates the homogenization process and ultimately leads to a cast of uniform properties.

Although the process of solidification is highly complex and requires in principle a set of PDEs, the derivation of the concentration distribution in the dendrite is apparently amenable to a simple treatment. The literature on the subject proposes the following differential equation for the description of the process:

$$(x_L - x_S^*)df_S = (1 - f_S)dx_L \tag{2.20}$$

where f_S is the local weight fraction of the solid, x_L is the liquid-phase weight fraction of the component that crystallizes out, and x_S^* is the equilibrium weight fraction at the solid–liquid interface.

Although superficially this expression resembles a mass balance, it is not clear how it is arrived at. It does not contain time or distance as an independent variable, variables that appear when we model a compartment or a 1-D pipe. We must also rule out a cumulative balance, which is always algebraic in form. The question then arises whether a new type of mass balance formulation is required to cover this case. Fortunately, this does not turn out to be the case. The three mainstay formulations — compartmental, 1-D distributed, and cumulative — are able to cover this case as well.

We assume the liquid to be well mixed with a uniform concentration of x_L. This is a reasonable assumption, as the size of the pool is quite small, typically 50 μm to 200 μm in width. The solid phase, on the other hand, cannot be considered uniform because solid-phase diffusivities are several orders of magnitude smaller. Concentrations in the dendrite will consequently vary in the lateral direction, with the local interfacial concentration x_S^* in instantaneous equilibrium with the uniform liquid pool at all times (Figure 2.10c). Drawing an envelope around the liquid pool one obtains the following mass balances:

TOTAL MASS BALANCE

$$\text{Rate of total mass in} - \text{Rate of total mass out} = \frac{d}{dt}(\text{total contents})$$

$$0 - R = \frac{d}{dt}L \tag{2.21a}$$

COMPONENT MASS BALANCE

$$\text{Rate of solute in} - \text{Rate of solute out} = \frac{d}{dt}(\text{solute content})$$

$$0 - x_S^* R = \frac{d}{dt}x_L L \tag{2.21b}$$

where R = rate of solidification (kg/s), and L is the total mass of liquid in the pool (kg).

As expected, the time variable t makes its appearance, as does the unknown variable R, the rate of solidification. Neither of these appears in Equation 2.20, thus bringing the proceedings to a seeming impasse.

Some thought will reveal that both of these undesirable variables may be eliminated by simply *dividing* the two equations as suggested previously (see Equation 2.10). We obtain

$$x_S^* = \frac{d(x_L L)}{dL} \tag{2.21c}$$

or, in expanded form,

$$x_S{}^* = x_L + L\frac{dx_L}{dL} \tag{2.21d}$$

This is beginning to look much more like the desired expression (Equation 2.20), and one final step will bring us to that goal. We introduce a cumulative mass balance that serves to convert liquid mass L to the solid weight fraction f_S:

$$L + S = M_{Tot} \tag{2.21e}$$

where S = mass of solid, and M_{Tot} = total mass (liquid pool + half the dendrite width). Consequently,

$$\frac{S}{M_{Tot}} = f_S = 1 - \frac{L}{M_{Tot}} \tag{2.21f}$$

and Equation 2.21d becomes, after some rearrangement,

$$(x_L - x_S{}^*)df_S = (1 - f_S)dx_L \tag{2.20}$$

which is the desired result.

Comments

Two features stand out in the treatment of this problem:

1. We reinforced our confidence in the three basic mass balance formulations at the algebraic and ODE levels: the compartmental, 1-D distributed, and cumulative balances. They are vindicated as a comprehensive tool kit at this level of modeling.
2. We see here again the near-miraculous reduction of a highly complex process to manageable proportions. This was accomplished by assuming the liquid pool to be well mixed and in equilibrium with the solid interface, both reasonable assumptions. We followed this up with the neat "trick" of dividing the two mass balances, thereby eliminating both the independent variable t and the unknown R. This gave us the desired relation between the solid-phase concentration and the fraction solidified.

The analysis used here is not confined to liquid–solid systems. It can be applied to any process in which a transfer of mass takes place between two phases that are in constant equilibrium with each other. It was first applied by Rayleigh to analyze equilibrium batch distillation and to derive the attendant concentration changes as a function of the fraction distilled. This case, as well as other batch processes of the same type, are considered in Chapter 7, dealing with equilibrium stage processes.

Equation 2.20 as it stands cannot be integrated directly because it contains more than two variables (i.e., f, x_L, and x_S^*). A second equation will be required, which is given by the following equilibrium relation:

$$x_S^* = f(x_L) \tag{2.22a}$$

Many binary alloys have phase diagrams in which the ratio x_S^*/x_L can be approximated as a constant K, termed the *partition coefficient*. Combining Equation 2.20 and Equation 2.22a, we obtain

$$[(K-1)x_L]df = (1-f_S)dx_L \tag{2.22b}$$

which, after separating variables and formal integration, yields the result

$$-(K-1)\int_{x_L0}^{x_L} \frac{dx_L}{x_L} = \int_0^r \frac{df_S}{1-f_S} \tag{2.22c}$$

and, hence,

$$\frac{x_L}{x_L^0} = (1-f_S)^{K-1} \tag{2.22d}$$

Because the fraction of solid formed, f_S, also equals the fractional distance z/L, we can write

$$\frac{x_L}{x_L^0} = (1-z/L)^{K-1} \tag{2.22e}$$

and for the solid-phase mole fraction x_S^*,

$$\frac{x_S^*}{x_L^0} = K(1-z/L)^{K-1} \tag{2.22f}$$

This last expression is important. It represents the *distribution* of x_S^* after solidification has ceased and provides the starting condition for the homogenization process. Any PDE model of that process will have to incorporate Equation 2.22f as an initial condition to arrive at a solution.

2.5 The General Conservation Equations

The method we described for setting up PDEs has a cumbersome feature attached to it. It needs to be repeated each time there is a change in geometrical configuration or in the process conditions. A switch from rectangular to cylindrical coordinates, for example, requires a new balance to be made, as does the inclusion of reaction terms.

This drawback can be overcome by formulating the mass balances in a generalized vectorial form, using the symbolism of vector calculus. These symbols, or *operators* as they are termed, arise in a natural way in the formulation of generalized transport equations. They are at first sight forbidding, and the beginner will probably be best served by regarding them as a convenient shorthand, without delving into their deeper origins. The symbol $\nabla^2 u$, for example, which is termed the *Laplacian* of u (and pronounced "del square u"), is shorthand for a collection of *second-order* partial derivatives. The symbols ∇u and $\nabla \bullet \mathbf{v}$, ("del u" and "del dot v") serve the same purpose for combinations of *first-order* partial derivatives. For example, in rectangular coordinates, del dot of the velocity vector **v** is synonymous with the sum of the first-order derivatives of the velocity components. Thus,

$$\nabla \bullet \mathbf{v} = \frac{\partial v_x}{\partial x} + \frac{\partial v_y}{\partial y} + \frac{\partial v_z}{\partial z} \tag{2.23a}$$

Similarly, we have for the Laplacian,

$$\nabla^2 u = \frac{\partial^2 u}{\partial x^2} + \frac{\partial^2 u}{\partial y^2} + \frac{\partial^2 u}{\partial z^2} \tag{2.23b}$$

Thus, both del dot and del square tend to be scalar expressions. Del u, on the other hand, is a vector.

Some thought will lead us to the conclusion that del dot terms will likely arise in flowing systems, while the Laplacian will most probably appear in the description of diffusion processes. This is indeed the case.

The use of these operators in the formulation of mass balances leads to the following generalized conservation equations. We have, for the component mass balance,

$$\mathbf{v} \bullet \nabla C_A \quad + \quad D\nabla^2 C_A \quad \pm \quad r_A \quad = \quad \frac{\partial C_A}{\partial t} \tag{2.24a}$$

Flow Diffusion Reaction Transient

and for the total mass balance, also known as the continuity equation,

$$\nabla \bullet \mathbf{v} = 0 \tag{2.24b}$$

Two restrictions apply to these expressions. First, they are confined to *incompressible flow* (i.e., systems in which density changes can be neglected, such as liquid flow or gas flow involving low pressure drops). Second, the formulation requires continuity of the concentration within the flow field. Systems in turbulent flow in which C_A undergoes an abrupt transition from linear gradient in the film to a constant value in the fluid core cannot be accommodated by these expressions. We must, in these cases, revert to the use of the classical shell balance.

To aid in the use of these equations, we compiled a "dictionary" of operator symbols, which provides a translation into scalar form for the three principal geometries (rectangular, cylindrical, and spherical). We can use this dictionary, Table 2.3, to extract several important subsidiary relations. For example, in the absence of flow and reaction, the general conservation equation becomes

$$D\nabla^2 C_A = \frac{\partial C_A}{\partial t} \tag{2.24c}$$

which is Fick's equation in three dimensions, with the Cartesian representation (see Table 2.2),

$$D\left[\frac{\partial^2 C_A}{\partial x^2} + \frac{\partial^2 C_A}{\partial y^2} + \frac{\partial^2 C_A}{\partial z^2}\right] = \frac{\partial C_A}{\partial t} \tag{2.24d}$$

This expression is identical to Equation 2.18d, which was derived by means of a shell balance. Its counterpart for heat conduction, known as Fourier's equation, is given by

$$\alpha\left[\frac{\partial^2 T}{\partial x^2} + \frac{\partial^2 T}{\partial y^2} + \frac{\partial^2 T}{\partial z^2}\right] = \frac{\partial T}{\partial t} \tag{2.24e}$$

The Graetz problem can be accommodated in similar fashion. Here the transient and reaction terms are dropped, and we obtain

$$\mathbf{v} \bullet \nabla C_A = D\nabla^2 C_A \tag{2.24f}$$

TABLE 2.3

Dictionary of Vector Operators

∇u
(1) *Cartesian*
$(\nabla u)_x = \frac{\partial u}{\partial x} \qquad (\nabla u)_y = \frac{\partial u}{\partial y} \qquad (\nabla u)_z = \frac{\partial u}{\partial z}$
(2) *Cylindrical*
$(\nabla u)_r = \frac{\partial u}{\partial r} \qquad (\nabla u)_\theta = \frac{1}{r}\frac{\partial u}{\partial \theta} \qquad (\nabla u)_z = \frac{\partial u}{\partial z}$
(3) *Spherical*
$(\nabla u)_r = \frac{\partial u}{\partial r} \qquad (\nabla u)_\theta = \frac{1}{r}\frac{\partial u}{\partial \theta} \qquad (\nabla u)_\varphi = \frac{1}{r\sin\theta}\frac{\partial u}{\partial \varphi}$
$\nabla \bullet \underline{v}$
(1) *Cartesian*
$\nabla \bullet v = \frac{\partial v_x}{\partial x} + \frac{\partial v_y}{\partial y} + \frac{\partial v_z}{\partial z}$
(2) *Cylindrical*
$\nabla \bullet v = \frac{1}{r}\frac{\partial}{\partial r}(rv_r) + \frac{1}{r}\frac{\partial v_\theta}{\partial \theta} + \frac{\partial v_z}{\partial z}$
(3) *Spherical*
$\nabla \bullet v = \frac{1}{r^2}\frac{\partial}{\partial r}(r^2 v_r) + \frac{1}{r\sin\theta}\frac{\partial}{\partial \theta}(v_\theta \sin\theta) + \frac{1}{r\sin\theta}\frac{\partial v_\varphi}{\partial \varphi}$
$\nabla^2 u$
(1) *Cartesian*
$\nabla^2 u = \frac{\partial^2 u}{\partial x^2} + \frac{\partial^2 u}{\partial y^2} + \frac{\partial^2 u}{\partial z^2}$
(2) *Cylindrical*
$\nabla^2 u = \frac{1}{r}\frac{\partial}{\partial r}\left(r\frac{\partial u}{\partial r}\right) + \frac{1}{r^2}\frac{\partial^2 u}{\partial \theta^2} + \frac{\partial^2 u}{\partial^2 z}$
(3) *Spherical*
$\nabla^2 u = \frac{1}{r^2}\frac{\partial}{\partial r}\left(r^2\frac{\partial u}{\partial r}\right) + \frac{1}{r^2\sin\theta}\frac{\partial}{\partial \theta}\left(\sin\theta\frac{\partial u}{\partial \theta}\right) + \frac{1}{r^2\sin\theta}\frac{\partial^2 u}{\partial \varphi^2}$

The dot product on the left is composed by the rules of vector algebra. In other words, it equals the sum of the vector component products. Setting $v_r = \frac{\partial C_A}{\partial \theta} = \frac{\partial^2 C_A}{\partial z^2} = 0$ and using the tabulations of Table 2.3, we obtain

$$v_z \frac{\partial C_A}{\partial z} = D \frac{1}{r} \frac{\partial}{\partial r}\left(r \frac{\partial C_A}{\partial r} \right) \qquad (2.24g)$$

which is in agreement with Equation 2.19b.

Finally, when all but the reaction and transient forms are omitted, we are led to the result

$$\pm V r_A = V \frac{dC_A}{dt} \qquad (2.24h)$$

This will be recognized as a mass balance for a batch reactor of volume V.

Illustration 2.10: Laplace's Equation, Steady-State Diffusion in Three-Dimensional Space: Emissions from Embedded Sources

Steady-state diffusion was considered in some detail in Chapter 1 at the elementary level of one-dimensional transport. In Cartesian space, the operative expression was Fick's law (Equation 1.4), which can also be written in the equivalent form

$$\frac{d^2 C}{dx^2} = 0 \qquad (2.25a)$$

The extension to three Cartesian dimensions is given by

$$\frac{\partial^2 C}{\partial x^2} + \frac{\partial^2 C}{\partial y^2} + \frac{\partial^2 C}{\partial z^2} = 0 \qquad (2.25b)$$

or in a generalized vectorial form by

$$\nabla^2 C = 0 \qquad (2.25c)$$

This is the classical and much-studied expression known as *Laplace's equation*. It can be obtained from Fick's equation by omitting the transient term, or from the general conservation equation (Equation 2.24a) by omitting transient, reaction, and flow terms.

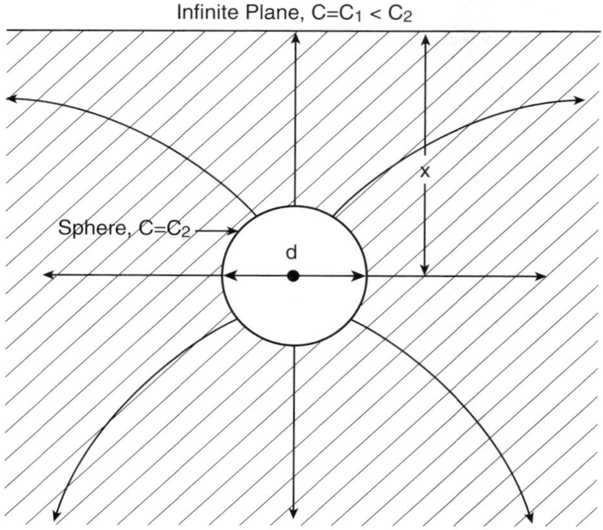

FIGURE 2.11
Diffusion from a sphere embedded in a semi-infinite medium.

A host of solutions of this problem are known from the analogous case of heat conduction and are easily adapted by substituting concentration and diffusivity for temperature and thermal diffusivity. Many of these solutions, particularly those dealing with finite geometries, are forbidding in form and difficult to apply in practice. The intent here is to draw the reader's attention to some simple solutions, which are useful in environmental and biological contexts.

The geometry we consider first is a semi-infinite medium bounded by a plane surface. An object of finite dimensions (sphere, disk, cylinder) is embedded at a distance x from the surface and is assumed to have a constant surface concentration C_2. Solute diffuses into the surrounding space and ultimately reaches the bounding surface, which is held at a constant concentration C_1 (Figure 2.11). C_1 is often near zero due to dispersion into a flowing fluid or the atmosphere.

Examples of such embedded objects are underground deposits of a toxic or benign nature and medical implants (see Illustration 2.12). The situation described in Illustration 1.4 — helium storage in an abandoned salt mine — is another example. There is also a strong heat transfer presence in this area which considers such problems as heat losses from underground steam lines and the potential for the freezing of water mains (see Practice Problem 2.12).

The primary information obtained from Laplace's equation comes in the form of the three-dimensional concentration distributions, which are of no direct practical use. It is common practice to convert these results into an equivalent rate equation that is of much greater interest. For conduction, it takes the form

TABLE 2.4

Shape Factors for Embedded Objects for Use in $N = SD\Delta C$

Embedded object	Shape factor
A. Infinite Medium	
1. Sphere, diameter d	$2\pi d$
2. Disk, diameter d	$4d$
3. Horizontal cylinder, length L, diameter d, ends sealed	$\frac{2\pi L}{\ln(4L/d)}$
B. Semi-infinite Medium (Figure 2.11)	
4. Sphere	$\frac{2\pi d}{1-d/4x}$
5. Disk, $x > 0.5\,d$	$\frac{4d}{1-d/5.67x}$
6. Disk, $x = 0$	2d
7. Horizontal cylinder	$\frac{2\pi L}{\ln(4x/d)}$
C. Bounded Medium	
8. Concentric spheres	$2\pi d(1 + d/2\delta)$
9. Concentric cylinders	$\frac{2\pi L}{\ln(1+2\delta/d)}$

Notes: x = distance from center to bounding; d = inner diameter; δ = distance between surfaces plane.

$$q(\mathrm{J/s}) = kS(T_2 - T_1) \tag{2.26}$$

and for diffusion,

$$N(\mathrm{mol/s}) = DS(C_2 - C_1) = (R_C/A)\Delta C \tag{2.27}$$

These are, of course, precisely the type of simple expressions one wishes to have on hand because it allows us to calculate flux from the embedded object to its surroundings. Equation 2.27 requires knowledge only of the bounding concentrations and the diffusivity. The geometry of the system is accounted for through the so-called shape factor S, which has the dimensions of length and is extensively tabulated in standard handbooks of heat transfer. A short version is given in Table 2.4.

One notes from these listings that objects embedded in an infinite medium have the lowest shape factors, while in bounded media (concentric spheres and cylinders) the values are at their maximum. This is in line with physical reasoning. The two cases also provide useful *upper* and *lower* limits to the emission rates. We will make use of these limits in Illustration 2.12 to estimate the performance of controlled-release implants.

Objects embedded in a semi-infinite medium bounded by a plane occupy an intermediate position. Their shape factors fall between the two limits mentioned, again in line with physical reasoning. Item 6 represents the special case of a disk embedded at the surface and is used to determine emission rates *into* a semi-infinite medium.

Let us use these tabulations to reexamine the helium storage problem of Illustration 1.4. For distances from the surface x much larger than the cavity dimension, we have, with $C_1 \approx 0$,

$$N = DS(C_2 - C_1) \approx D\, 2\pi d\, C_2 \tag{2.28a}$$

or, equivalently,

$$N = D\, 4\pi r\, p/RT \tag{2.28b}$$

This expression is identical to Equation 1.8c, obtained by a spherical shell balance, and represents diffusion into an infinite medium.

When distance x is reduced, the result begins to be affected by the proximity of the surface plane, but not by much. Suppose, for example, that the center of the storage cavity is at a distance of only 100 m from the surface. We then have

$$N = D[2\pi d/(1 - 100/400)]C_2 \tag{2.28c}$$

that is, the losses increase by 33%.

The use of a spherical shell balance within an infinite medium is thus justified at large depths but begins to break down in the vicinity of the planar surface. This is in agreement with physical reasoning.

Illustration 2.11: Lifetime of Volatile Underground Deposits

Consider a shallow dump, 10 m in diameter and buried at a depth $x = 5$ m. In one case, it contains, dispersed in it, 100 mol of a toxic substance with a low vapor pressure of 10^{-2} mmHg = 1.3 Pa. In the second case, the substance is the carcinogen benzene with a vapor pressure of 95 mmHg = 1.25×10^4 Pa. Setting aside removal by water seepage and bacterial action, what would be the lifetime of these toxins?

Item 5 of Table 2.4 provides a reasonable shape factor for this case. A rough estimate for diffusivity can be obtained from the expression $D = D_c\, \varepsilon/4$, where D_o is the free space diffusivity = 10^{-5} m²/s and ε is the soil void fraction, which is set equal to 0.3. We then have, from Equation 2.27b,

$$N(\text{moles}/\text{s}) = D_o S C_2 = D_o(\varepsilon/4)\frac{4d}{1 - d/5.67x}\, p/RT \tag{2.29a}$$

$$N = 10^{-5}(0.3/4)\frac{4 \times 10}{1 - 10/5.67 \times 5}p/8.31 \times 300 \tag{2.29b}$$

$$N = 1.8 \times 10^{-8}\ p\ (\mathrm{Pa}) \tag{2.29c}$$

and for the lifetime of the toxins (100 mol),

$$t = 5.6 \times 10^{9}/p(\mathrm{Pa}) \tag{2.29d}$$

There results, for the low-volatility toxin,

$$t = 5.6 \times 10^{9}/1.3 = 4.3 \times 10^{9}\ \mathrm{s} = 136\ \mathrm{yrs} \tag{2.29e}$$

and for benzene,

$$t = 5.6 \times 10^{9}/1.25 \times 10^{4} = 4.5 \times 10^{5}\ \mathrm{s} = 5.2\ \mathrm{days} \tag{2.29f}$$

Comments

The figures used in arriving at these results have a good deal of elasticity. For example, lowering D — never an easy quantity to estimate — by a factor of ten increases the lifetime of the deposits by the same amount. Adsorption of the toxins onto the soil lowers the equilibrium partial pressure, with similar consequences. What the illustration attempts to do is convey a sense of the *scale* of these events and draw attention to the tools available to do this.

Illustration 2.12: Performance of a Controlled-Release Device: Bracketing the Solution

Controlled-release devices, whether worn externally as "patches" or implanted surgically, serve to convey medication to the body at a preset, steady rate. The device often consists of a *suspension* of the soluble substance encapsulated in a permeable membrane. As material is released and passes through the membrane into the body, it is constantly replenished by dissolution of the suspension so that the concentration at the interface remains essentially constant at the saturation level, C_s. This is the "upstream" value of the diffusional driving force $\Delta C = C_s - C_1$. The term C_1 represents the concentration attained in the nearby capillaries that convey the substance into the bloodstream or some intended target. Compared to C_s, C_1 is usually assumed to be negligible if the consumption or dilution rate is high or if one seeks to establish the maximum possible release rate.

Whether it comes from an external source or an internal implant, the released substance encounters a complex three-dimensional geometry with ill-defined receiving boundaries. The concentrations and flux that result are described by Laplace's equation, but a rigorous solution of it would be a

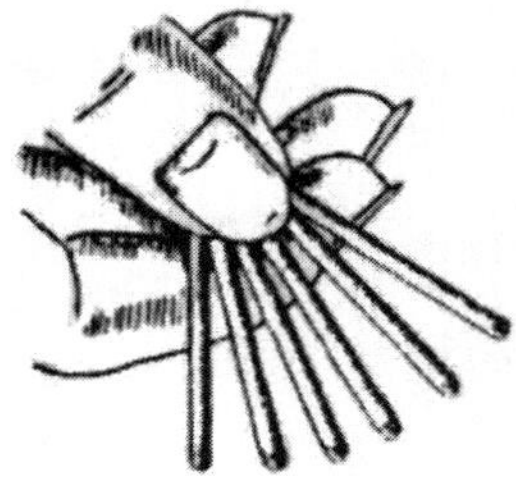

FIGURE 2.12
Norplant capsules.

doomed enterprise. The shape factors listed in Table 2.4 provide a possible way out by allowing us to establish, at the very least, *upper and lower bounds* to the release rate ("bracketing the solution").

Consider the following practical example. Norplant is a widely used implant that releases the contraceptive agent progestin into the bloodstream. It comes in the shape of thin cylinders of about 2 mm diameter and 3 cm length (see Figure 2.12). Typically, six such units are implanted above the elbow, with a combined lifetime of 5 years.

To do a sample calculation, we set the saturation concentration at 10 μmol/cm³ and use a typical tissue diffusivity of $D = 10^{-6}$ cm²/s. Using Item 3 of Table 2.4, we obtain, for the minimum release rate,

$$N_{\text{Min}} = \frac{2\pi L}{\ln(4L/d)} DC_s = \frac{(2\pi)3}{\ln(4\times 3/0.2)} 10^{-6}\times 10^{-5} \tag{2.30a}$$

$$N_{\text{Min}} = 4.6\times 10^{-11} \text{ mol/s} \tag{2.30b}$$

Note that this is the minimum to be expected in the *absence* of any intervening *sinks* due to consumption or blood flow. To establish the upper bound, we assume that "sink" ($C_1 = 0$) to be uniformly distributed at a distance $\delta = 2$ mm from the cylinder surface, an inspired, but reasonable, guess. Using Item 9 of Table 2.4, we then obtain

$$N_{\text{Max}} = \frac{2\pi L}{\ln(1+2\delta/d)} DC_s = \frac{(2\pi)3}{\ln(1+2\times 0.2/0.2)} 10^{-6}\times 10^{-5} \tag{2.30c}$$

$$N_{\text{Max}} = 17.1\times 10^{-11} \text{ mol/s} \tag{2.30d}$$

Thus, even though we have succeeded in bracketing the solution, the "bracket" is fairly wide: a factor of 17.1/4.6 = 3.7. Nevertheless, these limits provide useful guidelines in dimensioning the implant and choosing an appropriate solvent in the initial design stages. None of the devices are put to use without further extensive testing.

Practice Problems

2.1 Modeling Batch Distillation

A ternary mixture is to be distilled in the batch still shown in Figure 2.7. Identify the number and type of model equations required for a complete description of the process.

2.2 Modeling an Oil Spill on Land

An accidental oil spill onto porous soil has occurred. The initial layer thickness h_o gradually diminishes as oil seeps into the ground in the vertical direction z.

a. What are the dependent and independent variables?
b. How many equations are required to describe the process, and what are they?
c. Identify the auxiliary relation.

2.3 Compartmental Modeling in Pharmocokinetics

The subject of pharmocokinetics covers the study of the fate of an injected drug in the human or animal body. A typical exercise involves the monitoring of the drug concentration in the blood after injection and extracting from the resulting time history relevant kinetic or transport parameters. A favorite device for the interpretation of the experimental findings is the one-compartment model. After the one-shot injection of the drug, the drug concentration declines exponentially in the manner shown in the accompanying diagram. That curve, determined experimentally, is usually described by a first-order rate law, which is incorporated in a mass balance around the compartment. We have

$$\text{Rate of drug – in Rate of drug out} = \frac{d}{dt}\text{contents}$$

$$0 - k_e CV = \frac{d}{dt}VC$$

with the solution

$$\ln C/C_o = -k_e t$$

where k_e is the so-called elimination-rate constant, which can be obtained from a semilog plot of the experimental concentration-time data. It is often

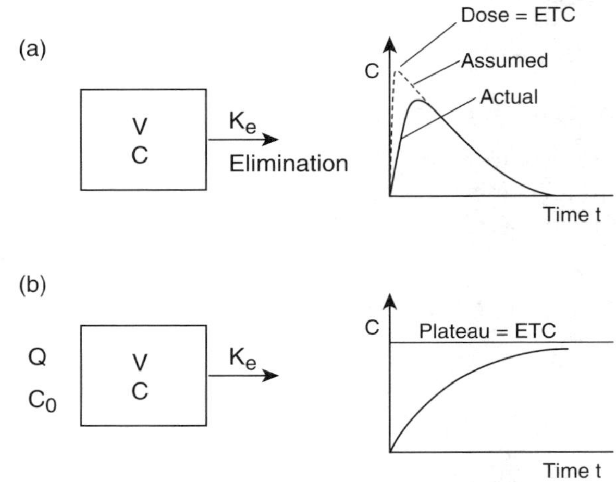

FIGURE 2.13
One-compartment models and the time course of drug concentration: (a) clearance following injection and (b) steady infusion.

reported in terms of the half-life of the drug $t_{1/2} = \ln 2/k_e$ (i.e., the time required to reduce its concentration to one-half the initial level). Note that the injected dose does not enter the mass balance directly but appears instead as the initial concentration C_o, which is obtained by extrapolation of the data to time $t = 0$ (Figure 2.13a). The term k_e can then be used to determine the time course of drug concentration during its steady infusion, or administration, to a patient. Infusion follows the time course shown in Figure 2.13b. It reaches a plateau value termed the *effective therapeutic concentration* (ETC) when infusion and elimination rates are exactly in balance.

Consider the case of a patient who has undergone major surgery and requires a slow, intravenous infusion to control hemorrhage. The plan is to infuse the drug at the rate of 1 g/h, and the drug is known to have a half-life of 3.9 h. If the drug is effective only above 0.048 g/L, how long does it take to reach this level after the start of infusion? Assume a fluid volume of 50 L.

Answer:

3.25 h

2.4 The Streeter–Phelps Model of River Pollution

Figure 2.14 shows the oxygen distribution — the so-called "oxygen sag curve" — that results in a river downstream from a steady inflow of pollutant.

a. Give a physical explanation for the shape of the oxygen sag curve and relate it to Equation 2.6j.

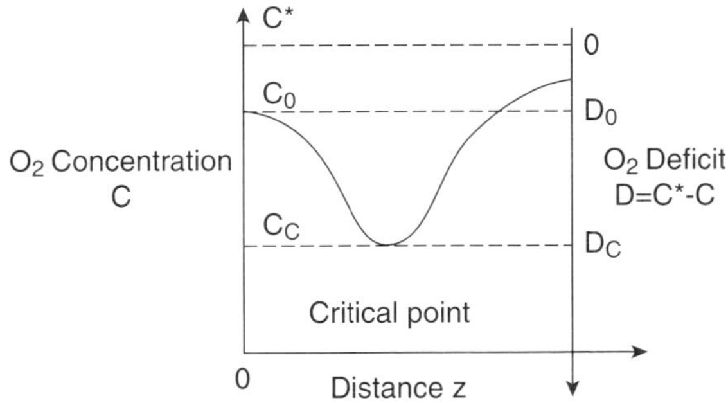

FIGURE 2.14
The Streeter–Phelps model. Dissolved oxygen profiles in a river with a steady influx of a pollutant.

b. How many model equations are required to describe the process, and what are they?

The full solution of the model appears in Chapter 5, Equation 5.11m.

2.5 Solute Release from a Tubular Wall into a Stream in Turbulent Flow

A solute is released from a tubular wall into a fluid in turbulent flow. Two cases are to be considered:

a. The substance is released at a constant rate N (mol/m^{2}s) all along the length of a tube. This case arises in some medical applications.
b. The solute concentration at the wall is constant and equal to its equilibrium solubility C^*. This leads to the type of profile development we saw in Figure 2.9a.

Derive for both these cases the variation in the axial direction of the bulk concentration C_b of the solute.

Answer:

a. $C_b = \dfrac{4N}{vd}x$

b. $(C^* - C_b)/C^* = \exp(4k_c x/vd)$

2.6 Clearance of a Contaminated River Bottom Sediment

A toxic substance is released into a river at a constant concentration and over a time interval Δt. The contaminant is adsorbed onto the bottom sediment and, after contamination has ceased, is gradually cleared into the river. It is desired to calculate the time necessary for the sediment to be cleared of 95% of its contaminant. Assume the concentration in the sediment and the water to be uniform in the vertical direction but not in the direction of flow.

a. Is the process an unsteady one or can it be considered to be at a steady state?
b. What are the dependent and independent variables for the system?
c. Describe the space over which the mass balance has to be applied.
d. Describe the resulting model equations without actually deriving them.

2.7 Performance of a Dryer

A batch of wet solids holding a total of 10 kg water is to be dried by passing hot air over it. The air enters with a humidity of 0.03 kg H_2O/kg air and leaves with a humidity of 0.10 kg H_2O/kg air. The (dry) air flow rate is constant at 0.05 kg/s. Calculate the time necessary to evaporate 95% of the moisture. (*Hint*: Make a cumulative balance.)

2.8 The Countercurrent Heat Exchanger

The countercurrent heat exchanger is the heat transfer analogue to the countercurrent gas scrubber described in Illustration 2.4. Cold fluid enters the central tube at a flow rate of F_c kg/s countercurrent to a hot fluid, which enters the concentric shell on the right and with a flow rate F_h kg/s. Heat is transferred to the cold fluid at a rate $q = UA\Delta T$, causing a change in the enthalpy H_c and, consequently, the temperature T_C of the cold fluid. Enthalpy is related to temperature through the expression $H = FC_p\,(T - T_{ref})$.

a. Make an integral heat balance over the entire exchanger. The resulting expression corresponds to Equation 2.11b obtained for the countercurrent gas scrubber.
b. Derive the thermal counterparts to the differential mass balances (Equation 2.12d and Equation 2.12e) but do not solve.

2.9 Unsteady Diffusion with a Chemical Reaction

Consider the case of a substance diffusing into a porous solid slab (see Illustration 2.7) in which it irreversibly bound to the solid substrate in accordance with a first-order rate law. Modify the result obtained in Illustration 2.7 to describe this process.

2.10 Unsteady Diffusion from a Sphere into a Solution of Finite Constant Volume

Soluble material is leached from a slurry of porous spherical particles by suspending them in a well-stirred liquid solvent of constant volume. Show that the model for this system is made up of a single PDE for the spheres and an ODE for the stirred tank. What are the relevant boundary and initial conditions?

2.11 A Total Mass Balance: The Continuity Equation in Three Dimensions

Show that the total mass balance for a flowing system with and without chemical reactions in a rectangular coordinate system is given by

$$-\left[\frac{\partial}{\partial x}\rho v_x + \frac{\partial}{\partial y}\rho v_y + \frac{\partial}{\partial z}\rho v_z\right] = \frac{\partial \rho}{\partial t}$$

where ρ = density, and v_x, v_y, v_z are the velocity components in the three directions. (*Hint*: Consider a cube with sides Δx, Δy, Δz.)

2.12 Diffusion and Reaction in a Spherical Catalyst Pellet

Use the general conservation Equation 2.24a and Table 2.2 to derive the differential equation for uniform diffusion and reaction in a spherical catalyst pellet.

2.13 Equivalence of Models and Solutions

We speak of the equivalence of two models when they are described by two sets of differential equations and boundary conditions that are *identical in form* but differ in the *physical significance of the variables*. Fick's equation and Fourier's equation are equivalent in this sense, provided they have the same form of boundary and initial conditions.

Recognizing equivalence has the advantage that one can use known solutions of one model to obtain an immediate solution of the other, needing only to make a suitable translation of the physical variables and parameters.

Consider the following two processes:

a. A solute diffuses radially out of a porous cylindrical particle into a well-stirred medium with negligible solute content but finite mass transfer resistance. The particle is initially uniformly loaded.
b. A liquid solution carrying a reactant is fed into a tubular reactor coated with a catalyst. The reactant diffuses to the wall where it undergoes a first-order reaction. Flow is laminar, but radial variations in velocity are to be neglected as a first approximation.

Identify the model equations and boundary conditions, and provide a short "dictionary" translating the variables from one model into the other.

2.14 Use of Separation Factors in Solidification Processes

The equilibrium partition coefficient K used in Illustration 2.9 does not in most cases provide an exact representation of the phase equilibrium. A more realistic representation is given by the expression

$$x_s{}^* = \frac{x_L}{(\alpha - 1)x_L + \alpha}$$

where α is the so-called separation factor, defined as

$$a = \frac{x_L(1 - x_s{}^*)}{x_s{}^*(1 - x_L)}$$

with $\alpha > 1$. At low values of x_L, $\alpha = 1/K$.

Solve the differential Equation 2.20. (*Hint:* $\int dx/x(a + bx) = -1/a[\ln{(a + bx)/x}]$.)

2.15 More on Shape Factors

Derive the shape factor S for two concentric spheres (Item 8 of Table 2.4).

2.16 A Heat Transfer Analogue: Freezing of an Underground Water Main

Water at 5°C is to flow through an underground water pipe of 100 m length and 10 cm internal diameter at the rate of 0.01 m^3/s. The surface is exposed to temperatures as low as 12°C for prolonged periods. What is the minimum allowable distance x from the surface if the water temperature is to drop by no more than 1°C? What is the result if a drop of 2° is allowed? Comment on the results. ($k = 2 \times 10^4$ kW/msK.)

Answer:

3.8 m ($\Delta T = 1$°C)

2.17 Design of a Controlled-Release Device

A newly developed medication is to be incorporated into an implant device. A particular solvent showed solubility for the substance (500 Da) of 20 g/L. Is this sufficient to produce the target delivery rate of 10^{-7} g/s using the device of Illustration 2.11?

2.18 More about Transdermal Patches

Yet another newly developed drug is to be used in an externally worn transdermal patch (see Figure 1.15) of no more than 3 cm in diameter. The weighted average diffusivity between the patch reservoir and the target blood capillaries is estimated at 5×10^{-7} cm^2/s. The highest saturation capacity attained among a variety of solvents tested was 10 g/L. Is this likely to be sufficient to deliver the desired dose of 1 mg/day? (*Hint*: Consult Table 2.4.)

3

Diffusion through Gases, Liquids, and Solids

In Chapter 1, the notion of diffusional processes was introduced in which mass transfer takes place by molecular motion only and is proportional to the concentration gradient of the diffusing species. This proportionality is enshrined in Fick's law of diffusion, and this introductory chapter was used to acquaint the reader with some simple applications of that law (see Illustrations 1.4 and 1.5). In this chapter and the one that follows, the intent is to amplify and expand the material on diffusion that was presented in Chapter 1.

We examine in some detail the important diffusion coefficient, which, when viewed mathematically, is the proportionality constant in Fick's law. It is also a material property that depends on the nature of the diffusing species, the matrix through which diffusion takes place, as well as on temperature and, in the case of gases, on pressure. We consider in some detail the diffusivities of gases within gases, within liquids, and within solids, and the diffusivities associated with the interdiffusion of liquids and solids. These coefficients are of considerable practical importance in various engineering disciplines, in materials processing, and in the biological and environmental sciences.

3.1 Diffusion Coefficients

3.1.1 Diffusion in Gases

The mechanisms by which diffusion in gases takes place are depicted schematically in Figure 3.1a. Gas molecules move in space in random motion with an average velocity u, repeatedly undergoing collision with other moving gas molecules, which causes them to be deflected into a new direction. The average distance traveled by a molecule is referred to as the *mean free path* λ, where λ is of the order 10^{-7} m at atmospheric pressure and is a direct measure of the diffusivity of a substance. It varies inversely with pressure and is some three orders of magnitude higher than that of a liquid molecule. A selected list of diffusivities of gases and vapors in air appears in Table 3.1.

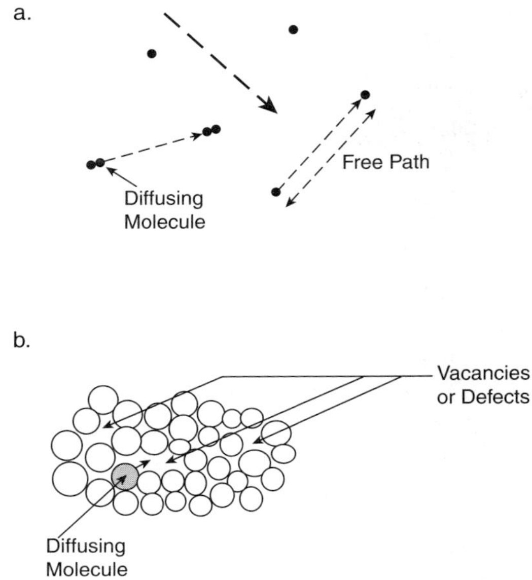

FIGURE 3.1
Diffusional mechanisms in (a) gases and (b) liquids and solids.

TABLE 3.1
Diffusivities in Air (P = 1 atm, T = 25°C)

Diffusing Species	Diffusivity (cm^2/s)
Hydrogen	0.78
Helium	0.70
Ammonia	0.22
Water	0.26
Oxygen	0.20
Ethanol	0.14
Acetic acid	0.12
Benzene	0.090
Toluene	0.086
n-Hexane	0.080
Carbon tetrachloride	0.083
Chlorotoluene	0.065
DDT	0.047
2,4,2′,4″-Tetrachlorobiphenyl (a PCB)	0.052
Mercury	0.13

A first simple expression for the diffusivity of gases was derived some 150 years ago based on the kinetic theory of gases. In this theory, the molecules are regarded as point entities that undergo elastic collisions with each other without the intrusion of intermolecular attractive or repulsive forces. This simple model led to the expression

$$\mathrm{D} = \frac{1}{3} u\lambda \propto \sqrt{\frac{T}{M}} / \frac{n}{V} \tag{3.1}$$

where the average velocity u varies directly with the square root of absolute temperature, and the mean free path is inversely proportional to the molar density n/V. This expression correctly predicts the order of magnitude of D (see Practice Problem 3.1) and the dependence on gas pressure and molar mass M, but it gives a much less accurate representation of the variation with temperature.

This first attempt at a prediction of D was followed by a series of more elaborate theories, which took account of the finite size of the gas molecules, as well as the effect of intermolecular forces. Probably the most popular among current prediction methods is that due to Fuller, Schettler, and Giddings, who proposed the following expression for the calculation of gas diffusivities:

$$D_{AB} = \frac{1.00 \times 10^{-3} T^{1.75} (1/M_A + 1/M_B)^{1/2}}{P_T \left[(\Sigma V_A)^{1/3} + (\Sigma V_B)^{1/3} \right]^2} \tag{3.2}$$

Here D_{AB} is in units of square centimeter per second (cm²/s), T is the absolute temperature (K), P_{T} is the total pressure in atmospheres, and V are the atomic and molecular volume contributions. These are empirical constants that correspond approximately to the molar volume of the substances in cubic centimeter per mole (cm³/mol). They have been tabulated, and a partial list for use with organic molecules appears in Table 3.2.

Illustration 3.1: Diffusivity of Cadmium Vapor in Air

Cadmium vapor is a toxic substance whose diffusivity in air is not readily available in the literature. It is desired to calculate its diffusivity at its boiling point of 1038 K and a pressure of 1 atm. Because an empirical atomic volume is not available, we use the reported value for its *liquid* volume of 14 cm³/mol for an atomic weight of 112.4. We obtain, using Equation 3.2,

$$D_{AB} = \frac{10^{-3}(1038)^{1.75} \left(\frac{1}{112.4} + \frac{1}{29} \right)^{1/2}}{1.0 \left[14^{1/3} + 20.1^{1/3} \right]^2}$$

$$D_{AB} = 1.6 \text{ cm}^2/\text{s}$$

TABLE 3.2

Atomic and Molecular Volume Contributions for Diffusivity Calculations

Species	Volume V (cm³/mol)
A. Gases (Fuller, Schettler, and Giddings method)	
C	16.5
H	1.98
O	5.48
N	5.69
Cl	19.5
Aromatic ring	20.2
Air	20.1
B. Liquids (Wilke–Chang Method)	
C	14.8
H	3.7
O	7.4
O in high esters and others	11.0
O in acids (OH)	12.0
Cl (terminal)	21.6
6-numbered ring	15

A data point for the diffusivity of Cd in N_2 at 273 K is available for comparison. Its value is 0.15 cm²/s, and we obtain, by applying a temperature correction in line with Equation 3.2,

$$D_{AB} = 0.15 \times \left(\frac{1038}{273} \right)^{1.75} = 1.55 \text{ cm}^2 / \text{s}$$

This is in agreement with the calculated value of 1.6 cm²/s.

Comments

Some remarks are in order regarding the *magnitude* of gas-phase diffusivities. We note from Table 3.1 that at 25°C and a pressure of 1 atm, most diffusivities cluster around a value of 0.1 cm²/s. This includes metal vapors, as well as medium-sized organic molecules such as DDT and the PCBs. The reason for this lies in the relatively weak dependence of D_{AB} on molar volume and mass and the limited number of gaseous or volatile substances available. The larger organic molecules such as polymers, proteins, and carbohydrates which would lead to low diffusivity values do not exist in the vapor phase. Thus, gas diffusivities lower than 0.01 cm²/s are unlikely to be encountered. An upper ceiling is provided by the lightest molecules hydrogen and helium, which have a mutual diffusion coefficient of 1.35 cm²/s at 25°C and 1 atm.

A reasonable order of magnitude estimate can therefore be arrived at in most cases by starting with a value of $D_{AB} = 0.1\ cm^2/s$ and applying temperature or pressure correction factors in accordance with Equation 3.2. Applying this procedure to the cadmium vapor of Illustration 3.1, we obtain a value of

$$D_{Cd-Air} = 0.1\left(\frac{1038}{273}\right)^{1.75} = 1.04\ cm^2/s$$

which is of the correct order of magnitude.

3.1.2 Diffusion in Liquids

Liquid densities exceed those of gases at normal atmospheric pressures by a factor of about 1000. These differences are reflected in the intermolecular distances that exist in the two phases. In gases under standard conditions, these distances are some three orders of magnitude greater than the molecular dimensions. Liquid molecules are by contrast closely packed, with intermolecular distances of the same order as the molecular size.

Gas molecules spend most of their time in transit between collisions and are only modestly affected by intermolecular forces. In liquids, these forces are the dominant factor that determines the mobility of the molecules. They are notoriously difficult to quantify, and as a consequence the prediction of liquid diffusivities has lagged behind theories describing the motion of gas molecules.

The Stokes–Einstein equation, one of the earliest theoretical expressions for liquid diffusivities, viewed the diffusion process as a hydrodynamic phenomenon in which the thermal motion of the molecules is resisted by a Stokesian drag force. This theory, along with subsequent modification by Sutherland and Eyring, established the following proportionality for the diffusion coefficient:

$$D_{AB} \propto \frac{T}{\mu_B V_A^{1/3}} \tag{3.3a}$$

This relation, which is most successful for large molecules ($V_A > 500\ cm^3/mol$), states that diffusivity varies inversely with the viscosity of the solvent and the molecular dimension of the diffusion molecule. This agrees with our intuitive grasp of the process.

Modern theories consider the diffusing particle to be contained in a cage with dimensions that are constantly fluctuating. Local fluctuations in density periodically open up holes or vacancies large enough to allow the particle to diffuse out of the cage (Figure 3.1b). Although this view has led to some progress in the quantification of the diffusion process, we are at present still constrained to using semi-empirical expressions for the prediction of diffusivities. One such relation, proposed by Wilke and Chang, has been reasonably successful in predicting diffusion coefficients of small molecules in aqueous and organic systems at normal temperatures. It has the form

$$D_{AB}(\text{cm}^2/\text{s}) = 7.4 \times 10^{-8}(\varphi M_B)^{1/2}\frac{T(K)}{\mu_B(cp)V_A^{0.6}} \tag{3.3b}$$

where M_B and V_A (cm^3/mol) are the molar mass of solvent and volume of the diffusing species, and φ is an empirical coefficient with a value of 2.6 for water and 1.0 for unassociated solvents. For organic solutes, V_A is composed of atomic and ring contributions, a partial list of which is given in Table 3.2B. An example of the application of Equation 3.3 is given in Practice Problem 3.4.

Diffusivities in molten salts and metals are even more difficult to predict, and here one often resorts to an Arrhenius-type relation to express the strong temperature dependence of the diffusion coefficient, which is concealed in the viscosity of Equation 3.3a and Equation 3.3b:

$$D_{AB} = D_0 \exp(-E_a/RT) \tag{3.4}$$

This equation also finds use as a correlation for diffusion coefficients in solids. Tabulation of D_0 and the activation energy E_a for various species can be found in the pertinent literature.

Given the uncertainties of current prediction methods for liquid diffusivities, it is clearly preferable to use measured values of D_{AB}. Table 3.3 lists diffusion coefficients in water at 25°C of a variety of solutes, including gases (Part A) and ions (Part C), and solutes of a biological or toxic nature (Part D and Part E). In Table 3.4 we reproduced a small selection of diffusivities in liquid metals and molten salts at different temperatures.

The remarkable feature that emerges from these tabulations is the relatively small numerical range of the diffusivities for a wide variety of different substances. Most coefficients cluster around a value of 10^{-5} cm^2/s, with the lighter and smaller solutes (H_2 and He) exceeding this benchmark by a factor of 5 to 7, while very large molecules of a biological origin fall below this value by factors of 2 to 20. Diffusivities in molten metals and salts, which are quite different in nature from normal liquids, likewise fall in the range 10^{-5} to 10^{-4} cm^2/s, the higher values due mainly to the higher temperatures involved. A rough initial estimate of $D = 10^{-5}$ cm^2/s when liquid diffusivities are not known will therefore not be too far off the mark. The corresponding average value for gases is, as was seen, 10,000 times higher at $D = 10^{-1}$ cm^2/s. However, because gases are much less dense than liquids and consequently have lower concentration gradients, the diffusion rates are only about ten times higher than the corresponding values in liquids.

TABLE 3.3

Diffusivities in Water (T = 25°C)

Diffusing Species	Diffusivity (cm²/s × 10⁵)	Molar Mass
A. Gases		
Hydrogen	4.8	
Helium	7.3	
Methane	1.8	
Ammonia	2.0	
Carbon monoxide	2.17	
Oxygen	2.42	
Nitrogen	2.0	
Hydrogen chloride	3.1	
Carbon dioxide	2.0	
Sulfur dioxide	1.7	
B. Liquids		
Methanol	1.28	
Ethanol	1.24	
Acetic acid	1.26	
Acetone	1.28	
Benzene	1.02	
C. Ions		
H^+	9.3	
OH	5.3	
NH_4^+	2.0	
Na^+	1.3	
Mg^{2+}	0.71	
Cl	2.0	
K^+	2.0	
Fe^{3+}	0.60	
Cu^{2+}	0.71	
NO_3	1.9	
SO_4^{2}	1.1	
D. Biological Substances		
Urea	1.4	60
Glucose	1.3	
Oxygen in blood	1.4	32
Oxygen in muscle tissue	1.7	32
Lysozyme (egg white)	1.0	14,000
Hemoglobin	0.69	68,000
Fibrinogen	0.2	330,000
Tobacco mosaic virus	0.044	40,000,000

TABLE 3.3 (continued)

Diffusivities in Water (T = 25°C)

Diffusing Species	Diffusivity ($cm^2/s \times 10^5$)	Molar Mass
E. Environmentally Toxic Substances		
Chlorine	1.45	
Chlorobenzene	0.91	
1,2,4-Trichlorobenzene	0.76	
Mercury	2.9	
DDT	0.49	
2,4,2′,4-Tetrachlorobiphenyl (a PCB)	0.55	

TABLE 3.4

Diffusivities in Liquid Metals and Salts

T°C	Diffusing Species	Melt	D (cm^2/s) × 10^5
1270	Fe	Fe (4.6%C)	10.0
40	Hg	Hg	2.0
600	Zn	Zn	5.0
450	Sn	Pb	2.0
906	Na^+	NaCl	14.2
933	Cl	NaCl	8.8
328	Na^+	$NaNO_3$	2.0
328	NO_3	$NaNO_3$	1.36

To demonstrate the use of liquid-phase diffusivities, we turn to the somewhat unusual case of ion migration in an electrolytic process and the electrical current that results from it.

Illustration 3.2: Electrorefining of Metals: Concentration Polarization and the Limiting Current Density

The final processing in the production of high-purity metals is often carried out electrolytically and is referred to as electrorefining. In this process, the metal to be refined, such as copper or silver, has a typical initial purity of 95 to 99%, and the aim is to reduce the impurity level to less than 0.1%. Conventional purification processes are often either inadequate or too expensive for this purpose. In electrorefining, the impure metal (e.g., copper) is placed in an electrolytic bath as an anodic plate paired with a cathode on which the purified metal is deposited electrolytically. The electrolyte typically consists of an aqueous solution of a salt of the metal to be purified (e.g., copper sulfate), and the electrolytic cell is composed of an array of closely spaced alternating cathodes and anodes. A sample electrode pair and the configuration of the electrolytic cell are shown in Figure 3.2a.

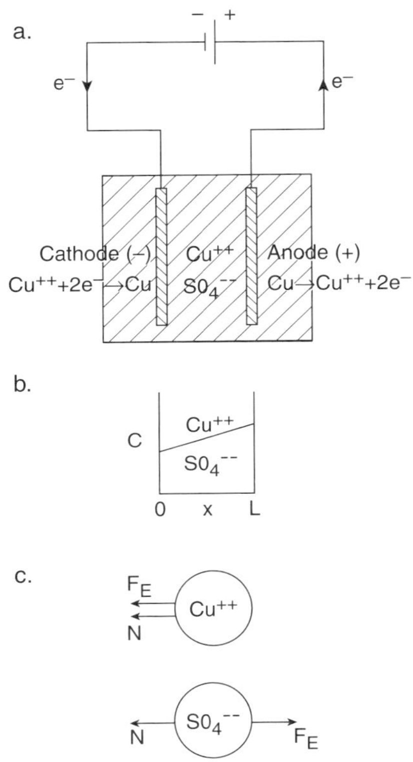

FIGURE 3.2
Electrorefining and concentration polarization: (a) sample electrode pair and electrolytic bath; (b) concentration gradients; and (c) flux due to diffusion and electrostatic potential.

When a potential is applied to the electrodes, two processes take place. At the anode, the metal, along with its impurities, dissolves as positive cations into the electrolytic bath. These ions migrate to the negative cathode where they are discharged as metal. The impurities, such as iron, usually require a higher potential to be deposited and, consequently, remain in solution.

As a result of the applied voltage and the attendant migration of ions, a concentration gradient develops within the bath which ultimately leads to a linear steady-state profile as shown in Figure 3.2b. A distinction must be made between the behavior of the cations, here exemplified by Cu^{++}, and the negative counterions represented by SO_4. For the former, the transport rate is made up of two components that act in the *same* direction (Figure 3.2c): diffusional transport with a rate N, and transport due to the applied electrical potential, represented by the rate F_E. We can then write

$$N_{\text{Tot}} / A = D_{\text{Cu}^{++}} \frac{dC_{\text{Cu}^{++}}}{dx} \quad + \quad F_E / A \tag{3.4a}$$

Diffusional flux   Electrostatic flux

For the anions, the two transport mechanisms act in opposite directions (Figure 3.2c). SO_4 ions migrate toward the cathode by diffusion, as do the copper ions, but they are repelled there and driven back into solution by the negative charge of the electrode. The result is that there is no net flux of the anions. Diffusional transport is exactly balanced by electrostatic transport, and we can write

$$0 = D_{SO_4^{--}} \frac{dC_{SO_4^{--}}}{dx} - F_E / A \tag{3.4b}$$

We note that $C_{SO_4^{--}} = C_{Cu^{++}}$, and for electrical neutrality to be maintained, one must have $D_{SO_4^{--}} = D_{Cu^{++}}$. Adding Equation 3.4a and Equation 3.4b to eliminate the electrostatic flux, we obtain

$$N_{Tot} / A = 2D_{Cu^{++}} \frac{dC_{Cu^{++}}}{dx} \tag{3.4c}$$

On integration by separation of variables, there results

$$N_{Tot} / A = 2 \frac{D_{Cu^{++}}(C_{Cu^{++}}|_L - C_{Cu^{++}}|_0)}{L} \tag{3.4d}$$

The maximum flux occurs when the concentration at the cathode has dropped to zero. Noting that the concentration at the anode is then twice the initial electrolyte concentration, $C^o_{Cu^{++}}$, one can write

$$(N_{Tot} / A)_{Max} = 4 \frac{D_{Cu^{++}} C^o_{Cu^{++}}}{L} \tag{3.4e}$$

Let us apply this equation to a particular condition. Suppose that the electrodes are spaced 1 cm apart, and the initial concentration of copper sulfate is 0.1 *M*. Using the diffusivity for Cu^{++} listed in Table 3.3C, we then obtain

$$(N_{Tot} / A)_{Max} = \frac{4 \times 0.71 \times 10^{-5} \times 10^{-4}}{1} \tag{3.4f}$$

$$(N_{Tot} / A)_{Max} = 2.84 \times 10^{-9} \text{ mol/cm}^2\text{s} \tag{3.4g}$$

Comments

Equation 3.4e gives a good semiquantitative description of the behavior of electrolytic cells and their counterpart, the galvanic cells or batteries. It shows that there is a limit to the rate of electrolytic deposition or production which cannot be exceeded by increasing the applied voltage. Similarly, there is a limiting maximum current that can be drawn from batteries. These limits are imposed by the diffusional processes, which accompany all electrochemical processes, be they galvanic or electrolytic in nature. Equation 3.4e also shows the need for close electrode spacing in order to achieve high production rates and high value of electrode area *A*.. In car batteries, for example, the plate spacing is of the order of 1 mm, and the number of electrode pairs is of the order 100.

Deviations from Equation 3.4e occur mainly as a result of the concentration dependence of the diffusivities. This is the case particularly at high electrolyte concentration where electrostatic interactions reduce the ion mobility below the values listed in Table 3.3C. The latter apply to conditions at infinite dilution.

Equations of the form of Equation 3.4e can also be used to calculate the current associated with electrochemical processes. To do this, we use the conversion factor given by the so-called Faraday number F:

$$F = 96{,}520 \text{ C/mol electrons} \tag{3.5}$$

where C denotes coulombs or ampere-seconds.

Let us use this conversion and Equation 3.4e to calculate the current associated with the electrorefining of copper. We assume the unit to be composed of 20 cathodes and 20 anodes, each 1 × 1 m in dimension and spaced $L = 2$ cm apart. Diffusivity of the copper ions is 0.71×10^{-5} cm²/s (see Table 3.3C), and the concentration C is set at 10^3 mol/cm^{-3}. One obtains for the current i

$$i = 2F \times A \times D_{\text{Cu}^{++}} \times C / L \tag{3.6a}$$

$$\text{i} = 2 \times 96{,}520 \times (20 \times 100 \times 100) \times 0.71 \times 10^{-5} \times 10^{-3}/2 \times 10^{-2} \tag{3.6b}$$

This yields

$$i = 137\ A \tag{3.6c}$$

Note that while currents in these operations are considerable, the applied voltage is quite small. Dissolution and deposition potentials almost exactly cancel each other, and the only voltage drop that occurs is due to the Ohmian resistance of the electrolyte solution. This rarely amounts to more than a fraction of a volt.

Practice Problem 3.4 deals with the calculation of the size of an electrorefining plant.

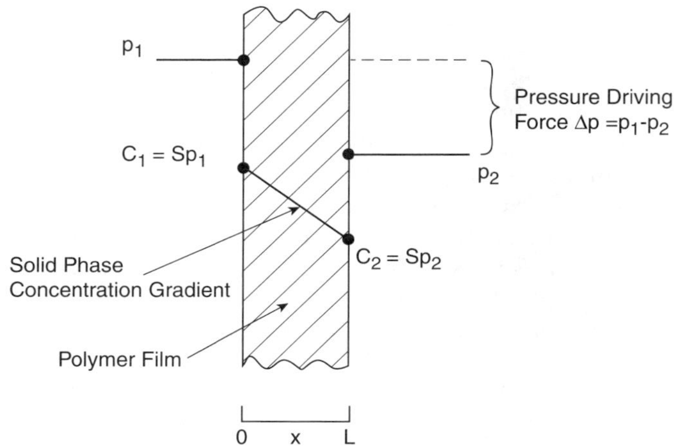

FIGURE 3.3
External gas pressures and internal solid-phase concentration in the diffusion of gases through polymers.

3.1.3 Diffusion in Solids

In diffusion through solids, several distinct cases arise which depend on the nature of the diffusing species and of the solid medium. The diffusing species can be gaseous or liquid in form, or, surprisingly, can also be a solid. For the solid medium, a distinction is made between consolidated media, such as polymers, and those that have a porous structure. We limit ourselves here to a discussion of three important cases of transport through solids, which we take up in turn: diffusion of gases through polymers and metals and through porous media, and the interdiffusion of solids.

3.1.3.1 Diffusion of Gases through Polymers and Metals

The diffusion of gases through polymers and similar consolidated media is viewed as a three-step process (Figure 3.3):

1. At the high-pressure interface, the gas dissolves or condenses in the solid matrix.
2. Following dissolution or condensation, the gas diffuses along a solid-phase concentration gradient dC/dx in accordance with Fick's law to the low-pressure interface.
3. On arrival at the low-pressure gas–solid boundary, the dissolved gas is desorbed or released to the gaseous medium.

In Steps 1 and 3, the gas is assumed to be in instantaneous equilibrium with the neighboring solid matrix. The solid-phase concentration C is related to the gas pressure by a linear relation, given as

$$C = Sp \tag{3.7}$$

where S is termed the solubility of the gas in the solid. The units used for C are somewhat unconventional and are expressed as cubic centimeter gas at STP per cubic centimeter solid. Units for pressure p are either in atmospheres (atm) or pascal (Pa), both of which are currently still in use. As a consequence of this dual usage, the solubility S is reported either in units of cubic centimeter (STP) per cubic centimeter pascal (cm^3 (STP)/cm^3 Pa) or cubic centimeter (STP) per cubic centimeter atmosphere (cm^3 (STP)/cm^3 atm).

A dual approach is also used in the formulation of the diffusional rate laws. In the first version, Fick's law is used with concentrations as defined above — that is, we have

$$N(\text{cc STP / s}) = -DA\frac{dC}{dx} \tag{3.8a}$$

where D is the diffusion coefficient in square centimeter per second (cm^2/s).

The second and preferred version uses a pressure gradient and leads to the rate law:

$$N(\text{cc STP / s}) = -PA\frac{dp}{dx} \tag{3.8b}$$

Here P is the so-called permeability, which is most commonly expressed in units of cubic centimeter (STP) times centimeter per square centimeter second (cm^3 (STP) × cm/cm^2s) or cubic centimeter (STP) times centimeter per square centimeter second pascal (cm^3 (STP) × cm/cm^2s Pa). Several other additional units are still in current use, and the translation from one set of units to another is a frequent necessity. To ease this task, provided in Table 3.5 is a listing of the most commonly required conversion factors.

Permeability is related to the Fickian diffusivity D through the solubility S. This is seen by substituting Equation 3.7 into Equation 3.8a and comparing the result with Equation 3.8b. We obtain the relation

$$P = DS \tag{3.9}$$

which can be used to calculate D from independent measurements of permeability and the equilibrium solubility S.

In contrast to the diffusivities in gases and liquids, which cluster around values of 10^{-1} and 10^{-5} cm^2/s, the diffusion coefficients and permeabilities we encounter here are not confined to a narrow numerical range. They are strongly material dependent and range over several orders of magnitude. This is shown in Table 3.6, which lists values of P and S for six common gases and vapors in several commercial polymers.

TABLE 3.5

Conversion Factors for Gas Permeabilities in Solids

From	Multiplication Factors to Obtain P in		
	$\frac{cm^3\ cm}{cm^2\ s\ cm\ Hg}$	$\frac{cm^3\ cm}{cm^2\ s\ Pa}$	$\frac{cm^3\ cm}{m^2\ day\ atm}$
$\frac{cm^3\ cm}{cm^2\ s\ cm\ Hg}$	1	7.5×10^{-4}	6.57×10^{-10}
$\frac{cm^3\ mm}{cm^2\ s\ cm\ Hg}$	10^{-1}	7.5×10^{-5}	6.57×10^{-9}
$\frac{cm^3\ cm}{cm^2\ s\ atm}$	1.32×10^{-2}	9.87×10^{-6}	8.64×10^{-8}
$\frac{in^3\ mil}{100\ in^2\ day\ atm}$	9.82×10^{-12}	7.37×10^{-15}	6.45×10^{-1}
$\frac{cm^3\ cm}{cm^2\ s\ Pa}$	1.33×10^{-3}	1	8.75×10^{-13}
$\frac{cm^3\ mil}{cm^2\ day\ atm}$	3.87×10^{-14}	2.9×10^{-17}	2.54×10^{-3}

TABLE 3.6

Permeabilities and Solubilities of Gases in Polymers

Polymer		H_2	He	O_2	CO_2	N_2	H_2O
Polyethylene	P	7.4	3.7	2.2	9.5	0.73	68
(low density)	S	1.6	0.054	0.47	2.5	0.23	
Polyethylene	P		0.86	0.30	0.27	0.11	9.0
(high density)	S		0.028	0.18	0.22	0.15	
Polystyrene	P	17	14	1.9	7.9	0.59	1350
(biaxially oriented)	S						
Polyvinyl chloride	P	1.3	1.5	0.034	0.12	0.0089	206
(PVC-unplasticized)	S	0.26	0.055	0.29	4.7	0.23	870
Polyvinyldene chloride	P		0.233	0.0038	0.0022	0.00071	7.0
(Saran)	S		(30°C)	(30°C)	(30°C)	(30°C)	
Polytetrafluoroethylene	P	7.4	9.0	3.2	7.5	1.0	6.8 (38°C)
(Teflon)	S	4.9	1.1	2.1	9.2	1.2	
Polychloroprene	P	10		3.0	19.0	0.88	683
(Neoprene G)	S	0.29		0.74	8.2	0.36	
Cellulose hydrate	P	0.0046	0.00038	0.0016	0.19	0.0024	18,900
(Cellophane)	S						
Vulcanized rubber	P	34		15	98	5.3	
	S	0.40		0.69	8.9	0.35	

Notes: $T = 25°C$ unless otherwise indicated. P in [cm^3 STP × cm/cm^2 s Pa]10^{13}. S in [cm^3 STP/cm^3 Pa]10^6.

Permeabilities typically vary over the range 10^{-17} to 10^{-11} cm^3 STP × cm/cm^2 s Pa, values for $D = P/S$ over the somewhat narrower range of 10^{-8} to 10^{-5} cm^3/s, with no apparent relation to species and material properties.

Even odder behavior is observed when one compares the mobility of different gases in the same material. Large molecules with a high molar mass, which one would intuitively expect to be less mobile than their lighter counterparts, can, in fact, have considerably higher permeation rates. Thus, in polyvinyl chloride (PVC) water vapor with a molar mass 4.5 times that of helium and twice its molecular size, nevertheless has a permeability over 100 times that of helium. This reinforces the notion that the movement of gas molecules through polymers is a highly complex process that does not exhibit the simple inverse relation to molecular mass and size seen in diffusion through gases and liquids. Much more elaborate theories, which are still in a state of development, are required to quantify this process.

The following example provides some practice in the use of solubilities and permeabilities.

Illustration 3.3: Uptake and Permeation of Atmospheric Oxygen in PVC

Consider a PVC sheet 0.1 mm thick with a one-sided area of 1 m^2. Its density is 1.1 g/cm^3, and it is exposed to atmospheric air at 100 kPa. We wish to calculate (1) the uptake of oxygen from the air and (2) the daily rate of permeation which prevails when one face of the sheet is in contact with the atmosphere and the other with pure nitrogen at the same total pressure of 100 kPa.

1. Uptake of oxygen

We start with Equation 3.7 and convert from C (cc STP O_2/m^3) to mass m_{O_2} of oxygen. This yields

$$m_{O_2} = S \times p_{O_2} \times V_{PVC} \times M_{O_2} / V_{O_2} \tag{3.10a}$$

Here p_{O_2} equals the partial pressure of oxygen in the air (Pa), V_{PVC} is the volume of the sheet (cm^3), and M_{O_2} and V_{O_2} are the molar mass and volume (STP) of the oxygen, respectively.

Using the values for solubility given in Table 3.6, we obtain

$$m_{O_2} = 0.29 \times 10^{-6}\,\text{cm}^3\,\text{STP/cm}^3\,\text{Pa} \times 21{,}000\,\text{Pa}$$

$$\times(100^2 \times 0.01)cm^3 \times 32\,g\,/\,cm^3\,/\,22{,}410\,cc\,/\,mole \tag{3.10b}$$

and, consequently,

$$m_{O_2} = 8.7 \times 10^{-4}\ \text{g} \tag{3.10c}$$

that is, approximately 1 mg.

2. Permeation rate of oxygen

The relevant expression for this case is Equation 3.8b, which after insertion of the appropriate numerical values yields

$$N(\text{cc STP/day}) = 0.034 \times 10^{-13}\ \text{cc STP cm/cm}^2\ \text{s Pa}$$

$$\times 10^4\ \text{cm} \times (21{,}000\ \text{Pa}/0.01\ \text{cm})$$

$$\times 3{,}600\ \text{s/hr} \times 25\ \text{hr/day} \tag{3.10d}$$

$$N = 6.17\ \text{cc STP/day} \tag{3.10e}$$

or, equivalently,

$$N'(\text{g/day}) = \frac{6.17}{22{,}410} 32 = 8.8 \times 10^{-3}\ \text{g/day} \tag{3.10f}$$

Comments

The permeation rate of atmospheric oxygen in PVC is seen to be quite low, in spite of the considerable area of the sheet in question and its very small thickness. PVC would consequently be a good packaging material in cases where atmospheric oxygen is to be excluded.

Illustration 3.4: Sievert's Law: Hydrogen Leakage through a Reactor Wall

Permeation of gases through metals initially follows the same mechanism that applies to diffusion through polymeric materials — that is, the gas dissolves or condenses in the solid metal matrix. On dissolution, however, a number of diatomic gases such as hydrogen, oxygen, and nitrogen undergo *dissociation* into their component atoms according to the following scheme:

$$\frac{1}{2} G_2\,(\text{g}) \underset{\leftarrow}{\overset{\rightarrow}{}} G\,(\text{s}) \tag{3.11a}$$

As a consequence of this process, the equilibrium solid-phase concentration C is no longer related linearly to gas pressure but depends instead on the *square root* of p. The linear relation (Equation 3.7), $C = Sp$, must consequently be replaced by the expression

$$C = Kp^{1/2} \tag{3.11b}$$

where K is the equilibrium dissociation constant for the reaction (Equation 3.11a). A corresponding change must be effected in the definition of the permeability. Whereas previously we defined P as $P = DS$ (Equation 3.9), we now write

$$P = DK \tag{3.11c}$$

where P has new units (cm^3 STP cm/cm^2 s atm$^{1/2}$).

Expression 3.11b is known as Sievert's law and can be used to calculate the permeation rate of diatomic gases through metals.

The example considered here involves a tubular reactor made of steel that is to be used in the high-pressure hydrogenation of hydrocarbon vapor. The vessel is 3 m long, has an internal radius r_i of 5 cm, and is to be operated at a pressure of 100 atm and a temperature of 450°C. The permeability P at this temperature has a value of 8.4×10^{-6} cm^3 (STP) cm/cm^2 s atm$^{1/2}$.

We wish to calculate the wall thickness required to keep hydrogen losses within reasonable bounds. The reader may recall that a similar configuration and process was considered in Illustration 1.5 to derive the diffusion rate through a hollow cylinder. That rate was given by the expression

$$N = D2\pi L \frac{(C_i - C_o)}{\ln r_o / r_i} \tag{1.9c}$$

where the subscripts i and o refer to inside and outside conditions.

For the case at hand, we replace C by pressure using Equation 3.11b and eliminate D by introducing the permeability P given by Equation 3.11c. This leads to the result

$$N[\text{cm}^3\,(\text{STP})/\text{s}] = P2\pi L \frac{\left(\sqrt{p_i} - \sqrt{p_o}\right)}{\ln r_o / r_i} \tag{3.11d}$$

Assuming negligible external hydrogen pressure ($p_o = 0$), we obtain

$$N[\text{cm}^3\,(\text{STP})/\text{s}] = 8.4 \times 10^{-6}\, 2\pi 300 \frac{\sqrt{100}}{\ln r_o / r_i} \tag{3.11e}$$

The resulting diffusion rates as a function of external radius r_o are tabulated below.

N[cm³ (STP)/s]	r_o (cm)	$r_o - r_i$ (cm)
1.7	5.5	0.5
0.87	6	1
0.33	8	3
0.23	10	5
0.18	12	7

We note that, after an initial rapid drop, the diffusion rate tapers off asymptotically with an increase in wall thickness $r_o - r_i$ due to the logarithmic dependence on external radius. A wall thickness of 0.5 cm is clearly insufficient because it leads to substantial hydrogen losses at a level of 1.7 cc (STP)/s. On the other hand, tripling the wall thickness from 5 cm to the inordinately high value of 15 cm merely brings about a reduction from 0.23 to 0.11 cc (STP)/s. A wall thickness of 5 cm therefore appears to be a reasonable compromise value. Ventilation will nevertheless have to be provided to prevent a dangerous buildup of hydrogen.

Illustration 3.5: The Design of Packaging Materials

The use of packaging materials is all-pervasive. Virtually every product that reaches the stores and the consumer is packaged at some stage during its passage from the manufacturing plant to its ultimate destination. Packaging in crates, cardboard boxes, or sacks and bags eases handling and provides protection from damage during transportation.

Additional packaging may be provided to prolong shelf life. Food items are often packaged to maintain freshness and prevent the loss of moisture. In other cases, the opposite result is desired: Atmospheric moisture is the enemy and has to be excluded by a suitable protective barrier. This arises in the packaging of moisture-sensitive items such as electronic components. Both instances require a barrier with low water permeability.

Consider the case of a moisture-sensitive item that has to be kept in an atmosphere of less than 10% relative humidity. This means that the partial pressure of moisture in the air cannot exceed 10% of the saturation vapor pressure. In order to protect the item against accidental excursions of the humidity of the surrounding air, it is proposed that it be packaged in an appropriate material. Inspection of Table 3.6 shows that high-density polyethylene has a suitably low permeability to water vapor. We now stipulate that the thickness of the packaging material should be sufficient to protect the item against accidental exposure to 95% humidity air of 1 h duration.

It is not immediately clear at the outset how this information is to be obtained. Because the permeation process is unsteady, a good way to start is to set up an integral unsteady moisture balance, in which the rate of permeation into the package is balanced by the change in moisture content of the interior air. Some further thought will then reveal that the thickness L being sought resides in the permeation gradient $\Delta p/L$. Consequently, we can write

$$\text{Rate of moisture in} - \text{Rate of moisture out} = \begin{array}{c}\text{Rate of change}\\ \text{of moisture}\\ \text{contents}\end{array}$$

$$PA\frac{p_o - p_i}{L} - 0 = \frac{d}{dt} n_{H_2O} \qquad (3.12a)$$

where p_o and p_i are the constant exterior and varying interior partial water pressures, and n_{H_2O} equals the mol of water vapor contained in the package. Because we have two dependent variables in p_i and n_{H_2O}, a second expression relating the two is required. This relation is provided by the ideal gas law:

$$n_{H_2O}(\text{mol}) = \frac{p_i V}{RT} \qquad (3.12b)$$

Combining Equation 3.12a and Equation 3.12b and introducing the molar conversion factor then yields

$$PA(p_o - p_i) / 22.410\ L = \frac{V}{RT}\frac{dp_i}{dt} \qquad (3.12c)$$

where the factor 22.410 is used to convert cm³ (STP) contained in the permeability to units of mole.

Integrating by separation of variables, we obtain

$$\frac{(PA)(RT)}{22,410\ VL}\int_0^t dt = \int_{p_i^o}^{p_i^f} \frac{dp_i}{p_o - p_i} \qquad (3.12d)$$

and, consequently,

$$\frac{(PA)(RT)t}{22,410\ VL} = \ln\frac{p_o - p_i^o}{p_o - p_i^f} \qquad (3.12e)$$

where the superscripts o and f denote the initial and final moisture content of the interior air. The desired packaging thickness is obtained by solving for L:

$$L = \frac{(PA)(RT)t}{22,410\ V\ \ln \dfrac{p_o - p_i^o}{p_o - p_i^f}} \tag{3.12f}$$

We set temperature at 25°C, area-to-volume ratio *A/V* at 2, and the initial interior humidity at zero. Using a permeability of 9×10^{-13} cm^3 STP cm/cm^{2}s PA listed in Table 3.6, we obtain

$$L = \frac{(9 \times 10^{-13}) \times (2) \times (8.314 \times 10^6) \times (298) \times (3600)}{22,410 \times \ln \dfrac{0.9\ p_{sat}}{0.9\ p_{sat} - 0.1\ p_{sat}}} \tag{3.12g}$$

$$L = 0.0061 \text{ cm} \tag{3.12h}$$

Comments

This thickness of less than 0.1 mm is not unduly large. In fact, we could extend the exposure time to 10 h without exceeding acceptable thickness limits.

3.1.3.2 Diffusion of Gases through Porous Solids

The permeation process considered in the previous section was somewhat unusual. It required the prior dissolution of the gas in the solid matrix before it could make its way through the medium by a process of solid-phase Fickian diffusion. A more conventional permeation process occurs when the solid involved is porous or is composed of loosely packed particles. Here the permeating gas enters the solid through the pore openings and continues its way through the porous passages, all the while remaining in the gas phase. Diffusion is strictly Fickian, and no penetration of the solid matrix, per se, takes place.

To describe this process, two factors need to be taken into account. The first is the reduction in cross-sectional area available for diffusion, which reduces the diffusion coefficient by a factor equal to the void fraction ε of the solid. Here ε is expressed as the ratio of open cross-sectional area to total cross-sectional area and ranges from 0 to 1. The second factor is due to the tortuous nature of the porous pathway that is often associated with these media. The gas molecules are made to zigzag their way through the solid, rather than going straight through, occasionally coming to a complete halt in so-called dead-end pores (Figure 3.4a and Figure 3.4b). The net effect is to lengthen the diffusional pathway and consequently reduce the effective diffusion rate. This lengthening of the path and the effect of dead-end pores is accounted for through the so-called tortuosity factor τ, which has a value greater than unity. Both of these effects are lumped into the diffusion coefficient, resulting in an effective and reduced diffusivity D_e — that is, we have

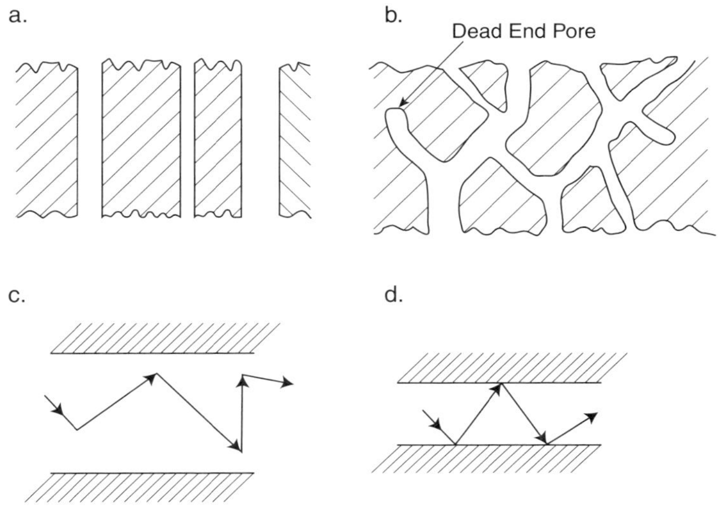

FIGURE 3.4
Diffusion in a porous solid: (a) straight-through pores; (b) tortuous pores with branching; (c) molecular diffusion; and (d) Knudsen diffusion.

$$D_e = \frac{D\varepsilon}{\tau} \tag{3.13a}$$

where D remains the ordinary diffusivity applicable to free space. D_e is used in conjunction with Flick's law, which retains its original form — that is, we have

$$N / A = -D_e \frac{dC}{dx} \tag{3.13b}$$

Void fractions for many porous media typically vary over the range 0.1 to 0.5, with 0.3 being a good average value. Tortuosity τ has a range of 1.5 to 10, with occasional excursions to higher values. A value of $\tau = 4$ gives a good initial estimate in the absence of precise data. The combined effect of the reduction in cross-sectional area and the lengthening of the diffusional pathway is to reduce the free-space diffusivity by approximately one order of magnitude. Thus, the diffusion of oxygen in air, previously set at 0.2 cm^2/s (Table 3.1), now drops by a factor of ten to approximately 0.02 cm^2/s.

We have up to this point assumed that permeation in the pores follows the mechanism of gaseous diffusion in free space (i.e., that the gas molecules undergo repeated collisions with each other as they progress through the medium, Figure 3.4c). However, when pore diameter drops below the value of the mean free path λ, the diffusional mechanism undergoes a change to what is termed *Knudsen diffusion*. Here the impeding collisions no longer occur between gas molecules, but rather between gas molecules and the wall of the pore. The pathway is still random, but it now zigzags between the walls of the pores, which deflect the gas molecules into a new direction (Figure 3.4d). Equation 3.2, which was given for diffusivity in free space, no longer applies. Although Knudsen diffusion still varies directly with temperature, and inversely with molar mass M, it now also depends on the radius of the pore. The relevant expression is given by

$$D_K(\text{cm}^2/\text{s}) = 9700\, r_p\, (T/M)^{1/2} \tag{3.14}$$

where r_p is the pore radius in centimeters, and T is the absolute temperature in degrees Kelvin. Knudsen diffusivities can be several orders of magnitude smaller than molecular diffusivities and are the controlling transport coefficient in many diffusional processes through porous media.

To illustrate diffusion through porous media, we consider three examples. The first deals with diffusion into and out of a leaf, which takes place through tiny, straight-through pores in the underside of leaves. This illustration and the practice problem that accompanies it introduce the reader to the fascinating world of plant physiology and the biophysical processes that sustain plant life, as well as our own. In the second example, we consider diffusion in a porous catalyst pellet. Here the pore structure is no longer "straight-through" but becomes tortuous and often gives rise to Knudsen diffusion because of the small dimension of the pores. Transport into and out of catalyst particles plays an important role in determining the overall performance of the catalyst and is taken up in more detail in Chapter 4.

The third example also deals with Knudsen diffusion, but here the mechanism is exploited to achieve a separation of gases, that of the uranium isotopes $U^{235}F_6$ and $U^{238}F_6$. The process was first used on a large scale during World War II but is now largely of historical interest, having been overtaken by the centrifuge process described in Illustration 1.3. It provides a glimpse of the desperate measure taken at the time, at huge cost, and also contains some marvelous examples of the application of chemical engineering principles.

Illustration 3.6: Transpiration of Water from Leaves: Photosynthesis and Its Implications for Global Warming

The principal transport processes in a leaf, apart from the conveyance of nutrients, are the uptake of carbon dioxide from the air and the release of oxygen and water vapor to the atmosphere. The latter process is referred to as transpiration.

The cell structure of a typical leaf is shown in Figure 3.5a and Figure 3.5b. Transport takes place mainly through openings termed *stomatal pores* (*stoma*: Greek for *mouth*), which are concentrated at the underside of the leaf. Stomata

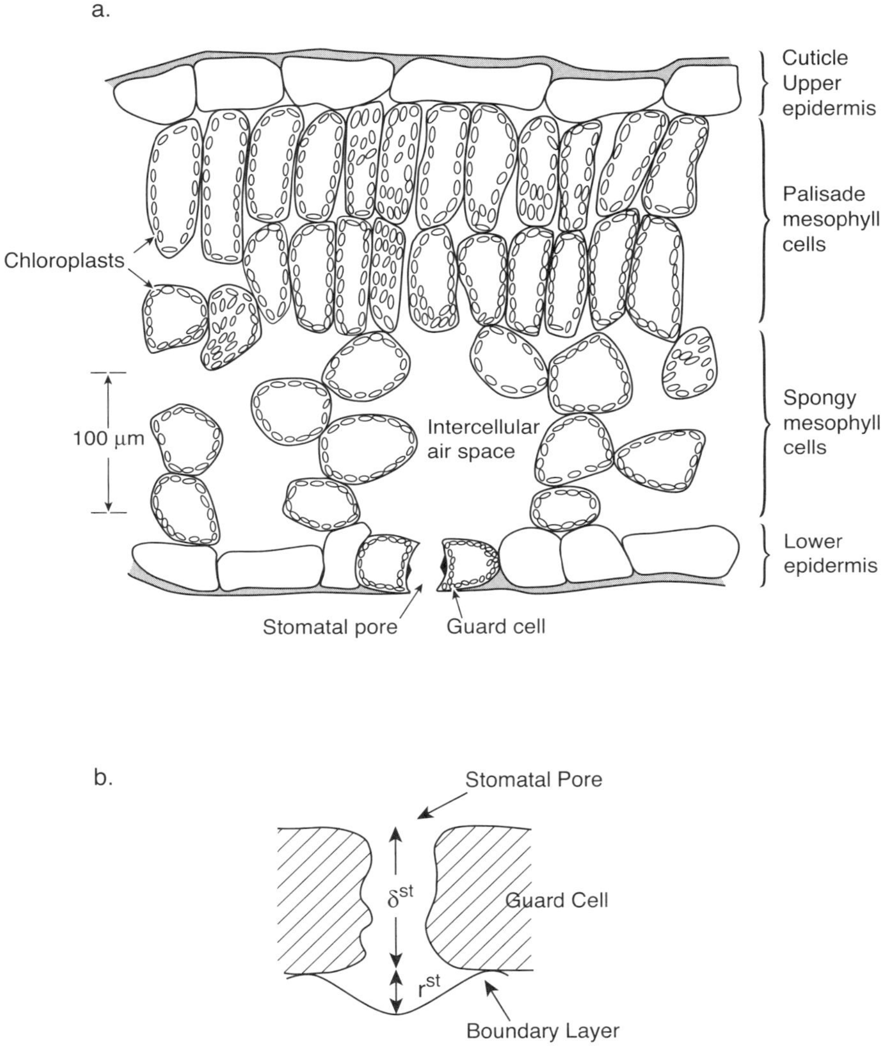

FIGURE 3.5
Mass transport in a leaf: (a) leaf cell organization and (b) transport in a stomatal pore.

have a typical length of 10 to 20 μm, an average radius of 5 to 10 μm, and cover a fractional area ranging from 0.002 to 0.02 (0.2 to 2%). The remainder of the leaf surface is covered by a layer termed *cuticle*, which is essentially impermeable to gases (Figure 3.5a).

The stomatal pores are flanked on either side by guard cells that control the size of the pore opening by expanding or contracting in response to external stimuli. In the dark, and at low external humidities, the guard cells are triggered to expand, partially closing the stomatal pores. Under these conditions, transport of gases through the pores is reduced or ceases. The movement of the guard cells, which causes the losses, is brought about by a change in osmotic pressure of the cell fluids.

The interior of the leaves contains the *mesomorphic cells*, which are arranged in either a loosely packed, "spongy" configuration or in the denser "palisade" form. The intervening spaces are taken up by intercellular air. These mesomorphic cells contain smaller cells termed *chloroplasts*, which in turn carry chlorophyll, the principal substance responsible for photosynthesis (i.e., the conversion of carbon dioxide into organic compounds and oxygen). This process is addressed in greater detail in Practice Problem 3.7.

In the present illustration, we consider the transport of water vapor from the interior of the leaf through the stomatal pores into the surrounding atmosphere. To obtain an assessment of the maximum possible moisture loss, we assume the interior of the leaf to be saturated with water vapor.

In principle, the external resistance will depend on wind conditions that vary with time as well as with location. Extensive studies have shown that this resistance can be accounted for in an approximate fashion by adding a distance equal to the average stomatal radius, r^{st}, to the length of the pore, σ^{st}. The total effective length then becomes $^{st} = \sigma^{st} + r^{st}$ (Figure 3.5b).

Let us use these concepts to calculate the water loss from a typical garden or city tree of modest size. We assume a stomatal length and radius of $\sigma^{st} = 15 \times 10^{-6}$ m and $r^{st} = 7.5 \times 10^{-6}$ m, with a fractional open area of 1% ($\varepsilon = 0.01$), and we set the leaf area at 25 cm^2 and the leaf population at 10,000 leaves per tree. External humidity is assumed to be 50% at a temperature of 25°C, with a corresponding saturation water vapor p_{sat} pressure of 23.8 mmHg. Diffusivity D of water vapor in air is 2.6×10^{-5} m^2/s (Table 3.1). The relevant transport equation is then given by

$$N = D\varepsilon A \frac{\Delta C}{\Delta x} = D\varepsilon A \frac{C_{sat} - 0.5C_{sat}}{\sigma^{st} + r^{st}} \tag{3.15a}$$

We first calculate the saturation concentration C_{sat}, which is given by

$$C_{sat} = \frac{n_{sat}}{V} = \frac{p_{sat}}{RT} = \frac{(23.8/760)1.01 \times 10^5}{8.31 \times 298} = 1.28\ \text{mol/m}^3 \tag{3.15b}$$

Substitution of this and other numerical values into Equation 3.15a then yields

$$N = 2.6 \times 10^{-5} \times 0.01 \times 10^4 \times 25 \times 10^{-4} \frac{1.28 - 0.5 \times 1.28}{15 \times 10^{-6} + 7.5 \times 10^{-6}} \tag{3.15c}$$

$$N = 0.185\ \text{mol/s} \tag{3.15d}$$

This is the maximum possible moisture loss under the stipulated conditions. Actual losses may be considerably lower due to less than saturation values in the interior and due to evaporative cooling, which lowers the interior water vapor pressure and hence diminishes the driving force.

Comments

The first impression of this model is its striking simplicity in the face of fairly complex circumstances. The use of an equivalent resistance for the boundary layer equal to the radius of the stomata is not immediately transparent but becomes more reasonable upon closer scrutiny of the configuration involved. Because no water vapor emanates from the leaf except at the stomata, the humidity will be constant over the impermeable portion of the leaf. Concentration cannot change abruptly to the level prevailing at the pore mouth, and one must therefore expect that contours in the form of humps will develop around the opening (Figure 3.5b). Based on these concepts, early workers postulated an equivalent resistance equal to the average pore radius. This was an inspired approximation that was later confirmed in more elaborate studies. These more refined results did not result in substantial changes in the original approximation. The use of the equivalent length r^{st} is now firmly entrenched in the literature.

In Practice Problem 3.8, we will examine the reverse process of CO_2 uptake by the leaf. The transport of carbon dioxide into the leaf interior is a more complex phenomenon involving several resistances as well as the reaction steps of photosynthesis. It is nevertheless possible to arrive at some simple results that illuminate this hugely important process with implications for global warming. Practice Problem 3.11 balances these results against the daily emissions of an average car.

Illustration 3.7: Diffusivity in a Catalyst Pellet

Catalysts almost invariably consist of porous particles with a substantial internal surface area, which typically varies from a few square meters per gram to a few tens or even 100 m^2/g. For a reaction to occur, the reacting species must diffuse into the interior of the particle to reach the reactive surface. Conversely, the product gases, once formed, must in turn diffuse out in order to maintain a steady state. Both the diffusional process and the local reaction rate play a role in determining the overall rate of conversion of reactants. When diffusion is fast, which is the case for small particles with large pores, the reaction becomes the rate-determining step. Conversely, when the particle is large and the pores are small, the rate of diffusion becomes the dominant factor. This interplay of diffusion and reaction will be examined in greater detail in Chapter 4 (Illustration 4.9). It is shown there that the impeding effect of diffusional resistance can be expressed in terms of an effectiveness factor E, which varies in value from zero to one. When diffusion is fast, reaction is the dominant process, and E approaches one. Conversely, low diffusion rates and fast reactions lead to small values of the effectiveness factor.

The evaluation of E requires knowledge of both the diffusivity and the reaction rate constants. Our task in the present illustration is to estimate the diffusivity of a reacting species, given certain physical parameters of the catalyst particle.

Let us consider the diffusion of oxygen in a silica-alumna cracking catalyst with an average pore radius of 24 Å = 24×10^{-8} cm and a void fraction ε of 0.3 cm^3/cm^3. The tortuosity factor is not known, and we consequently use an average value of $\tau = 4$. Because pore radius is much smaller than the mean free path, which is of the order of 10^5 cm at atmospheric pressures, it is suspected that Knudsen diffusion may be operative. We turn to Equation 3.14 and obtain, for a reaction temperature of 420 K,

$$(D_K)_{pore} = 9700 \times 24 \times 10^{-8} \, (420/32)^{1/2} \tag{3.16a}$$

$$(D_K)_{pore} = 8.5 \times 10^{-3} \text{ cm}^2/\text{s} \tag{3.16b}$$

This value is more than ten times lower than the molecular diffusivities of 0.20 cm^2/s. We conclude therefore that Knudsen diffusion is indeed the operative mode of diffusion. To calculate the effective diffusivity D_e for the entire particle, we draw on Equation 3.13a and obtain

$$D_e = D_{pore} \, \varepsilon/\tau = 8.5 \times 10^{-3} \times 0.3/4 \tag{3.16c}$$

$$D_e = 6.4 \times 10^{-4} \text{ cm}^2/\text{s} \tag{3.16d}$$

Illustration 3.8: Uranium Isotope Separation by Knudsen Diffusion

The process for gas separation by diffusion used in the Manhattan Project during World War II relies on the inverse relation of the Knudsen coefficient D_K with molar mass M. It requires the use of a porous barrier or membrane, operated at a sufficiently low pressure to allow Knudsen diffusion to be the dominant transport mechanism. In addition, the downstream product is usually rapidly pumped away in order to minimize back-diffusion ($C_2 \approx 0$). We can then write, for the U^{235} enrichment process,

$$(N/A)_{235} = \left(D_K \frac{C_1}{L}\right)_{235} = \left(D_K \frac{y_1 P_T}{RTL}\right)_{235} \tag{3.17a}$$

and similarly for the heavy species,

$$(N / A)_{238} = \left(D_K \frac{y_1 P_T}{RTL} \right)_{238} \tag{3.17b}$$

where L = membrane thickness, and the mole fraction $y^{238} = 1 - y^{235}$. Division of the two equations then yields

$$\frac{N^{235}}{N^{238}} = \frac{y_1^{235}}{1 - y_1^{238}} \left(\frac{M^{238}}{M^{235}} \right)^{1/2} \tag{3.17c}$$

At steady state, the same ration of N applies at the downstream end (y_2) and can be expressed in terms of the enriched downstream mole fraction y_2^{235}:

$$\frac{N^{235}}{N^{238}} = \frac{y_2^{235}}{1 - y_2^{235}} \tag{3.17d}$$

We then have, by division of Equation 3.17c and Equation 3.17d,

$$\alpha = \frac{y_2^{235} / 1 - y_2^{235}}{y_1^{235} / 1 - y_1^{235}} = \left(\frac{M^{238}}{M^{235}} \right)^{1/2} \tag{3.17e}$$

α will be recognized as the same "separation factor" as was used to characterize the gas centrifuge process (Illustration 1.3). For the uranium hexafluoride compounds in question, α becomes

$$a = \left(\frac{352}{349} \right)^{1/2} = 1.0043 \tag{3.17f}$$

This value is vastly inferior to that obtained by the centrifuge method (1.18, see Illustration 1.3). It implies that extensive staging is required to separate the isotopes by diffusion. The actual number of stages ultimately used in the production unit (Figure 3.6) amounted to more than 4000. Of particular interest to chemical engineers is the fact that the *minimum* stage requirement, which applies under conditions of *no product withdrawal*, was calculated by an equation derived some 10 years earlier by Fenske for multistage *distillation* at total reflux (*Ind. Eng. Chem.* 24, 482 [1932] see Illustration 7.10).

Comments

The outline given above gives a mere hint of the difficulties that had to be overcome. The preparation of the membrane barriers with the required pore

FIGURE 3.6
The Oak Ridge, Tennessee, gaseous diffusion plant, built 1943–1944 at a cost of $500 million.

size presented a huge initial challenge. Each of the 4000-plus stages was serviced by two compressors and one cooler, and all operational parts had to be leakproof and resistant to the highly corrosive hexafluoride gas. Some thirty mathematicians were kept occupied for 2 years to work out the difference-differential equation that would allow them to calculate the approach to steady state, which took several months. The power plant ultimately installed was rated at 250 MW, one hundred times the requirement of the modern centrifuge process.

Readers who question the morality of the entire enterprise should keep in mind the uncertainties and pressures of the times. From an engineering perspective, it is interesting to speculate whether a crash development program would have led to a viable centrifuge process and possibly have changed the course of the war. This was certainly considered, and the decision to opt for the membrane process must have been among the most difficult of the project.

3.1.3.3 Diffusion of Solids in Solids

The diffusion of solid ions, atoms, or molecules through solid matrices, while seemingly not possible, can and does in fact take place. It does so by a mechanism akin to that which described liquid-phase diffusion: Local

TABLE 3.7

Approximate Diffusivities of Solids in Solids

Diffusion Species	Solid Matrix	D (cm²/s)	T (K)
C	Fe	6×10^{-6}	1667
C	Fe + 2% Cr	5×10^{-5}	1667
C	Fe + 1% Mn	4×10^{-4}	1667
B	Ge	1×10^{-16}	1000
B	Ge	3×10^{-14}	1110
B	Ge	4×10^{-13}	1176
Si	Si	7×10^{-15}	1429
Si	Si	3×10^{-12}	1667

density fluctuations lead to a momentary opening or vacancy into which a neighboring particle can displace itself (see Figure 3.1b). Thus, a diffusional flow occurs, which follows Fick's law, as do more conventional diffusional processes.

A short compilation of important diffusion coefficients is presented in Table 3.7. Several features are of note here. The first is the strong dependence on temperature that we see in the diffusivities of boron in germanium, and the self-diffusion of Si. A 10 to 20% increase in absolute temperature can lead to a rise in diffusivity of several orders of magnitude. The temperature dependence is clearly exponential and is commonly expressed by an Arrhenius-type relation:

$$D = D_o \exp(-E_a/RT) \tag{3.18}$$

where E_a is an activation energy, and D_o is a preexponential factor. The implication here is that the solid-phase diffusion is to be viewed as an activated process in which the diffusing particle has to attain a threshold activation energy E_a before displacement can take place.

A second feature seen in Table 3.7 is the strong influence of impurities on the magnitude of diffusivities. An impurity level of only 1% can result in a dramatic increase in the diffusion coefficient of several orders of magnitude (see the diffusion of carbon in iron). The reason for this lies in the local change in packing of the molecules with an attendant increase in the probability of a vacancy opening up. While the magnitude of solid–solid diffusivities generally falls below values of 10^{-10} cm²/s, much higher levels comparable to those that prevail in liquids are also encountered. An example of this is seen in the diffusion of carbon in solid iron (Table 3.7).

The reader may be puzzled as to the relevance of solid–solid diffusional processes. The low values that pertain and the universal nature of the system would appear to make this a topic of mainly academic interest. In fact, the diffusion of solids within solids is of high importance in a number of disciplines. In metallurgy, solid-phase diffusion is induced to desegregate local accumulations of alloy components and to relieve stresses caused during casting. This is implemented by maintaining the cast form just below its softening point for prolonged periods of time. The resulting diffusional process is unsteady and uses as its initial condition the concentration profile that was derived in Illustration 2.9.

Solid-state diffusion also plays an important role in the manufacture of semiconductors. To produce the junctions needed in these devices, a dopant such as boron is deposited on the surface of the semiconductor crystal (e.g., silicon or germanium) and is subsequently made to diffuse into the interior. This process, termed *drive-in diffusion*, is again carried out at elevated temperatures. An analysis of it and some relevant calculations appear in Chapter 4, Practice Problem 4.7. The diffusivity required in these calculations is derived in the short illustration below.

Illustration 3.9: Diffusion of a Dopant into a Silicon Chip: Temperature Effect on Transport Rates

The strong effect of the exponential Arrhenius relation on transport rates is often not fully realized. Even a minor increase in temperature can bring about a substantial increase in diffusion rates. Table 3.7 gave some indication of this effect.

As a further demonstration, consider the following example. Suppose that at 1000 K a dopant concentration is found to penetrate a required distance in 1 *h*. It is desired to reduce the time to 1 *min* (i.e., increase diffusivity 60-fold). The ratio of the old to new diffusivity is immediately obtained from Equation 3.18:

$$\frac{\exp(-E_a / RT)}{\exp(-E_a / RT)} = 60 \tag{3.19a}$$

For a value of E_a = 357 kJ/mol (boron in silicon), we obtain

$$\frac{\exp(-3.57 \times 10^5 / 8.31T)}{\exp(-3.57 \times 10^5 / 8.31 \times 10^3)} = 60 \tag{3.19b}$$

and, hence, solving for T,

$$T = \frac{3.56 \times 10^5 / 8.31}{3.56 \times 10^5 / 8.31 \times 10^3 - \ln 60} \tag{3.19c}$$

$$T = 1105\ K \tag{3.19d}$$

Thus, a relatively small increment in T of about 100 K (10% of the previous value) has brought about a 60-fold increase in the diffusion rate. Evidently this felicitous property of the Arrhenius relation allows us to manipulate transport rates to a considerable extent without resorting to exotic temperature levels.

Practice Problems

3.1 A Fermi Problem: Estimation of Gas Diffusivities

Make an order of magnitude estimate of the diffusion coefficient in simple gases other than He and H_2, given that the velocity of sound at atmospheric pressure is of the order 300 m/s and the liquid to gas densities for most gases at atmospheric pressure are in the approximate ratio of 1:1000.

(*Hints*: Sound propagates approximately at the same speed as that of molecular motion. The diameter of a typical small molecule is of the order of 1 Å = 10^8 cm.)

3.2 Pressure Dependence of Diffusivity in Gases

a. Explain why molecular diffusivity varies inversely with pressure.
b. Why is Knudsen diffusivity independent of pressure?
c. Is Knudsen diffusion more likely to occur at high pressures or low pressures?
d. Given that the mean free path of a gas molecule varies inversely with pressure and is approximately 10^{-7} m at 1 atm, calculate the pressure that separates Knudsen from molecular diffusion for (1) a 100-Å diameter pore and (2) a 1000-Å pore. 1 Å (Ångstrom) = 0.1 mm. What happens when these threshold values are exceeded?

3.3 Flow of Gases at Low Pressures

Pressures as low as 10^{-10} atm are routinely attained in research laboratories.

a. What is the mean free path of a gas molecule at this pressure?
b. The conduits used to evacuate vessels are surprisingly large (many centimeters in diameter). Give a reason for this.
c. What is the nature of gas flow at this pressure?

3.4 Estimation of the Diffusivity of DDT in Water

Use the Wilke–Chang equation (Equation 3.3) to estimate the diffusion coefficient of DDT in water at 25°C. The viscosity of water at this temperature is 0.894 cp (centipoises). Compare the result with the value tabulated in Table 3.3E. (*Hint*: The formula for DDT is given by 1-trichloro-2,2 bis(*p*-chlorophenyl)ethane. It contains two phenol rings and five chlorine atoms.)

3.5 Electrorefining of Copper

Copper is to be refined electrolytically at the rate of 10 kg/h using a cell with 2 cm plate spacing and a 1 *M* $CuSO_4$ electrolyte.

a. What is the minimum electrode area required to carry out the process?
b. If the electrode dimensions are 2 m × 2 m × 1 cm, what is the minimum length of the electrolytic cell?

Answer:

a. 309 m^2

3.6 Diffusivity in Polymers

Although permeability is the most commonly employed transport coefficient for polymers, occasional use is also made of diffusivities.

a. Calculate the range of diffusivities for the substances listed in Table 3.6. How do they compare with the diffusivity of gases in liquids?
b. Repeat the calculation for Illustration 3.3, Part 2, using a diffusivity and solid-phase concentration instead of permeability and a partial-pressure driving force.

3.7 Performance of Saran Wrapping

A package is made of three ears of corn by placing them on a Styrofoam® tray and enclosing them with Saran Wrap® 0.01 mm thick. The exposed area is 400 cm^2. It is desired to keep moisture losses below 1% of the total water content estimated at 100 g per package for a shelf life of 10 days. Does the wrapping meet these requirements? Assume a constant internal saturation vapor pressure of 23.8 mmHg (25°C) and zero external moisture.

3.8 Photosynthesis: CO_2 Uptake by an Average Tree

Photosynthesis involves the uptake by plants of carbon dioxide from the surrounding air and its conversion to organic compounds and oxygen:

$$CO_2 + H_2O + other \xrightarrow[synthesis]{Photo-} Organic\ Compounds + Oxygen$$

TABLE 3.8

Representative Values of Resistances for CO_2 Diffusion into Leaves

Component	Resistance (sm^{1})
Crops, open stomata	170–830
Trees, open stomata	500–2500
Mesophyll cells	
Estimation	600
Measurements	200–800
Chloroplasts	
Estimation	<500
Measurements	<400

It is the largest synthetic process on earth and results in the fixing of approximately 7×10^{13} kg carbon per year. Transport of CO_2 into the leaves is considered to be the rate-determining process in this chain of events and involves a series of steps represented as a sum of resistances (i.e., we use the principle of additivity of resistances introduced in Chapter 1). The Fickian rate process is then expressed in the form

$$N / A = \frac{\Delta C}{\Sigma R}$$

where ΔC is the concentration driving force, and ΣR represents the sum of resistances. In a leaf, the major resistances are three in number and are due to (1) the stomata, (2) the mesophyll cells, and (3) the chloroplasts (see Figure 3.5a). The range of values of resistances is tabulated in Table 3.8. We wish here to calculate the maximum daily CO_2 uptake by the modest-sized tree used in Illustration 3.6. To this end, we choose the lower values of the resistances listed in Table 3.8 and set the chloroplast resistance at 200 sm^{1}. We further neglect the carbon dioxide produced by the tree by respiration to maintain its own life. This is usually less than 10% of the total carbon dioxide uptake. Because photosynthesis is rapid compared to the transport process, the carbon dioxide concentration within the chloroplasts is set at zero. These are the approximations we use to calculate the carbon dioxide uptake.

Answer:

15 mol CO_2/day

3.9 Choice of a Packaging Material

Packaging materials often come in the form of laminates composed of individual film barriers, each one of which is designed to meet a particular requirement. These include the reduction of light or heat transmission and the retention or exclusion of certain gases and vapors. Most packaged foods continue, after packaging, to undergo certain metabolic changes, including respiration. They are, in a sense, still "alive." These changes are usually undesirable, and packaging materials must be designed to reduce these effects to a minimum. Consider the case of respiration, which is analogous to breathing in humans and has to be minimized as far as possible. Which of the materials listed in Table 3.5 would give the best performance in this respect?

3.10 Performance of a Weather Balloon

A weather balloon filled with helium is to rise to a height of 3 km to collect data. The temperature and pressure at this level are at 270 K and 70 kPa. The balloon is made of Saran Wrap® half a millimeter thick and is assumed to be at 1 atm pressure.

a. What is the diameter of the balloon? (No mass transfer principles here. Use Archimedes law.)
b. The balloon is to lose no more than 1% of its contents per hour in order to maintain its approximate position. Is this requirement met?
c. Comment on possible alternative materials.

Answer:

a. 1.88 m

3.11 Carbon Dioxide Emissions from Cars: How Many Trees Will Compensate the CO_2 Output of One Vehicle?

The task in this example is to compute the daily production of carbon dioxide by a vehicle driven an average of 100 km daily (36,500 km per year) with a gasoline mileage of 15 km/L. Assume the fuel to be pure octane (C_8H_{18}) with a specific gravity of 0.8, which is quantitatively burned to carbon dioxide. Compare the calculated result with that given for Practice Problem 3.8.

3.12 On U^{235} Enrichment

Identify two factors that lead to much higher separation factors by the centrifuge method than by membrane diffusion.

3.13 More on the Temperature Dependence of Solid-Solid Diffusion

Diffusion of solids within solids usually takes place at elevated temperatures of the order of 1000 K. The activation energies required to trigger the process are typically in the range 100 to 500 kJ/mol.

Show that at the level of 1000 K, a temperature rise of 100 K enhances the rate of diffusion by the following factors:

Activation energy (kJ/mol)	100	200	300	400	500
Enhancement factor EF	2.99	8.92	26.6	79.5	237

4

More about Diffusion: Transient Diffusion and Diffusion with Reaction

The process of diffusion is central to much of what we describe as mass transfer. It manifests in a variety of ways and needs to be looked at repeatedly, even at this introductory level, in order to grasp its full ramifications. This was done by first introducing, in Chapter 1, the notion of the rate of diffusion, enshrined in Fick's law, and to the linear driving force mass transfer rates that derive from it. In Chapter 2 we demonstrated how these rate expressions are incorporated into mass balances leading to models of various mass transfer processes. In Chapter 3, diffusivities, the all-important proportionality constants in Fick's law, were examined at both the molecular and macroscopic levels.

The subject of diffusion is returned to in this chapter, and an examination takes place of what happens when the rate of diffusion varies with both time and distance (Section 4.1) and when diffusion occurs simultaneously with a chemical reaction (Section 4.2). These are more advanced topics, which in the case of Section 4.1 lead to partial differential equations, notably Fick's equation given in Chapter 2 (Equation 2.18c). We do not attempt to solve it here, which would merely distract us from the main task, so we confine ourselves instead to a presentation of the more important results in either analytical or graphical form. These results are then used to solve a range of practical problems, a task that is far from trivial in spite of the appearance it gives of applying a set of convenient "recipes." The discussion in Section 4.2 is confined to steady-state processes in which the state variable varies only with distance. Hence, no partial differential equations arise here. We do, however, have to deal with ordinary differential equations that sometimes require going beyond the elementary separation-of-variables technique seen in previous chapters by using the so-called *D*-operator method. This procedure is outlined in Appendix A1 at the end of the text.

4.1 Transient Diffusion

In transient diffusion, the concentration of a species varies, as we have seen, with both time and distance. The underlying process of diffusion may take place in isolation or it may be accompanied by a chemical reaction, by flow, or by both reaction and flow. These more complex cases are not taken up here, and we limit ourselves instead to the consideration of purely diffusive processes. Furthermore, with one or two exceptions, the treatment is confined to a *single* spatial coordinate represented by the Cartesian x- (or z-) axis or by the radial variable r. The last is used in formulating diffusion in a sphere, or in the radial direction of a cylinder. Fick's equation for these three cases can be deduced from the general conservation equation (Equation 2.24a) and Table 2.3. The three cases are as follows:

Cartesian coordinate in one dimension

$$\frac{\partial C}{\partial t} = D\frac{\partial^2 C}{\partial x^2} \tag{2.18a}$$

Spherical coordinate

$$\frac{\partial C}{\partial t} = D\left[\frac{\partial^2 C}{\partial r^2} + \frac{2}{r}\frac{\partial C}{\partial r}\right] = D\left[\frac{1}{r^2}\frac{\partial}{\partial r}\left(r^2\frac{\partial C}{\partial r}\right)\right] \tag{2.18b}$$

Cylindrical coordinate (radial direction)

$$\frac{\partial C}{\partial t} = D\left[\frac{\partial^2 C}{\partial r^2} + \frac{1}{r}\frac{\partial C}{\partial r}\right] = D\left[\frac{1}{r}\frac{\partial}{\partial r}\left(r\frac{\partial C}{\partial r}\right)\right] \tag{2.18c}$$

The alternative formulations given on the right are sometimes found useful in the solution of steady-state problems.

Associated with these equations are a host of boundary and initial conditions. One can, for example, specify the concentration on the surface, or the flux prevailing there. Alternatively, there may be a film resistance at the surface, in which case the rate of mass transfer must be equated to the rate of diffusion at the surface.

Initial conditions may also be complex. Suppose, for example, that we wish to solve Fick's equation for transient diffusion into a sphere. The simplest case here is to assume that the sphere is initially "clean" (i.e., contains no solute). But what if it is not? We would then have to specify an initial concentration distribution $C_{t=0} = f(r,\theta,\varphi)$, and this distribution would have to be entered into the solution process as an initial condition. Evidently there are an infinite number of such distributions; hence, Fick's equation for this case will have an unlimited number of solutions.

Yet another complication arises if the diffusion process is triggered by a mass *source*. A puff of smoke emanating from a chimney, for example, constitutes such a source. If it lasts for only an instant, we speak of an *instantaneous source*. If the emanation persists, we speak of a *continuous source*. The existence of such sources must be incorporated into the Fickian model and leads to what we term *solutions of source problems*. Because of their importance, particularly in an environmental context, we will address this topic separately in some detail.

This short discussion serves to alert you to the dismaying fact that Fick's equation has an unlimited number of solutions that are intimately linked to the initial and boundary conditions and to the presence of sources. To keep our deliberation to manageable proportions, we propose to address the following limited but highly important and instructive cases:

1. Diffusion due to instantaneous and continuous sources emitting into infinite and semi-infinite domains — Both point sources as well as area sources of infinite extent are addressed. These are referred to as *Source Problems*.
2. Transient diffusion in a semi-infinite domain, into a solid bounded by parallel planes, in a sphere, and in the radial direction of a cylinder — The domain is assumed to be initially uniformly loaded or uniformly clean and to have a constant surface concentration. We term these *Nonsource Problems*. Occasional departures from the stated assumptions will be noted as they occur.

4.1.1 Source Problems

We start by considering a source located at the origin of a three-dimensional Cartesian coordinate system. At time $t = 0$, the source releases a substance of mass M_p kg. The release is instantaneous and leaves immediately (i.e., it is of infinitesimally short duration). Thereafter, the released substance diffuses into the three-dimensional infinite space surrounding the source, giving rise to time-dependent three-dimensional concentration profiles $C(x, y, z, t)$. The three coordinates x, y, z can be combined into a single radial distance r anchored at the origin, where $r^2 = x^2 + y^2 + z^2$. The resulting concentration profiles are given by the following expression:

INSTANTANEOUS POINT SOURCE EMITTING INTO INFINITE SPACE

$$C = \frac{M_p}{8(\pi D t)^{3/2}} \exp(-r^2 / 4Dt) \tag{4.1a}$$

where D is the diffusivity, and M_p is the emitted mass in kilogram or moles. In the corresponding case of a *heat* source, C is replaced by temperature T, M_p by amount of heat released q_p (J) divided by ρC_p, and D by thermal diffusivity α.

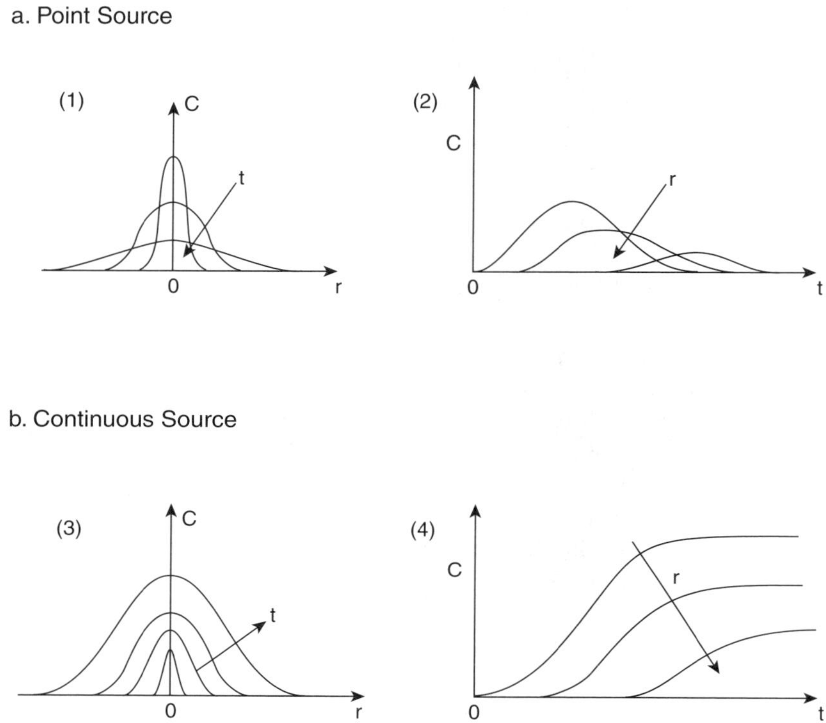

FIGURE 4.1
Concentration profiles and histories in source problems (infinite domain).

Equation 4.1a is termed the *fundamental solution to the point source problem*. All other solutions of such problems flow from it, in one way or another. Its derivation is not usually done in a formal mathematical way. It relies instead on three factors:

1. It correctly represents the physical profiles that have the shape of a Gaussian distribution or "bell curve" (Figure 4.1).
2. It satisfies Fick's equation (proof by substitution).
3. It satisfies the initial and boundary conditions.

For example, at the origin $r = 0$, Equation 4.1a reduced to the expression

$$C_{r=0} = \frac{M_p}{8(\pi D t)^{3/2}} \tag{4.1b}$$

that is, the concentration at the origin decays inversely with $t^{3/2}$ (Figure 4.1b (1)). If, in addition, one sets $t = 0$, one obtains

$$C_{r=0,t=0} = \infty \tag{4.1c}$$

that is, at the instant of release, the concentration at the origin is infinite.

Table 4.1 contains a number of additional solutions of both instantaneous and continuous source problems in infinite, semi-infinite, or half space. Let us examine two of these tabulated solutions.

The case of emissions into half space (see Item 2 of Table 4.1) can be obtained from the fully infinite case by arguing that the mass that would otherwise have gone into the lower half space is reflected upward, thus doubling the concentration. We obtain the following:

INSTANTANEOUS POINT SOURCE ON AN INFINITE PLANE EMITTING INTO HALF SPACE

$$C = \frac{M_p}{4(\pi Dt)^{3/2}} \exp(-r^2 / 4Dt) \tag{4.2}$$

The solutions for instantaneous *plane sources* (Items 3 and 4, Table 4.1) are similar in form, with the amount emitted, M_A, now being expressed in units of kilogram per square meter (kg/m^2).

The second major category to be considered is that of a *continuous source*. We would intuitively expect that the solution to this problem could in principle be obtained by integrating over time the result obtained for an instantaneous source. This is the case and leads to the emergence of a special type of integral, termed the *error function*, erf x, defined as

$$\text{erf } x = \frac{2}{\pi^{1/2}} \int_0^x \exp(-u^2)du \tag{4.3a}$$

The error function can be viewed as a partial area under the Gaussian distribution curve $(2/\pi^{1/2})\exp(-u^2)$ (Figure 4.2). It has the value 0 at $x = 0$, and a value of 1 at $x = \infty$. The latter case corresponds to the half area under the distribution curve. A related expression is the *complementary error function*, erfc x, defined as

$$\text{erfc } x = 1 - \text{erf } x \tag{4.3b}$$

and with the values erfc(0) = 1 and erfc(∞) = 0.

The error function integral has to be evaluated numerically and can be found tabulated in texts of diffusion or conduction, or in mathematical tables. An abbreviated listing appears in Table 4.2, and some important properties of the function are summarized in Table 4.3.

With these definitions in place, we can now proceed to present the solution to the continuous point source problem. It is given by (see Item 5 of Table 4.1)

TABLE 4.1

Source Problem Solutions

1. Instantaneous Point Source Emitting into Infinite Space

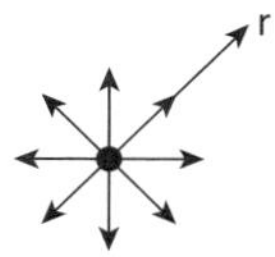

$$C = \frac{M_p}{8(\pi Dt)^{3/2}} \exp(-r^2/4Dt)$$

M_p in kg

2. Instantaneous Point Source on an Infinite Plane Emitting into Half-Space

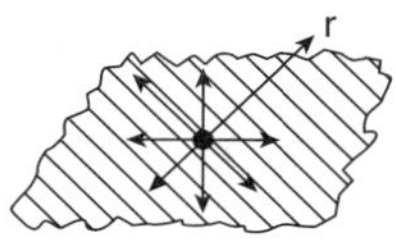

$$C = \frac{M_p}{4(\pi Dt)^{3/2}} \exp(-r^2/4Dt)$$

M_p in kg

3. Instantaneous Infinite Plane Source Emitting into Infinite Space

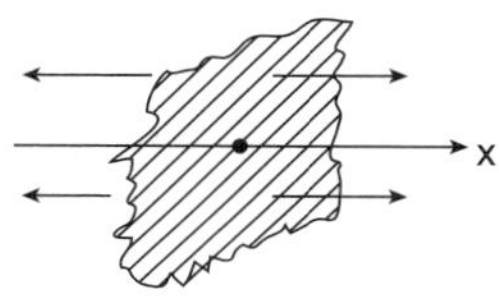

$$C = \frac{M_A}{2(\pi Dt)^{1/2}} \exp(-x^2/4Dt)$$

M_A in kg/m^2

4. Instantaneous Infinite Plane Source Emitting into Half Space

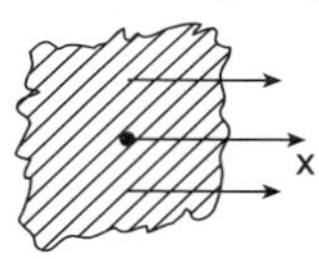

$$C = \frac{M_A}{(\pi Dt)^{1/2}} \exp(-x^2/4Dt)$$

M_A in kg/m^2

5. Continuous Point Source Emitting into Infinite Space

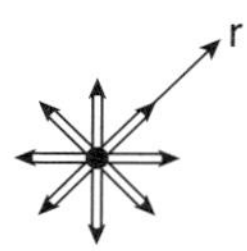

$$C = \frac{M_{CP}}{4\pi Dr} \operatorname{erfc} \frac{r}{2\sqrt{Dt}}$$

M_{CP} in kg/s

6. Continuous Point Source on a Semi-Infinite Plane Emitting into Half Space

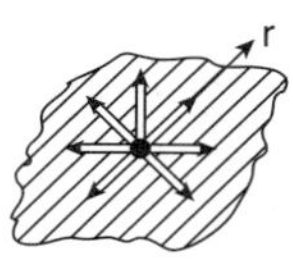

$$C = \frac{M_{CP}}{2\pi Dr} \operatorname{erfc} \frac{r}{2\sqrt{Dt}}$$

M_{CP} in kg/s

7. Continuous Infinite Plane Source Emitting into Half Space

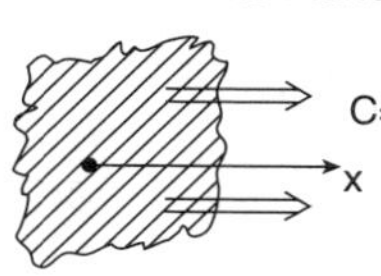

$$C = M_{CA} \left[\left(\frac{t}{\pi D}\right)^{1/2} \exp(-x^2/4Dt) - \frac{x}{2D} \operatorname{erfc} \frac{x}{2\sqrt{Dt}} \right]$$

M_{CA} in kg/m^2 s

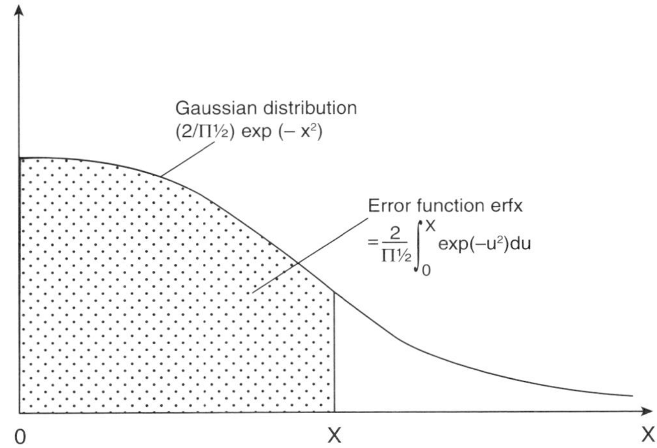

FIGURE 4.2
Gaussian distribution curve and the error function.

TABLE 4.2
Values of the Error Function

x	erf(*x*)
0	0
0.05	0.05637
0.1	0.11246
0.15	0.16800
0.20	0.22270
0.25	0.27632
0.30	0.32863
0.35	0.37938
0.40	0.42839
0.50	0.52050
0.60	0.60386
0.70	0.67780
0.80	0.74210
0.90	0.79691
1.0	0.84270
1.2	0.91031
1.5	0.96611
2.0	0.99532
2.5	0.999593
3.0	0.999978
∞	1.00000

TABLE 4.3

Properties of the Error Function

1. $\text{erf}(0) = 0$
2. $\text{erf}(\infty) = 1$
3. $\text{erf}(x) = \text{erf}(x)$
4. $1\ \text{erf}(x) = \text{erf}(x)$ (Complementary error function)
5. $\frac{d}{dx}\text{erf } x = \frac{2}{\sqrt{\pi}} e^{-x^2}$
6. $\int_x^\infty \text{erfc } x\ dx = \frac{1}{\sqrt{\pi}} e^{-x^2} - x \text{ erfc } x$
7. Approximation for small x: $\text{erf } x \cong 2\pi^{1/2}\, x$
8. Approximation for large x:

   $$\text{erfc } x \cong \pi^{-1/2} e^{-x^2}\left[\frac{1}{x} - \frac{2}{x^3} + \ldots\right]$$

CONTINUOUS POINT SOURCE EMITTING INTO INFINITE SPACE

$$C = \frac{M_{cp}}{4\pi Dr}\text{erfc}\frac{r}{2\sqrt{Dt}} \tag{4.4}$$

For emissions into half space, the concentration given by Equation 4.4 is again doubled (see Item 6 of Table 4.1). Solutions to continuous *plane* sources are, as before, similar in form (Items 7 and 8), with the emission rate M_{CA} now expressed in units of kilogram per square meter second (kg/m^{2}s). Table 4.1 also shows that concentrations due to *instantaneous* sources all contain *negative exponentials* (the distribution curve), while those due to *continuous sources* all contain *error functions* (area under distribution curve).

To obtain a better sense of the spreading concentration waves, we return to the concentration histories and profiles shown in Figure 4.1. In the case of instantaneous point sources, the pulse emitted at $t = 0$ gradually spreads about the origin and *diminishes* in strength (i.e., concentration decreases with the passage of time — see Part 1 of Figure 4.1). Continuous sources result in a similar spreading of concentrations about the origin, but here there is a continuous *increase* in the level, which ultimately reaches a plateau (see Part 4). Of interest is the appearance of *maxima* at a particular location of the domain. In the case of the instantaneous source, this maximum occurs at finite times (see Part 2), while the concentrations from continuous emissions attain their maximum only in the limit of $t \to \infty$ (see Part 4). These maxima will be addressed in Illustration 4.1 and in Practice Problems 4.1 and 4.2.

Illustration 4.1: Concentration Response to an Instantaneous Point Source: Release in the Environment and in a Living Cell

The problem considered here involves calculating the concentrations that arise when material released by an instantaneous point source diffuses into a semi-infinite domain. In the environment, this could come about, for example, when a tank containing pressurized liquefied gas is ruptured and releases the contents, which instantly evaporate into the surrounding air.

Let us set M_p at 1 kg (e.g., a small propane cylinder) and assume a typical molecular diffusivity in air of 10^{-5} m^2/s. We use the data to calculate the concentrations that result at a distance of $r = 100$ m from the source after 10, 100, and 1000 days, respectively. Substituting values into Equation 4.2, we obtain for $t = 10$ days,

$$C = \frac{1}{4(\pi \times 10^{-5} \times 3,600 \times 24 \times 10)^{3/2}} \exp\left(-\frac{100^2}{4 \times 10^{-5} \times 3,600 \times 24 \times 10}\right) = 0$$

and similarly for $t = 100$ and 1000 days. The results are tabulated below:

t (days)	C (kg/m³)
10	0
100	1.4×10^{-24}
1000	6.3×10^{-9}

The data show a very slow, gradual rise in concentration. A subsidiary question, which will be addressed before commenting on the results, focuses on the time it takes for the concentration to reach its *maximum value*. That maximum occurs, as was shown in Figure 4.1a, Part 2, at some finite point in time, which is *independent of the amount released*. This makes it a convenient quantity for establishing the time scale of these events.

We write Equation 4.2 in the following form

$$C(r,t) = \frac{u(t)}{v(t)} \tag{4.5a}$$

and require $\partial C/\partial t = 0$ or, alternatively,

$$(du/dt)v - u(dv/dt) = 0 \tag{4.5b}$$

We obtain

$$\exp(-r^2 / 4Dt)(r^2 / 4Dt)t^{3/2} - \exp(-r^2 / 4Dt)\frac{3}{2}t^{1/2} = 0 \tag{4.5c}$$

and, consequently,

$$(t)_{\text{Max}C} = \frac{1}{6}(r^2 / D) \tag{4.5d}$$

Substituting into Equation 4.2 yields the corresponding value for the maximum concentration:

$$C_{\text{Max}} = \frac{M_p}{4(\pi r^2 / 6)^{3/2}} \exp(-6/4) \tag{4.5e}$$

or

$$C_{\text{Max}} = 0.147\, M_p / r^3 \tag{4.5f}$$

The noteworthy feature of these results is the strong dependence on distance r and the fact that C_{Max} depends on amount released, M_p, but not at all on diffusivity, while the exact reverse is true for t_{Max}.

Substituting the numerical parameter values into Equation 4.5d and Equation 4.5f, we obtain

$$t_{\text{MaxC}} = \frac{1}{6}(100^2 / 10^{-5}) = 1.87 \times 10^8\,\text{s} = 2164\,\text{days} \tag{4.5g}$$

and

$$C_{\text{Max}} = 0.147 \times 1/100^3 = 1.47 \times 10^{-7}\ \text{kg/m}^3 \tag{4.5h}$$

Thus, it takes almost 6 years for the concentration to attain its maximum at the modest distance of 100 m. The good news is that by then the concentration has dropped to less than 1 ppm.

Evidently these numbers do not give a realistic picture of the actual process. Wind and other air currents would intervene to disperse the material at a much higher rate. Quantifying the effect of wind, even when it is unidirectional, is a more complex task, which is addressed in Illustration 4.3.

Let us next consider the instantaneous release from a point source within a living cell 50 μm in radius. Although the geometry here is finite, the initial stages of the diffusion process can be viewed as taking place into an infinite medium. If we assume $D = 10^{-10}$ m²/s, typical of a large protein, then the time it takes for the maximum concentration to reach $r = 10$ μm equals one half the value given by Equation 4.5d:

$$t_{\text{Max}} = \frac{1}{12}\frac{(10\times 10^{-6})^2}{10^{-10}} = 0.083 \text{ s} \tag{4.5i}$$

that is, attainment is near-instantaneous in spite of the much lower diffusivity. Note the vast difference in time scale for the two processes (6 years versus less than a second), which is entirely due to the quadratic dependence on *r*.

The lesson here, once again, is that molecular diffusion is extremely inefficient at distances exceeding a few centimeters but is highly effective at the micron level. Because most cellular processes take place at that level, the implications are considerable.

Illustration 4.2: Net Rate of Global Carbon Dioxide Emissions

The release of carbon dioxide into the atmosphere and its attendant effect on climate change have been widely publicized in recent years. A warning call came in the late 1980s and early 1990s when it was established that carbon dioxide concentrations in the air, then at a level of 370 ppm, were growing at a rate of 2 ppm per year. What we wish to do here is use this information to calculate the net rate of global carbon dioxide emissions (i.e., the amount in excess of what is consumed by plant life or taken up by water basins).

The tools required for this task reside in Item 7 of Table 4.1; this information describes the continuous emission from an infinite plane, here taken to be the surface of the earth, into a semi-infinite domain, the atmosphere.

We start by noting that at ground level, $x = 0$, the expression given there reduces to the form:

$$C = M_{CA}\left(\frac{t}{\pi D}\right)^{1/2} \tag{4.6a}$$

where M_{CA} is the rate of emission (in kg/m^{2}s), the unknown being sought here. The diffusivity D of carbon dioxide in air at 25°C is 2.2×10^{-5} m^2/s, the air density 1.18 kg/m^3.

We proceed as follows: Solving for t we obtain, for the interval of observation of 1 year,

$$t = \frac{C^2 \pi D}{M_{CA}^2} = 3{,}600 \times 24 \times 365 = 3.15 \times 10^7 \text{ s} \tag{4.6b}$$

The result can be recast in the following form:

$$M_{CA}^2 = \frac{\pi D(C_2{}^2 - C_1{}^2)}{t} = \frac{\pi D(C_2 + C_1)(C_2 - C_1)}{3.15 \times 10^7} \quad (4.6c)$$

where C_1 and C_2 are the concentrations at the start and the end of the year.

Before introducing numerical values, the data, given in parts per million (ppm) (m^3 CO_2/m^3 air) must be converted to kilogram per cubic meter (kg/m^3). This is done by multiplying parts per million values by the density of carbon dioxide, 1.79 kg/m^3. We obtain

$$M_{CA} = \left[\frac{\pi \times 2.2 \times 10^{-5}(370 + 372)(372 - 370)10^{-12}\ 1.79^2}{3.15 \times 10^7} \right]^{1/2} \quad (4.6d)$$

$$M_{CA} = 1.0 \times 10^{10}\ kg/m^2s \quad (4.6e)$$

The average radius of the earth is 6370 km. Hence, the averaged global amount of carbon dioxide emitted is given by

$$M_{Tot} = 1.0 \times 10^{10}\ 4\pi\ (6.37 \times 10^6)^2 \quad (4.7a)$$

$$M_{Tot} = 5.1 \times 10^4\ kg\ CO_2/s \quad (4.7b)$$

$$M_t = 5.1 \times 10^4\ kg\ CO_2/s = 1.5 \times 10^9\ tons/yr \quad (4.7c)$$

This is a *minimum* amount because we used the molecular diffusivity for CO_2. When D is increased hundred-fold to account for a mean turbulence at the ground level, M_t rises to 1.5×10^{10} tons/yr. This value corresponds roughly to the reduction targets set in the Kyoto Accord.

Comments

This is still a staggering amount, considering that it represents only the excess carbon dioxide produced in addition to what is removed by photosynthesis and other sinks. The gross rate of carbon dioxide production is thus considerably higher. Some measure of what plant life can accomplish to alleviate this effect was given in Practice Problem 3.8.

Illustration 4.3: Finding a Solution in a Related Discipline: The Effect of Wind on the Dispersion of Continuous Emissions

The problem of tracking the maximum concentration resulting from wind-borne continuous emissions is most conveniently solved by borrowing an equivalent solution from the literature on heat conduction. The definitive work on the topic by Carslaw and Jaeger (see Selected References) contains a section entitled "Moving Sources of Heat." Described in that section is a problem in which "... heat is emitted at the origin for times $t > 0$ at the rate q heat units per unit time and that an infinite median moves uniformly past the origin with a velocity U parallel to the x-axis" (p. 266). This situation corresponds to the case of a continuous point mass source emitting into half-space in which the air moves in the x-direction with a velocity U.

The complete solution is rather complex and is not reproduced. It does, however, reduce to a simple expression for the limiting steady-state case $t \to \infty$, which is of greater interest and which is given below in the form presented by Carslaw and Jaeger:

$$v = \frac{q}{4\pi KR} \exp\left[-U(R - x) / 2\kappa\right] \tag{4.8a}$$

where v = temperature (K), q = heat emitted (J/s), K = thermal conductivity (J/sm^{2}K), $R^2 = x^2 + y^2 + z^2$, and κ = thermal diffusivity (m^2/s).

Translation into the corresponding mass diffusion case is straightforward. We replace v by C (kg/m^3), q by M_{cp} (kg/s), and both K and κ by diffusivity D (m^2/s). The result is then given by

$$C = \frac{M_{cp}}{4\pi DR} \exp[-U(R - x) / 2D] \tag{4.8b}$$

and represents the maximum concentration attained at steady state in response to the continuous point source.

A quick inspection of this equation shows that the wind velocity has an enormous effect on the concentration response. With a modest wind velocity of U = 10 cm/s, an effective diffusivity of 10^{-3} m^2/s (average for all directions), and $(R - x)$ of the order 10^2 m, the exponential term immediately reduces to zero. Air movement would have to be reduced to the order of the diffusivity before any significant concentration levels arise.

The case of overriding interest here is that of the concentrations in the direction of the wind and along the positive x-axis. This would represent locations of maximum steady-state exposures. We have, in this case, $R = x$, and hence,

$$C_{\text{Max}} = \frac{M_{cp}}{4\pi Dx} \tag{4.8c}$$

We see here again the inverse dependence on distance x as was the case for the instantaneous point source, but this time it is linear, and the effective diffusivity enters the expression as well. It is now well established that such emissions originating in the southern part of the country can be airborne over long distances and affect the wildlife in Arctic regions.

FIGURE 4.3
Dispersion of a dye into a turbulent stream of water. The concentrations seen are the steady-state plateau values of a continuous source shown in Figure 4.1.

Before applying Equation 4.8c, an important question needs to be addressed: What value should be assigned to D? Evidently we are no longer dealing with molecular diffusivities but rather with what is termed "turbulent" or "eddy" diffusivities, D_e, which are several orders higher in magnitude. A sense of the scale of the events is provided by Figure 4.3, which shows the dispersion of a dye tracer in turbulent pipeline flow. Mixing in the axial direction is about one hundred times faster than in the radial direction.

Eddy diffusivity in the open atmosphere is highly variable and difficult to quantify, but a sense of its scale can be obtained from measured values of D_e in *large diameter* gas pipes (Sherwood et al., 1975, see Selected References). An approximate dependence on flow velocity, assumed to mimic our velocity, is given by the following relation, valid for $v > 0.1$ m/s:

$$\text{Axial } D_e \text{ (m}^2\text{/s)} \cong 0.1\, v \text{ (m/s)} \tag{4.8d}$$

Suppose a pollutant source in the south, or a cluster of such sources, produces toxic material at the rate of 100 g/s. If we use an eddy diffusivity of 10^{-3} m^2/s as an average over a long period of time, the maximum concentration attained at a distance of 1500 km will be

$$C = \frac{M_{Cp}}{4\pi Dx} \tag{4.8e}$$

$$C = \frac{10^{-1}}{4\pi 10^{-3} \times 1.5 10^{6}} \tag{4.8f}$$

$$C = 5 \times 10^{-6} \text{ kg/m}^3 \tag{4.8g}$$

or approximately 5 ppm for an air density of 1.3 kg/m^3. Note that this seemingly small amount is further enhanced by biomagnification in the animal food chain, leading to potentially harmful effects (see Illustration 5.7).

Comments

The actual transport of pollutants is considerably more complex. One mechanism, termed the *grasshopper effect*, consists of the repeated precipitation of the pollutant from the atmosphere and its revolatization. The simple model we provided here does not take these details into account. It does, however, set an upper limit to what can be expected if the emissions persist. This is important and useful information.

Locating solutions in related disciplines is not an easy matter. It requires a wide knowledge of what goes on elsewhere, but the results can be rewarding and often lead to new insights.

4.1.2 Nonsource Problems

Nonsource problems are by far the most prevalent type of problem involving diffusional processes, and almost all the material contained in standard monographs on diffusion is devoted to this topic. A parallel situation exists in the related field of heat conduction.

The variety of geometries and boundary conditions that can arise in these problems was already mentioned, and we proposed to limit ourselves to the semi-infinite, parallel plane, spherical, and cylindrical geometries subject to constant initial and surface concentrations. We start with the simplest of these geometries, the semi-infinite medium, and follow with a discussion of the other three principal geometries.

4.1.2.1 Diffusion into a Semi-Infinite Medium

Consider the case of the semi-infinite medium $x > 0$ in which the concentration C_o is uniform throughout and which is exposed at time $t \geq 0$ and at the position $x = 0$ to a constant surface concentration C_s. The solution to this problem can be given as a terse analytical expression that takes the following form:

$$\frac{C - C_s}{C_o - C_s} = \operatorname{erf} \frac{x}{2\sqrt{Dt}} \tag{4.9a}$$

For a medium initially devoid of solute, $C_o = 0$, the equation reduces to

$$\frac{C}{C_s} = 1 - \operatorname{erf} \frac{x}{2\sqrt{Dt}} = \operatorname{erfc} \frac{x}{2\sqrt{Dt}} \tag{4.9b}$$

Both of these expressions frequently appear in the literature. They contain, as did some source problems, an error function but lack the preexponential factor we have seen there. As a result, the concentration distributions that arise in this case are a function of only *one* dimensionless parameter, $x / 2\sqrt{Dt}$. From this, the following are deduced:

1. The distance of penetration of any given concentration is proportional to the square root of time.
2. The time required for any point to reach a given concentration is proportional to the square of its distance from the surface and varies inversely with the diffusivity.

We demonstrate the use of these simple relations with the following example.

Illustration 4.4: Penetration of a Solute into a Semi-Infinite Domain

Suppose a spill of a solvent has occurred (a) on land and (b) into a water basin. How much longer will it take a particular concentration to penetrate the same distance in water that it does in air? Because the concentration is the same in both cases, we can solve the problem without having to evaluate the error function. We write

$$\left[\frac{x}{2\sqrt{Dt}}\right]_{\text{air}} = \left[\frac{x}{2\sqrt{Dt}}\right]_{\text{water}} \tag{4.10a}$$

and, therefore,

$$t_{\text{water}} = \frac{D_{\text{air}}}{D_{\text{water}}} t_{\text{air}} \tag{4.10b}$$

Now, diffusion in air is of the order 10^{-5} m^2/s and that in water of the order 10^{-9} m^2/s (see Table 3.1 and Table 3.3). Consequently, the time of penetration in water is 10,000 times longer than that in air.

Consider next the following problem. If it takes a particular concentration 100 h to penetrate a distance of 1 m, how long will it take the same concentration to advance 10 m? Here again we are dealing with identical concentrations, which by virtue of Equation 4.9 lead to

$$\frac{x_1}{2\sqrt{Dt}} = \frac{x_2}{2\sqrt{Dt}} \tag{4.10c}$$

and, consequently,

$$t_2 = t_1 \left(\frac{x_2}{x_1} \right)^2 \tag{4.10d}$$

or

$$t_2 = 100 \left(\frac{10}{1} \right)^2 = 10^4 \text{ hrs} \tag{4.10e}$$

Hence, it takes 100 times longer to penetrate from a distance of 1 m to a distance of 10 m.

From these simple examples, it can be seen that the case of diffusion into a semi-infinite medium can yield rapid answers in a straightforward fashion. Furthermore, the geometry is not trivial. It can often be used to approximate finite geometries, particularly if the diffusion process is slow, as it is in liquid or solid media. Penetration will then be confined to short distances from the surface, at least initially, and the medium can consequently be regarded as semi-infinite for the short period under consideration. This case is taken up in some detail in Section 4.1.2.3.

In diffusion problems, we often seek to calculate the *cumulative* amount of material that has entered or left a medium, rather than a particular concentration level or the time it takes to attain that level. In the case of accidental spills, for example, it is often of greater interest to know the time required for *complete* evaporation or dissolution of the material, rather than the detailed concentration transients. To obtain this information, some mathematical manipulations of the distribution Equation 4.9 are required, which are discussed in Illustration 4.5.

Illustration 4.5: Cumulative Uptake by Diffusion for the Semi-Infinite Domain: The Higbie Equation

Cumulative uptake in general is derived in two steps: first one calculates the instantaneous *rate of diffusion* into the domain in question and then *integrate* that rate over time.

Calculation of the rate requires taking the derivative of the error function in Equation 4.9a. Let us consider the special (and usual) case of zero initial concentration $C_o = 0$. The need is then to evaluate

$$N / A = -D \left. \frac{\partial C}{\partial x} \right|_{x=0} = C_s \frac{\partial}{\partial x} \operatorname{erf} \left. \frac{x}{2\sqrt{Dt}} \right|_{x=0} \tag{4.11a}$$

Now from Item 5 of Table 4.1, we have

$$\frac{d}{dx}\,\text{erf}\,x = \frac{2}{\sqrt{\pi}}\exp(-x^2) \qquad (4.11b)$$

and, consequently,

$$\frac{\partial}{\partial x}\,\text{erf}\,\frac{x}{2\sqrt{Dt}} = \frac{2}{\sqrt{\pi}}\frac{1}{2\sqrt{Dt}}\exp(-x^2) \qquad (4.11c)$$

Combining Equation 4.11a and Equation 4.11c, we obtain for this intermediate step

$$N / A = -D\left.\frac{\partial C}{\partial x}\right|_{x=0} = \frac{DC_s}{\sqrt{\pi Dt}} \qquad (4.11d)$$

This is the expression that now has to be integrated with respect to time. We immediately obtain

$$M_t = 2C_s\left(\frac{Dt}{\pi}\right)^{1/2} \qquad (4.11e)$$

where M_t is the desired total mass per unit area which has diffused into the semi-infinite domain up to time t.

Note that M_t varies directly with the surface concentration C_s but only with the *square root* of time. In Practice Problem 4.6, use will be made of Equation 4.11e to estimate the time required for spilled solvent to evaporate into the (still) atmosphere.

The development just given can be used in another important way. We return briefly to Equation 4.11d and note that it can be cast in the following form:

$$N / A = k_C \Delta C = k_C C_s \qquad (4.11f)$$

where k_C is an equivalent mass transfer coefficient representative of the transient diffusion process. Comparison of the two equations yields the following expression for k_C:

$$\boxed{k_C = \left(\frac{D}{\pi t}\right)^{1/2}} \qquad (4.11g)$$

This equation, due to Higbie, was originally derived to describe mass transfer between rising gas bubbles and a surrounding liquid (*Tran. AIChE, 31*, 368 [1935]). It applies quite generally to situations where the contact time between the phases is short and the penetration (or depletion) depth is so small that transfer may be viewed as taking place from a plant to a semi-infinite domain. In Section 4.1.2.3 we will provide a quantitative criterion for this approach, which is also referred to as the Penetration Theory. It also describes both the short- and long-term behavior in diffusion between a plane and a semi-infinite space, and we used this property in Chapter 1, Table 1.4, to help us set upper and lower bounds to mass transfer coefficients and "film" thickness z_{FM}.

What the Higbie approach accomplishes, in essence, are two things: First, it translates the solution for diffusion in a semi-infinite space into the simple and familiar algebraic form of a mass transfer rate:

$$N / A = k_C C_s = \left(\frac{D}{\pi t} \right)^{1/2} C_s \tag{4.11h}$$

Second, it approximates diffusion in more complex geometries by the solution for the semi-infinite case.

What Higbie's solution does not do is provide the detailed *concentration profiles* that result in these cases. To obtain that information, one must return to the fundamental solution (Equation 4.9a).

4.1.2.2 Diffusion in Finite Geometries: The Plane Sheet, the Cylinder, and the Sphere

It was previously indicated that the diffusion into finite geometries leads to expressions of considerably greater complexity than was the case for the semi-infinite medium. The solutions typically take the form of infinite series, which not only are cumbersome to evaluate but also contain implicit parameters that cannot be conveniently extracted. It has become customary in these cases to represent the results graphically, which allows any one of several parameters to be read off with ease.

Figure 4.4 shows the concentration profiles that arise in a sheet of thickness $2L$ that is exposed to a surface concentration C_s at time $t = 0$ and contains an initial concentration C_o. We can use these plots to calculate, for example, the time required for a certain concentration C to reach a particular position x, or conversely, to calculate the prevailing concentration at a specified x after the lapse of time t. The position $x = 0$ is of particular interest, as it represents the midpoint of the sheet and is the farthest removed from the imposed surface concentration C_g. Note that the sheet is infinite in extent — that is, there are no concentration variations in the y and z directions. The finite three-dimensional case evidently leads to more complex distributions, which cannot be plotted conveniently.

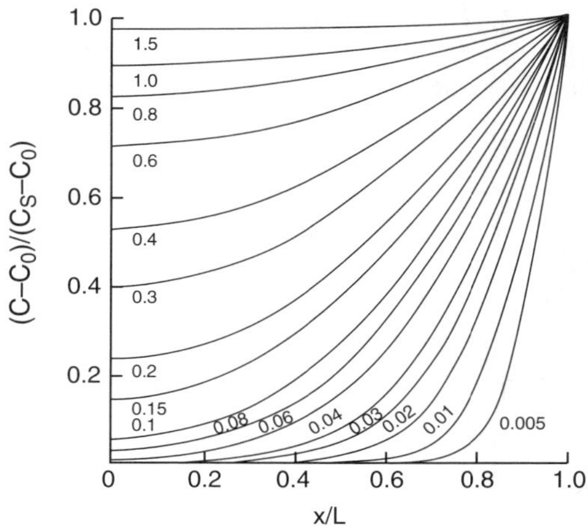

FIGURE 4.4
Concentration distributions at various times in the sheet $L < x < L$ with initial uniform concentration C_O and surface concentration C_S. Numbers on curves are values of the Fourier number $Fo_m = Dt/L^2$.

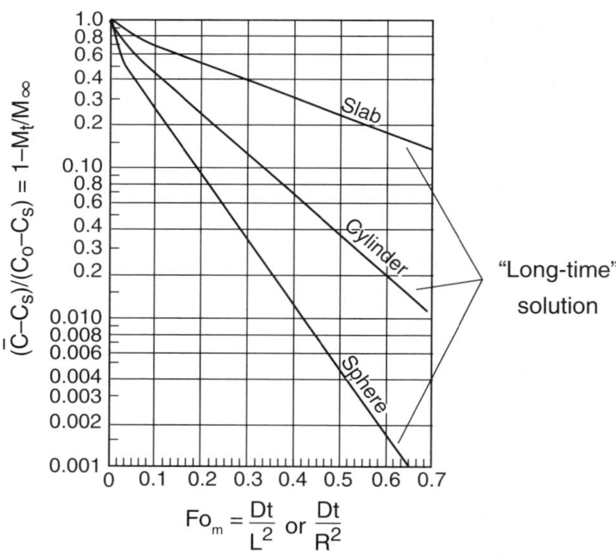

FIGURE 4.5
The relative change in average composition $\bar{C}$ for the basic shapes.

Figure 4.5 dispenses with the display of detailed profiles and presents instead the *average concentration* $\bar{C}$ in the medium after the lapse of time t. The ordinate of that plot, $\left(\bar{C} - C_s\right)/\left(C_o - C_s\right)$, can also be related to the mass fraction M_t/M_∞ taken up or released by each of these geometries up to a time t. Thus,

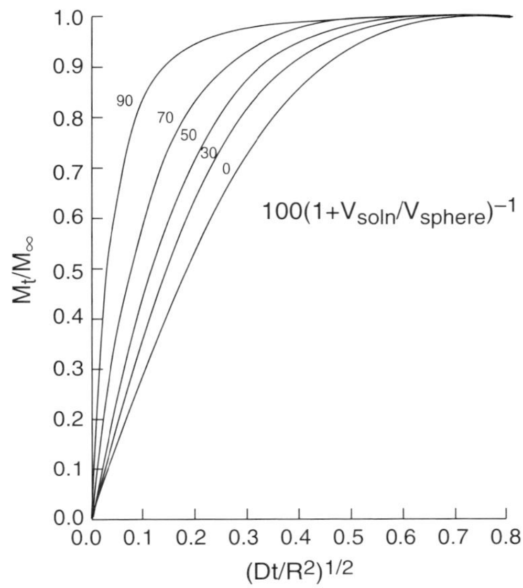

FIGURE 4.6
Fractional diffusional uptake and release in a sphere as a function of dimensionless time in a well-stirred solution of limited volume.

$$\frac{\text{Mass taken up or released up to } t}{\text{Mass taken up or released to } t \to \infty} = \frac{M_t}{M_\infty} = 1 - \frac{\bar{C} - C_s}{C_o - C_o} \tag{4.12}$$

where C_s is the constant surface or external concentration, and C_o is the initial concentration within each shape. C_s = constant implies that the external medium is *infinite in extent*.

When the external medium has a finite volume, V_{soln}, C_s will no longer be constant but will vary with time. The solution to this problem can still be expressed in terms of a fractional uptake or release and is shown in Figure 4.6 for the case of an immersed sphere or aggregate of spheres. It is used in the analysis of batch adsorption and leaching operations, among others.

The following three simple examples illustrate the use of these plots.

Illustration 4.6: Manufacture of Transformer Steel

To make a transformer steel with the proper hysteresis characteristics, it has to be loaded with silicon up to a certain prescribed content. This is to be accomplished by exposing a steel sheet that is low in silicon content and of 2 mm thickness to an atmosphere of $SiCl_4$ that dissociates to Si(g) and Cl_2(g). The silicon gas dissolves in the steel up to 3 wt% at equilibrium. The treatment is to be carried out at 1255 K. Silicon diffusivity in steel at this temperature is 8.2×10^{-13} m²/s.

The task here is to calculate the time necessary to achieve a target average concentration of 2.85 wt% silicon in the steel. We draw for this purpose on Figure 4.5 that has ordinate values that contain the average concentration $\bar{C}$ being sought.

We start by noting that one can set the initial concentration $C_o = 0$ in view of the low silicon content of the steel to be treated. The surface concentration C_s required for the ordinate value is taken to be 3 wt% — that is, it is assumed that at time $t = 0$, the steel surface is at equilibrium with the surrounding silicon atmosphere. We obtain, for the ordinate of Figure 4.5,

$$\frac{\bar{C} - C_s}{C_o - C_s} = \frac{2.85 - 3.0}{0 - 3.0} = 0.05 \tag{4.13a}$$

Linear extrapolation of the slab line of Figure 4.5 yields

$$\frac{Dt}{L^2} \cong 1.13 \tag{4.13b}$$

from which results

$$t = (1.13)\frac{(1 \times 10^{-3})^2}{8.2 \times 10^{-13}} = 13.8 \times 10^6 \text{ s} = 383 \text{ h} \tag{4.13c}$$

Thus, some 16 days are required for the target concentration to be attained. This is entirely due to the low diffusivity of silicon in the steel, which is some three orders of magnitude lower than typical diffusivities in liquids. To reduce the inordinate length of treatment, the operation will likely have to be carried out at a higher temperature. Because the diffusivity in solids varies exponentially with temperature (see Equation 3.17), a modest increase in the latter will quickly lead to substantially higher diffusivities and will thus shorten the duration of the treatment considerably.

Figure 4.5, which was used here to calculate time t required to achieve a certain average concentration, can also be put to the task of extracting diffusivities from experimental data.

Illustration 4.7: Determination of Diffusivity in Animal Tissue

To determine the diffusivity of carbohydrates in animal tissue, a sample specimen 1 mm in thickness is soaked in sugar solution for a lengthy period of time and is then mounted on a holder and placed in a large, well-stirred bath of water. After exposure for 1 h, the sample is removed, and the residual sugar content is determined. It is found to have dropped to 20% of the initial concentration. We wish to use the data to calculate the diffusivity of sugar within the tissue.

The problem again calls for the use of Figure 4.5, with the ordinate value now given by $\bar{C} / C_0 = 0.2$. The corresponding abscissa value is 0.54, so that

$$\frac{Dt}{L^2} = 0.54$$

and, consequently,

$$D = 0.54\ L^2/t = 0.54\ (0.5 \times 10^3)^2/3600$$

$$D = 3.8 \times 10^{11}\ \text{m}^2/\text{s}$$

Illustration 4.8: Extraction of Oil from Vegetable Seeds

Vegetable oils can be extracted from their parent seeds by contacting them with solvent in a way that approximates a well-stirred tank. External film resistance can in these cases often be neglected, and the principal events are confined to the interior of the seeds. Figure 4.6 can then be used to carry out pertinent calculations.

Suppose we desire to calculate the time required to extract 90% of the oil contained in oil-bearing vegetable seeds assumed to be spherical. The given data are as follows: $R = 0.25$ cm, $D = 5 \times 10^{-6}$ cm²/s, $V_{\text{Sol'n}}/V_{\text{Solids}} = 2$, so that $100\ (1 + V_{\text{Sol'n}}/V_{\text{Solids}}) = 33.3$.

From Figure 4.6, we obtain, for $M_t/M_\infty = 0.9$, $Dt/R^2 = 0.39$.

Hence, $t = 0.39\ R^2/D$, $t = 0.39\ (0.25)^2/5 \times 10^{-6} = 4.9 \times 10^3$ s $= 1.36$ h.

One common feature of these illustrations and the solution plots Figure 4.2 through Figure 4.4 is the appearance of the parameter Dt/R^2 or Dt/L^2. Known as the Fourier number, Fo_m, this dimensionless group is on par with the Reynolds number in importance and as a defining parameter. It appears in all solutions of Fick's equation and has its origin in its counterpart for heat transfer, Fo_h. The two groups are usually defined in the following generalized form:

For heat transfer (conduction):

$$Fo_h = \frac{\alpha t}{\ell^2} \tag{4.14a}$$

For mass transfer (diffusion):

$$Fo_m = \frac{Dt}{\ell^2} \tag{4.14b}$$

Hereℓs some characteristic length such as the radius of a sphere or cylinder, the half-width of a slab, or more generally the ratio of volume to surface area V/A.

The first use to which we wish to put the Fourier number is to provide a criterion for the validity of two important limiting solutions to Fick's equation. These apply, respectively, at "small" values of the Fourier number ("short-time solution") and "large" values ("long-time solution") of the same parameter. They are arguably the *most frequently used expressions* for transient diffusion or, when suitably translated, for transient conduction. Just how small or large Fo_m should be is discussed next.

4.1.2.3 Diffusion in Finite Geometries: The "Short-Time" and "Long-Time" Solutions

It was noted in Illustration 4.4 that if a diffusing solute penetrates only a short distance into the neighboring phase, either during uptake or release, the diffusion process can be viewed as taking place from a plane into a semi-infinite space. This is equivalent to "unfolding" the curved boundary and its adjacent region into a plane strip bordering on open space nearly devoid of solute (see Figure 4.7). The important consequence is that we can now use the simple results for diffusion between a plane and a semi-infinite space to calculate concentration distributions (Equation 4.9a) or cumulative uptake (Equation 4.11e), or to use Higbie's equation to determine equivalent mass transfer coefficients and film thickness.

The criteria for the validity of the short-time solution, obtained by comparison with the rigorous solutions of Fick's equation, for various geometries, are as follows:

For a curved space (sphere, cylinder, etc.):

$$Fo_m = \frac{Dt}{\ell^2} < 10^{-3} \tag{4.15a}$$

For a plane surface (slab):

$$Fo_m = \frac{Dt}{\ell^2} < 10^{-2} \tag{4.15b}$$

These criteria are conservative and assume a concentration change of only 1% at a distance of 0.1 R from the surface. They can be increased by a factor of five and still yield acceptable results.

The *"long-time" solution* has a different origin. It utilizes the fact that after a short "start-up" period, both cumulative uptake and release become linear

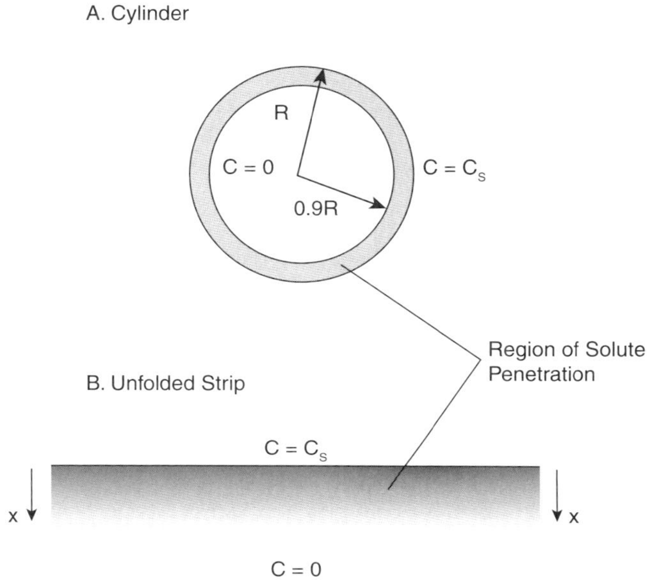

FIGURE 4.7
"Short-time" solution for diffusion into a cylinder: (a) near-surface penetration of diffusing species and (b) unfolding of near-surface region into a semi-infinite strip.

functions of the Fourier number on a semilog scale (see Figure 4.5). In other words, we can write

$$\log \frac{\overline{C} - C_s}{C_o - C_s} = \log(1 - M_t / M_\infty) = -m\frac{Dt}{\ell^2} \tag{4.16a}$$

where m equals 4.34 for the sphere, 2.50 for the cylinder, and 1.20 for the slab. The criterion for any of the quantities appearing in Equation 4.16a to be within 10% of the exact value is given by

Any geometry: $$Fo_m = \frac{Dt}{\ell^2} > 0.5 \tag{4.16b}$$

An equation identical in form to Equation 4.16a also arises if we describe the diffusion process by an ordinary differential equation (ODE) model using an equivalent (and empirical) "particle mass transfer coefficient" k_C. We have, for either uptake or release,

$$-k_C A(\overline{C} - C_s) = V\frac{d\overline{C}}{dt} \tag{4.16c}$$

TABLE 4.4

Diffusion in Finite Geometries

Solutions	Criterion for Validity
A. Short-Time Solution	
$(C - C_s) / C_o - C_s) - \text{erf}(x / 2\sqrt{Dt}$	Curved surface: $Dt/\ell^2 < 10^{-3}$
$M_t(\text{kg/m}^2) = 2C_s(Dt/)^{1/2}$ or $M_t(\text{kg/m}^2) = 2C_o(Dt/)^{1/2}$	Plane surface: $Dt/\ell^2 < 10^{-3}$
B. Long-Time Solution	**Error**
$(\bar{C} - C_s) / (C_o - C_s) = (1 - M_t / M_\infty)$	t to $M_t/M_\infty = 0.5$: +50%
$= \exp(2.303 mDt/^2)$	t to $M_t/M_\infty = 0.9$: +10%
	t to $M_t/M_\infty = 0.99$: +3%

Note: m = 4.3, 2.50, 1.20 for sphere, cylinder, slab.

TABLE 4.5

Relation between Particle Mass Transfer Coefficient and Diffusivity

Geometry	k_C	$k_C A/V = k_C a$
Sphere	3.33 D/R	10.0 D/R^2
Cylinder	2.88 D/R	5.76 D/R^2
Slab	2.77 D/L	2.77 D/L^2

and upon integration,

$$\log \frac{\bar{C} - C_s}{C_o - C_s} = \log(1 - M_t / M_\infty) = -\frac{k_C A}{2.303 V} t \tag{4.16d}$$

This approximation is widely and successfully used in engineering work, usually without any attempt to relate it to diffusional parameters. Equation 4.16a allows us to do that. A comparison with Equation 4.16d yields, for a sphere,

$$k_C = \frac{2.303(4 / 3\pi R^3)}{4\pi R^2}(4.35D / R^2) \tag{4.16e}$$

$$k_C = 3.33\ (d/R) \tag{4.16f}$$

and similarly for the other shapes.

In Table 4.4 and Table 4.5, we attempt to summarize this avalanche of equations in a rational way.

Illustration 4.9: Validity of Short- and Long-Time Solutions

To gain additional confidence in the use of these solutions, we return to Illustration 4.6 and make a comparison with the exact answer.

LONG-TIME SOLUTION

For the time necessary to achieve the desired average concentration $\bar{C}$ in the transformer steel, we have, from Table 4.4,

$$t = -\frac{\ell^2}{2.303\,mD} \ln(\bar{C} - C_s)/(C_o - C_s) \tag{4.17a}$$

$$t = -\frac{(1 \times 10^{-3})^2}{2.303 \times 1.2 \times 8.2 \times 10^{-13}} \ln(0.05) \tag{4.17b}$$

$$t = 1.32 \times 10^6 \text{ s} \tag{4.17c}$$

which is in agreement with the extrapolated value of Equation 4.13c. Note that the Fourier number has the value

$$Fo_m = \frac{Dt}{\ell^2} = \frac{8.2 \times 10^{-13} \times 1.3 \times 10^6}{(1 \times 10^{-3})^2} = 1.1 \tag{4.17d}$$

so that the validity criterion given in Table 4.4 is amply satisfied.

SHORT-TIME SOLUTION

This time interval was not addressed in Illustration 4.6, but some sample calculations will give us a sense of what happens during this initial period.

For the short-time solution to be valid, we must have (Table 4.4)

$$t < \frac{\ell^2}{D} 10^{-2} \tag{4.17e}$$

or

$$t < \frac{10^{-6}}{8.2 \times 10^{-13}} 10^{-2} = 1.2 \times 10^4 \text{ s} \tag{4.17f}$$

We set $t = 10^4$ s and use Equation 4.11g to obtain an approximate value of the depth of penetration ("film thickness") attained in that time:

$$z_{FM} = (\pi D t)^{1/2} = (\pi \times 8.2 \times 10^{13} \times 10^4)^{1/2} \tag{4.17g}$$

$$z_{FM} = 1.6 \times 10^4 \text{ m} = 0.16 \text{ mm} \tag{4.17h}$$

The short-time solution can then be used to obtain the near-surface concentration profile. For example, at a depth of 0.1 mm, we have (Table 4.4)

$$\frac{C - C_s}{C_o - C_s} = \text{erf}(10^{-4} / 2\sqrt{8.2 \times 10^{-13} \times 10^4}) = \text{erf}(1.1) \tag{4.17i}$$

and, hence, using the error function Table 4.3,

$$1 - \frac{C}{C_s} = 0.875 \tag{4.17j}$$

In other words, after a time interval of 10^4 s (~ 3 hrs), the concentration at a distance of 0.1 mm will have attained just under 13% of the surface concentration of the steel. This confirms the exceedingly slow progress of the diffusion process.

4.2 Diffusion and Reaction

Processes in which diffusion is accompanied by a chemical reaction arise frequently and in a variety of different contexts. All catalytic reactions, in which the catalyst resides within a porous matrix, are necessarily accompanied by diffusional transport of the reactants and products into and out of this catalyst particle. In noncatalytic gas–solid and liquid–solid reactions, diffusion occurs not so much within the solid particle but rather through a gas or liquid film, or through ash layers surrounding the reacting core. Here again, diffusion is coupled with reaction.

Reactions accompanied by diffusion also occur in fluid systems. The atmosphere is one vast reacting reservoir in which gaseous pollutants such as the nitrogen oxides (the famous Nox) or sulfur oxides (the equally famous Sox) diffuse into the air and undergo reactions with atmospheric oxygen. In gas–liquid systems, the liquid phase often contains a reacting component that interacts with the gas diffusing into it. Here, too, reaction is linked to diffusion.

Evidently, in each of the systems mentioned, a host of different reactions are possible. A reacting solid particle, for example, may involve the combustion of a fuel or the calcining of a solid (e.g., limestone) to calcium oxide and carbon dioxide. Reactants involved in catalytic reactions are almost infinite in their variety, as are those participating in atmospheric reactions.

In order to convey a flavor of these events without overwhelming the reader with a mass of details, we will limit ourselves to the following cases.

4.2.1 Reaction and Diffusion in a Catalyst Particle

In this example, we consider the diffusion of a reactant into a porous catalyst particle where it undergoes a first-order reaction. The model here is a second-order ODE, which is solved by the *D*-operator method given in the Appendix A1 and yields the concentration profile of the reactant within the particle. This is interesting but not immediately useful information for the design of catalytic reactors. To transform the result into a tool for engineering use, we derive the overall rate of reaction in the pellet with diffusional resistance and divide the result by the reaction rate that would prevail in the absence of a diffusional resistance. This ratio, known as the *catalyst effectiveness factor* E, can be viewed as the efficiency of the catalyst in converting the reactant to product. When diffusion is very fast, or the reaction slow, the interior reactant concentration will be nearly that prevailing at the surface. Catalyst efficiency will be approximately 100%, the maximum it can attain. With increasing diffusional resistance, the reactant concentration within the particle begins to fall below that prevailing at the surface, causing the reaction to slow. This is reflected in a lower effectiveness factor E, which is now below 1. We thus obtain a good sense, through the value of E, of how well the catalyst is performing, in coping with the diffusional resistance. This also enables us to take countermeasures to raise the efficiency, for example, by increasing porosity or reducing particle size, both of which have the effect of diminishing diffusional resistance.

There is a second important reason for introducing the concept of an effectiveness factor. In the ordinary course of events, concentrations within a catalytic reactor packed with catalyst particles will vary both axially in the direction of flow as well as radially within the catalyst pellets. The model mass balance for such a system would consequently lead to a partial differential equation (PDE). By using an effectiveness factor, we reduce the PDE to an equivalent set of two ODEs — one the pellet mass balance in the radial direction and the other the reactor mass balance in the direction of flow. The reaction rate, which previously varied in two directions $r_A(r, z)$, is now a function of the axial distance only. We replace $r_A(z, r)$ by $E\ r_{Ai}(z)$, where r_{Ai} is the so-called intrinsic reaction rate measured experimentally on a fine powder and excludes diffusional effects. The latter are lumped into the effectiveness factor that now acts as a fractional efficiency on the intrinsic rate r_{Ai}. This product of $Er_{Ai}(z)$ is used in the reactor mass balance.

4.2.2 Gas–Solid Reactions Accompanied by Diffusion: Moving-Boundary Problems

The topic addressed in this example is that of a solid particle that undergoes a continuous reaction with a gas, building up in the process a layer of porous solid product, which we denote by the general term *ash*. Reactant diffuses through a growing layer of ash to the surface of the shrinking particle where the reaction takes place. Such systems of two phases, in which the phase boundary undergoes a continuous movement due to some physical or chemical event, are referred to as *moving-boundary problems*. Examples of this type of behavior are numerous and important. In addition to reacting systems, moving-boundary problems arise in operations involving phase change, such as evaporation, condensation, freezing and melting, crystal growth and dissolution, metal or polymer casting, and the freeze-drying of foods.

The state variables in these processes, such as temperature or concentration, are, in principle, functions of both distance and time, leading to PDEs that are usually coupled and nonlinear. To reduce the model to a manageable set of ordinary differential and algebraic equations, the following assumptions are made:

1. The "core" contained by the moving front, such as a burning fuel particle, has uniform properties and can be treated as an unsteady compartment.
2. The movement of the front is sufficiently slow so that the transport gradients outside the core attain a quasi-steady state. This condition, which we encountered in Illustration 2.4, has the effect of eliminating time as a variable, with a consequent simplification of the model equation.
3. The processes involved — transport and reaction — are dominated by a rate-controlling slow step.

Thus, although both time and distance are retained as variables, distance (expressed through the changing size or mass of the core) becomes a dependent variable for the core-unsteady balance but is retained as an independent variable for the external, quasi-steady-state balance. Time is an independent variable as well, but it appears only in the core balance.

A systematic way of modeling these systems is to start with unsteady balances about the core, followed by a consideration of the quasi-steady-state process outside the moving boundary. It is good practice to keep track of the number of dependent variables that must ultimately be matched by the number of equations. These procedures will be demonstrated in Illustration 4.11.

4.2.3 Gas–Liquid Systems: Reaction and Diffusion in the Liquid Film

It is common practice in gas-absorption operations to employ a solvent that reacts with the solute being absorbed from the gas phase. The purpose of this practice is to promote the solute removal rate and to enhance the efficiency of the gas absorber. Acid gases such as H_2S and CO_2 are often contacted with solvents containing an alkaline component such as potassium or sodium hydroxide, or an ethanol amine. Conversely, the absorption of a basic solute such as ammonia can be promoted by reacting it with an acidic solvent.

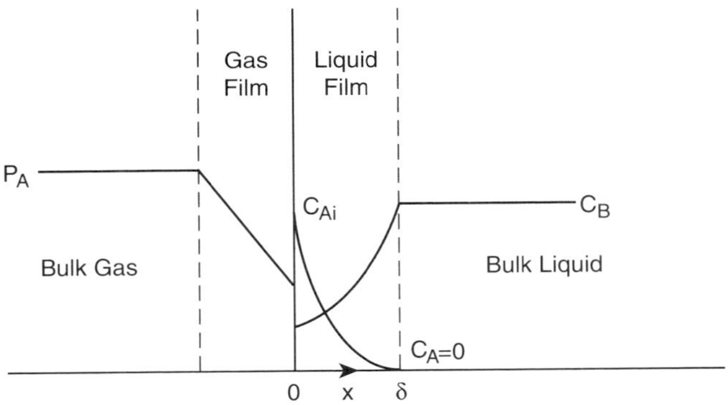

FIGURE 4.8
Concentration profiles in a gas–liquid reaction; fast first-order reaction in the liquid film, with reactant concentration dropping to zero at $x = \delta$.

The stoichiometry of these liquid-phase reactions can be represented in the general form

$$A(g) + bB() \rightarrow \text{products} \tag{4.18}$$

$B()$ may refer to pure liquid B or, more commonly, to B dissolved in a liquid solvent.

In Illustration 4.11 we take up the case of a first-order reaction taking place in the liquid, which is fast enough to cause the reactant concentration to drop to zero within a relatively short distance from the interface (i.e., within what is conventionally regarded as a liquid film).

The resulting concentration profiles are shown in Figure 4.8. We note here that because of the rapid reaction in the liquid film, the profiles are highly nonlinear and can no longer be approximated by a linear driving force. We must resort to a full diffusional mass balance. This is undertaken in Illustration 4.12.

Here again, as in the case of the catalyst particle, the resulting concentration profile does not yield information of immediate interest and is translated instead into a quantity of greater engineering usefulness. The quantity in question is the so-called enhancement factor E_h, which is defined as the *ratio of mass transfer with reaction to the mass transfer rate without reaction*. In contrast to the catalyst effectiveness factor, the value of E is above, rather than below, unity. This is because the reaction continuously removes reactant, thus sharpening its gradient and enhancing the mass transfer rate.

Illustration 4.10: Reaction and Diffusion in a Catalyst Particle: The Effectiveness Factor and the Design of Catalyst Pellets

We consider in this example the diffusion of a reactant A into a flat-plate catalyst particle where it undergoes a first-order reaction. The geometry, the

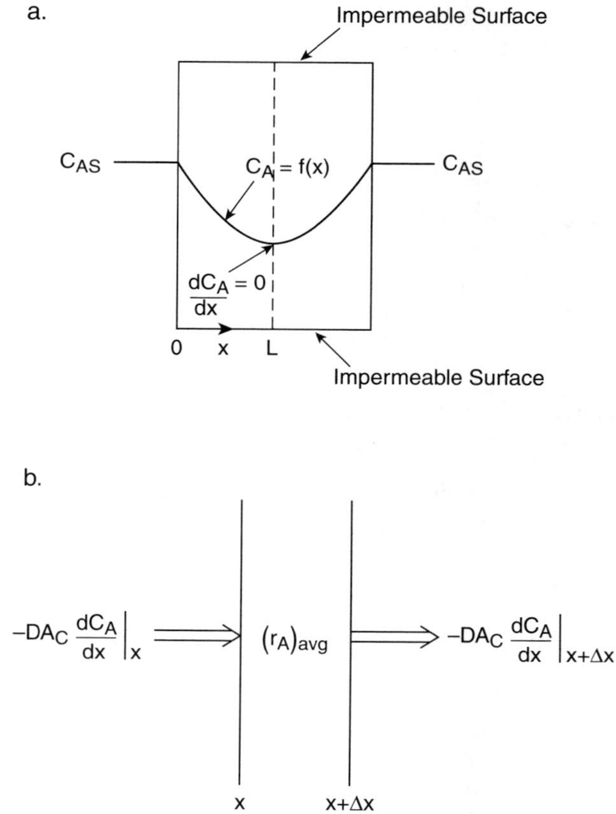

FIGURE 4.9
Diffusion and reaction in a flat-plate catalyst particle: (a) geometry and boundary conditions and (b) difference element.

associated boundary conditions, and the resulting concentration profile are shown in Figure 4.9a.

Diffusion is assumed to take place unidirectionally through the largest exposed area, while the edges of the plane are taken to contribute an insignificant amount of flow. The surfaces of the particle are maintained at a constant concentration C_{As}, leading to the parabolic concentration profile shown in Figure 4.9a. The distribution is symmetric about the midplane and has a vanishing derivative at that point.

These two conditions, prevailing at the surface and the center plane, respectively, constitute the boundary conditions (BCs) for the model equation.

Figure 4.9b shows the difference element over which the mass balance is taken. We obtain

$$\text{Rate of } A \text{ in} - \text{Rate of } A \text{ out} = 0$$

$$-D_e A_C \frac{dC_A}{dx}\bigg|_x - \left[\begin{matrix} -D_e A_C \dfrac{dC_A}{dx}\bigg|_{x+\Delta x} \\ +k_r (C_A)_{\text{avg}} A_C \Delta x \end{matrix}\right] = 0 \tag{4.19a}$$

where D_e is the effectiveness diffusivity.

Dividing by $A_C\Delta x$ and letting $\Delta x \to 0$, we obtain

$$d^2C_A/dx^2 - k_r C_A/D_e = 0 \tag{4.19b}$$

This equation can be cast in a convenient nondimensional form by defining

$$\psi = C_A/C_{As} \tag{4.19c}$$

$$z = x/L \tag{4.19d}$$

$$\varphi = L(k_r/D_e)^{1/2} \tag{4.19e}$$

Equation 4.19b then becomes

$$d^2\psi/dz^2 - \varphi^2\psi = 0 \tag{4.19f}$$

This expression now has to be solved for the dimensionless concentration ratio ψ. We draw for this purpose on the D-operator method given in Appendix A1, which yields the following solution:

$$\psi = C_1 \exp(\varphi z) + C_2 \exp(-\varphi z) \tag{4.19g}$$

$$\psi(0) = 1$$

$$\psi'(1) = 0 \tag{4.19h}$$

The integration constants obtained from these two conditions are

$$C_1 = \exp(\varphi)[\exp(\varphi) + \exp(-\varphi)]$$

$$C_2 = \exp(\varphi)/[\exp(\varphi) + \exp(-\varphi)] \tag{4.19i}$$

Upon introducing them into the solution (Equation 4.19g), we obtain

$$\psi = \frac{\exp[-\varphi(1-z) + \exp[\varphi(1-z)]}{\exp(\varphi) + \exp(-\varphi)} \tag{4.19j}$$

or, alternatively,

$$\frac{C_A}{C_{As}} = \psi = \frac{\cosh[\varphi(1-z)]}{\cosh\varphi} \tag{4.19k}$$

where we used the hyperbolic function cosh $\varphi = [\exp(\varphi) + \exp(\varphi)]/2$ listed in Appendix A2.

φ is called the Thiele modulus, after one of the pioneers in the field, and it expresses the combined effect of reaction rate constant k_r and effective diffusivity D_e on the concentration profile in the pellet. Large values of this parameter are associated with larger pellets and low diffusivities and imply a strong pore diffusion resistance. This means that the reaction rate outpaces the transport rate, which is unable to adequately replenish the consumed reactant. A sharp drop in reactant concentration results and may, in extreme cases, lead to a vanishing concentration at some location in the pellet. Small values of φ, on the other hand, imply a low reaction rate and high diffusivities. Concentration profiles under these conditions will be nearly flat, and the pellet operates near its maximum effectiveness.

We now turn to the task of converting the concentration profile (Equation 4.19h) into the more useful effectiveness factor E, which was defined as the ratio of the reaction rate with diffusional resistance to the reaction rate without diffusional resistance:

$$\mathrm{E} = \frac{\text{rate with diffusional resistance}}{\text{rate without diffusional resistance at } C_{As}} \tag{4.20a}$$

Because the rate of reaction in the pellet must, at steady state, equal the rate of supply of reactant, we have, alternatively,

$$E = \frac{\text{rate of diffusion at } z = 0}{\text{rate of reaction at } C = C_{As}} \tag{4.20b}$$

or

$$E = \frac{-D_e A_C (dC_A / dx)_{x=0}}{L A_C k_r C_{As}} \tag{4.20c}$$

where the derivative is obtained from Equation 4.19k.

The result, obtained after some manipulation, is given by

$$E = -\frac{1}{\varphi^2}\left[\frac{-\varphi \sinh\varphi(1-z)}{\cosh\varphi}\right]_{z=0} \tag{4.20d}$$

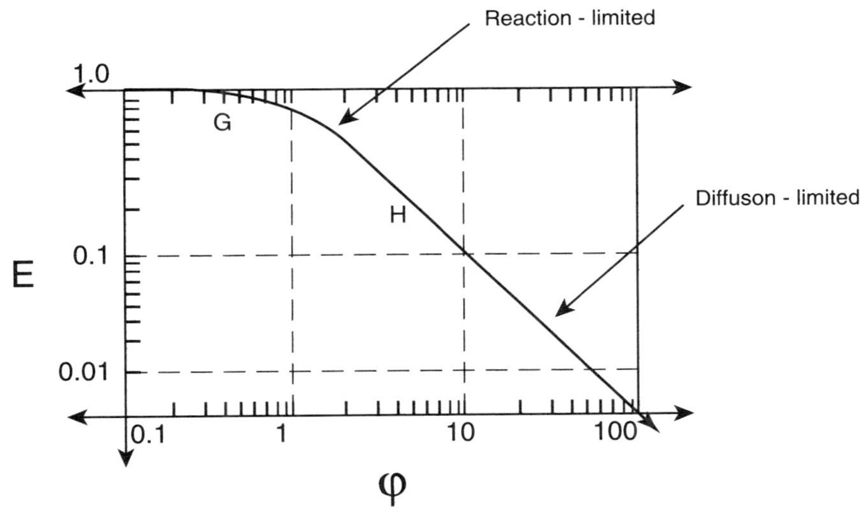

FIGURE 4.10
Catalyst particle effectiveness factor E as a function of the Thiele modulus for a flat-plate and isothermal first-order reaction.

or, alternatively,

$$E = (\tanh \varphi)/\varphi \tag{4.20e}$$

A plot of this relation is shown in Figure 4.10. The important fact that emerges from it is that in the range 0–0.5 the pellet is very nearly at its maximum effectiveness, $E = 1$. This region, termed *reaction-limited,* is precisely where we wish to operate, and we can use a threshold value of $\varphi = 0.5$ to establish some of the properties a good catalyst pellet should possess. Note that above values of $\varphi = 2$ the relation becomes inverse linear: $E = 1/\varphi$. This is the so-called *diffusion-limited region* in which diffusion is the dominant, rate-determining, and retarding process.

Suppose it is desired to carry out a gas-phase reaction in a reactor packed with catalyst pellets. A reasonable size for the pellets that balances low pressure drop and acceptable contact area is about 1 cm. Rate constants for many important gas-phase catalytic reactions are of the order 10^{-3} s^{-1}. The question then arises whether a typical pellet diffusivity D_e satisfies the criterion $\varphi \leq 0.5$ or whether it has to be adjusted, along with pellet size, to yield an acceptable effectiveness. We have the requirement

$$\varphi = L(k_r/D_e)^{1/2} \leq 0.5 \tag{4.20f}$$

A rough estimate of pellet diffusivity assuming molecular rather than Knudsen diffusion comes from the relation given in Chapter 3:

$$D_e = \frac{D\varepsilon}{4} \tag{4.20g}$$

where ε = porosity. If we set $\varepsilon = 0.4$ and use $D_e = 10^{-1}$ cm^2/s for the diffusivity of a typical gas, we obtain

$$\varphi = 0.5\ (10^{-3}/10^{-2})^{1/2} = 0.16 \tag{4.21}$$

Thus, our criterion is fully met. In fact, one can drop D_e by an order of magnitude or raise the rate constant by the same amount without violating the criterion. There is also some freedom to manipulate pellet size, if needed. All of this is revealed in simple fashion by an inspection of the Thiele modulus.

Illustration 4.11: A Moving Boundary Problem: The Shrinking Core Model

Concepts that must be applied in modeling the progress of a gas–solid reaction were previously introduced. They involve separating the system into a core, which reacts and continuously shrinks in size, and an external layer, which grows with time but is assumed to be at a quasi-steady state. This configuration is referred to as the *shrinking core model* and has associated with it the assumption of a quasi- or pseudo-steady state.

We illustrate these concepts with the following example: Suppose a solid particle undergoes a reaction according to the scheme

$$A(g) + bB(s) \rightarrow \text{Products} \tag{4.22}$$

with rates r_A and r_B related by the stoichiometry of the reaction as follows:

$$r_A = \frac{1}{b} r_B \tag{4.23}$$

The rate of reaction is assumed to be fast enough so that all the reactant arriving at the surface of the particle is instantaneously consumed. Hence, $(C_A)_{r=r_C} = 0$ (Figure 4.11). On the other hand, the movement of the core front is, because of its high density, sufficiently slow so that the external layer of solid products can be taken to be at steady state. The oxidative degradation of organic aerosols in the atmosphere is an example of such a reactor. The task is to determine the time dependence of the particle radius r_C and, hence, that of its mass.

We start, as recommended, with an unsteady core balance and write

$$\text{Rate of } B \text{ in} - \text{Rate of } B \text{ out} = \begin{matrix}\text{Rate of} \\ \text{change of } B\end{matrix} \tag{4.24a}$$

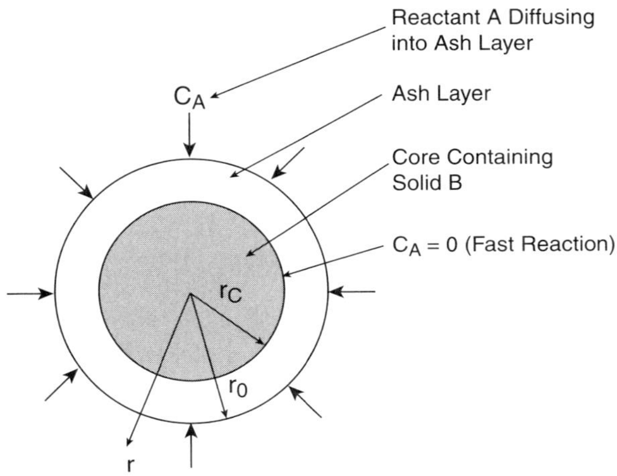

FIGURE 4.11
A reacting solid particle and its external ash layer.

$$0 - r_B = 4\pi r_C^{\,2} \rho_B \frac{dr_C}{dt}$$

where we placed the unknown reaction rate r_B in the "out" column and denoted particle density by ρ_B.

This is followed by a quasi-steady-state mass balance in the product layer which is composed as follows:

Rate of A in at r – Rate of A consumed at $r_C = 0$

$$4\pi r^2 D_e \frac{dC_A}{dr} - r_A = 0 \tag{4.24b}$$

where D_e is the effective diffusivity of the gaseous reactant A in the product layer. This last expression can be immediately integrated by separation of variables to yield

$$(4\pi D_e / r_A) \int_0^{C_A^o} dC_A = \int_{r_o}^{r_C} \frac{dr}{r^2} \tag{4.24c}$$

or

$$(4\pi D_e C_A^{\,o} / r_A) = \frac{1}{r_C} - \frac{1}{r_o} \tag{4.24d}$$

where r_o = initial radius of particle.

We pause at this stage to take a brief inventory. We have, in Equation 4.23, Equation 4.24a, and Equation 4.24d, three equations in the three dependent variables r_A, r_B, and the core radius r_C. The model is consequently complete, and one can proceed with the elimination of r_A and r_B to arrive at the desired solution, particle radius r_C as a function of time.

We obtain from the three aforementioned equations

$$-\frac{4\pi D_e b C_A^o}{1/r_C - 1/r} = r\pi r_C^2 \rho_B \frac{dr_C}{dt} \tag{4.24e}$$

which can again be integrated by separation of variables:

$$-\frac{D_e b C_A^o}{\rho_B} \int_0^t dt = \int_0^{r_C} \left(r_C - \frac{r_C^2}{r_o} \right) dr_C \tag{4.24f}$$

The final result is then

$$t = \frac{\rho_B r_o^2}{6bD_e C_A^o} \left[1 - 3\left(\frac{r_C}{r_o}\right)^2 + 2\left(\frac{r_C}{r_o}\right)^3 \right] \tag{4.24g}$$

Comments

The illustration demonstrates how the use of clever simplifying assumptions and some inspired modeling can reduce the complexities of a process to manageable proportions. The assumptions would be violated only if the progress of core consumption were extremely rapid, in which case events in the outer layer become both time and distance dependent, leading to a PDE.

The time required for the reaction to be complete is obtained from Equation 4.24g by setting $r_C = 0$. This leads to

$$t_{\text{Tot}} = \frac{\rho_B r_o^2}{6bD_e C_A^o} \tag{4.24h}$$

where t_{Tot} is seen to vary inversely with the effective diffusivity D_e and the external reactant concentration C_A^o.

This was anticipated on physical grounds, as both these factors increase the rate of reactant consumption. Somewhat less expected is the direct dependence on the square of the initial particle radius, r_o^2. It might have been argued that the time necessary for total consumption should vary with the mass of the particle (i.e., with the radius cubed). That this is not so is neatly revealed by the model and is because that the rate of consumption is dictated by, and hence proportional to, the surface area of the particle at any instant.

It will also be noted that knowledge of the reaction rates was not required in this instance. This is a consequence of the assumption that the rate of reaction is rapid and as a result ceases to affect the overall rate of the process. The rate-determining step here is the speed with which the core front recedes, which is very low given the high density ρ_b of the particle. This, together with the low values of D_e and $C_A{}^o$, leads to high values of the consumption time t_{Tot}.

Illustration 4.12: First-Order Reaction with Diffusion in a Liquid Film: Selection of a Reactive Solvent

We consider the situation shown in Figure 4.8, involving transport through a gas film to a liquid interface, followed by diffusion and reaction in the liquid film. The difference element over which the mass balance is taken is the same as that for the flat-plate catalyst pellet and leads to the same differential equation and the same boundary conditions:

$$\frac{d^2C_A}{dx^2} - k_r C_A / D_L = 0 \tag{4.25a}$$

$$CA(0) = CAi \tag{4.25b}$$

$$\frac{dC_A}{dx}(\delta) = 0 \tag{4.25c}$$

$$CA(\delta) = 0 \tag{4.25d}$$

Where the treatment of the two cases differs is in the ultimate goal of the solution. The catalyst pellet required us to derive the ratio of reaction rate with diffusion to reaction rate without diffusion. For the liquid film, the requirement is reversed: We compose the ratio of diffusion rate with reaction over diffusion rate without reaction. The results are exact inverses. We obtain the following:

For the catalyst pellet:

$$E = \frac{\tanh\varphi}{\varphi} \tag{4.20c}$$

For the liquid film:

$$E_h = \frac{\text{Ha}}{\tanh \text{Ha}} \tag{4.25e}$$

where E_h is the *enhancement factor* and $\text{Ha} = \delta(k_r/D_L)^{1/2}$ is the *Hatta number*, named after one of the pioneers in the field. For the case of no reaction, Ha = 0 and $E_h = 1$; that is, no enhancement occurs. Enhancement E_h rises above 1 when Ha > 0 and continues to increase with further increases in the reaction-rate constants k_r. Values of E_h can be read off Figure 4.7 by setting $\varphi = \text{Ha}$ and $E_h = 1/E$.

Details of the derivation of Equation 4.25e can be found in Practice Problem 4.22. The intent here is to use the enhancement factor to make a rational selection of a reactive solvent. Assume a typical turbulent flow film thickness of 0.1 mm (see Illustration 1.8) and a value of $D_L = 10^{-5}$ cm²/s, which is the commonly used order of magnitude of liquid-phase diffusivities (see Chapter 3). To find out at which value of Ha enhancement begins to exceed 1, we use the criterion Ha > 0.5 (i.e., the inverse of that proposed for the catalyst pellet, $\varphi < 0.5$). We obtain, for Ha = 0.5 ($E_h \approx 1$),

$$k_r = D_L\left(\frac{\text{Ha}}{\delta}\right)^2 = 10^{-5}\left(\frac{0.5}{10^{-2}}\right)^2 = 0.025\,\text{s}^{-1}$$

This is the threshold value of k_r which yields $E_h \approx 1$. Above that point, the following values apply:

k_r, s¹	2.5	10	25	100
E_h	5	10	50	100

Thus, for the reaction to have a significant effect on mass transfer rate, the rate constant has to be in excess of 0.025 s¹. This is, in fact, the range of many liquid-phase reaction rates. The requirements become less stringent with an increase in film thickness. For δ = 1 mm, for example, rate constants can be lowered by a factor of 100 to achieve the same result.

Comments

Some remarks are in order regarding the use of E. To apply the enhancement factor, we first must establish k_L in the absence of the reactive component, for example, in water devoid of NaOH. That value is then premultiplied by E to obtain the "effective" $(k_L)_{eff}$:

$$(k_L)_{eff} = E_h(k_L)_{\text{non-reactive}} \tag{4.25f}$$

Some typical E_h values are shown in Table 4.6.

When the reactive component is used in excess and the reaction is irreversible and rapid, the liquid film is reduced to the vanishing point. This particular and convenient limiting case is often used in the design of packed tower (see Illustration 8.3).

TABLE 4.6

Enhancement Factor for the Absorption of Carbon Dioxide in Various Solvents

Solvent	E_h
2*N* Potassium hydroxide	6.25
2*N* Sodium hydroxide	10.5
2*N* Ethanol amine	62.5

Illustration 4.13: Gas–Liquid Reactions Again: Biofiltration and the Biofilm

Bacteria and other microorganisms are increasingly being relied upon to deal with pollutants and toxins that are otherwise difficult to remove or dispose of. Bacterial action is coming into prominent use in soil remediation and has, of course, long played a key role in wastewater treatment (see Illustration 5.5). The enactment of more stringent air quality regulations in the 1990s led to the increased use of biodegradation for the removal of objectionable substances, in particular of the so-called volatile organic compounds (VOCs) and odors. The operation, which is now a major established technology, consists of passing the contaminated air through a *biofilter* consisting of beds of inert particulates (wood chips, activated carbon, perlite) which support a thin aqueous "biofilm" of microorganisms.

Superficially, at least, the process is identical to that described in Illustration 4.12: A gaseous component reaches a liquid interface, diffuses into the liquid, and undergoes an irreversible reaction (in the case of VOCs, to CO_2, H_2O, Cl, and SO_4). It differs, however, in two important respects: The kinetics of the reaction are generally more complex, and the liquid is immobile, stagnant, and composed of a thin layer of aqueous gels (the "biomass") of no more than a fraction of a millimeter thickness.

The kinetics of the reaction, which is enzymatic (i.e., catalytic), is described by the Michaelis-Menten or Monod equations, which take the form

$$r(\mathrm{kg\,m^{-3}h^{-1}}) = \frac{k_{\mathrm{Max}} S}{K_m + S} \tag{4.26a}$$

where S is the concentration of the so-called substrate — that is, of the reacting substance (g/m^3). The graph of this equation, shown in Figure 4.12a, shows two important limiting cases: At *low* concentrations of S ($K_m \gg S$), the rate is linear in S (i.e., *first order*), while at *high* values of S ($S \gg K_m$), the kinetics become independent of concentration (i.e., *zero order*). Physically, the latter case signals the fact that the catalytic sites of the enzyme are saturated, unable to receive any further substrate, and that the rate is consequently at a maximum. The latter case in particular is frequently seen in practice. The following are the expressions for the two limiting cases:

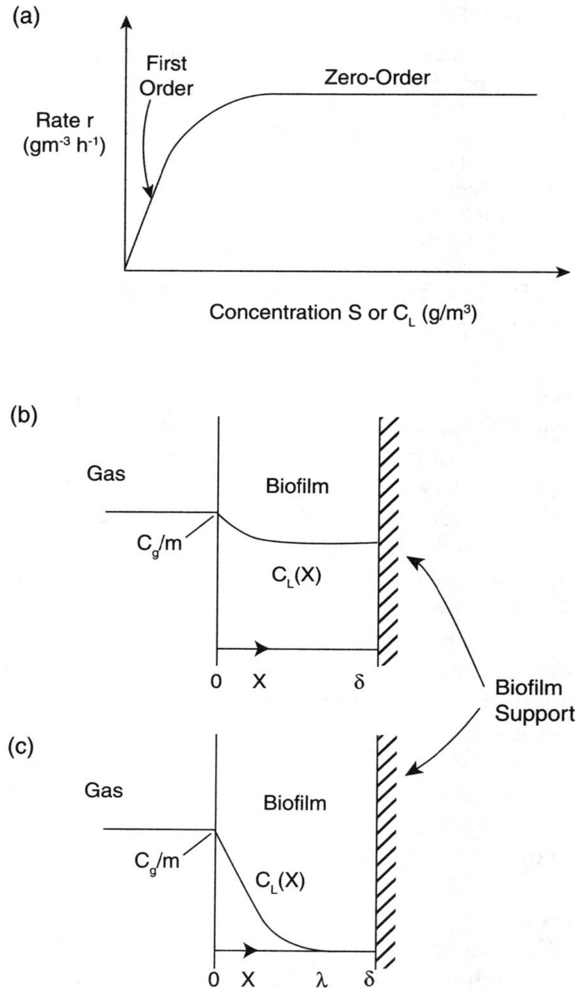

FIGURE 4.12
The Ottengraf biofilm model: (a) Michaelis-Menton kinetics; (b) reaction-limited process (zero order); and (c) diffusion-limited process.

$$r = k_{\text{Max}} S \text{ for } K_m \gg S \tag{4.26b}$$

and

$$r = k_{\text{Max}}/K_m = k_o \text{ for } S \gg K_m \tag{4.26c}$$

Just as in the case of synthetic catalysts, the process can be reaction limited (i.e., the reaction is the rate-determining step) or diffusion limited. Concentration profiles for these two cases within the biofilm are shown in Figure 4.12b

TABLE 4.7

Representative Values of m and k_o

Compound	Partition Coefficient (m)	Zero-Order Rate Constant k_o (g m³ h¹)
Ethanol	0.000257	88
Ethyl acetate	0.0055	110
Toluene	0.275	64
Hexane	74.1	65

and Figure 4.12c. Note that for the diffusion-limited case, both the gradient and the concentration C_L (substrate S) drop to zero at some interior point $x = \lambda$. The model for this case first presented by Ottengraf in 1983 is identical in form to that of the previous illustration (Equation 4.25a and Equation 4.25b) — that is, we have, for zero-order kinetics,

$$d^2C_L/dx^2 - k_o/D_e = 0 \tag{4.26d}$$

$$C_L\Big|_{x=0} = C_G / m \tag{4.26e}$$

$$(dC_L/dx)_{x=\lambda} = 0 \tag{4.26f}$$

$$C_L\Big|_{x=0} = 0 \tag{4.26g}$$

The film surface, $x = 0$, is assumed to be at equilibrium with the gas phase, expressed through a dimensionless partition or distribution coefficient m (g/m³)/(g/m³). m can be extracted from Henry's constants (see Table 6.1) or from activity coefficients (see Illustration 6.14). Some typical values appear in Table 4.7.

What we now wish to do is derive an effectiveness factor for the biofilm to help us establish conditions for maximum removal efficiency. The procedure is the same as that used for the catalyst pellet (Illustration 4.10). Equation 4.26d is integrated twice to obtain the concentration profile in the film $C_L = f(x)$, which is then differentiated to obtain the flux into the biofilm. The result, the derivation of which is found in Practice Problem 4.23, is given by

$$N = -D_e A \frac{dC_L}{dx}\bigg|_{x=0} = Ak_o\left(\frac{2C_g D_e}{k_o m}\right)^{1/2} \tag{4.26h}$$

which in turn yields the effectiveness factor

$$E = \frac{N}{k_o A\delta} = \frac{1}{\delta}\left(\frac{2C_g D_e}{k_o m}\right)^{1/2} \tag{4.26i}$$

This is a much simpler expression than that obtained for the catalyst pellet (Equation 4.20e) but differs from it by being a function of the external gas-phase concentration Cg. We can use the fact to set lower limits to the pollutant concentration Cg, which can be efficiently removed by biofiltration (i.e., $E \approx 1$). To exceed that threshold, we must have

$$Cg > \delta^2 \frac{k_o m}{2D_e} \tag{4.26j}$$

Two cases are considered: $m = 100$, representative of a highly hydrophobic solute such as hexane (Table 4.7), and, at the other end of the spectrum, $m = 10^{-3}$, characteristic of water-soluble VOCs. D_e is set at 10^{-10} m²/s (ten times lower than diffusivity in water), biofilm thickness δ at 0.1 mm, and k_o at 50 g/mol h, representative of a "good" biofilter. One obtains, for the two cases,

$$\begin{array}{ll} m = 100 & \\ (Hydrophobic) & Cg > (10^{-4})^2 \dfrac{50 \times 100}{2 \times 3600 \times 10^{-10}} = 70\,\text{g/m}^3 \approx 50{,}000\,\text{ppm} \end{array}$$

$$\begin{array}{ll} m = 10^{-3} & \\ (Hydrophilic) & Cg > (10^{-4})^2 \dfrac{50 \times 10^{-3}}{2 \times 3600 \times 10^{-11}} = 7 \times 10^{-4}\,\text{g/m}^3 \approx 0.5\,\text{ppm} \end{array}$$

Thus, although hydrophilic VOCs can be removed efficiently down to less than 1 ppm, their hydrophobic counterparts do not fare nearly as well and require much greater residence time or filter depth for their effective removal.

Practice Problems

4.1 More on Instantaneous Point Sources

Assuming an absence of wind, what is the effect of a tenfold increase in distance r from the emission source on:

a. The peak concentration?
b. The time of arrival of the peak?

What is the effect of a tenfold increase in diffusivity on:

c. The peak concentration?
d. Time of arrival of the peak?
e. What is the answer to these questions for any *arbitrary* concentration level?

4.2 More on Continuous Sources

a. What is the effect of a tenfold increase in distance from the emission source on the ultimate steady-state concentration?
b. Compare Equation 4.8e with Item 5 of Table 4.1 and comment.
c. Develop an expression for the concentration in the immediate vicinity of a continuous source after long exposure times.

Answer:

c. $$\frac{M_{Cp}}{4\pi Dr}\left(1-\frac{r}{\sqrt{\pi Dt}}\right)$$

4.3 Accident Response

Suppose 100 kg of a toxic substance with a tolerance level of 5 ppm by weight has been accidentally released over a short period of time to the atmosphere. Up to what distance from the site of the accident should evacuation measures be considered, assuming *random* air turbulence equivalent to a diffusivity $D = 10^{-4}$ m²/s?

4.4 Emissions from a Chimney

A 10-m-tall chimney emits a toxic substance at the rate of 10 g/s. Calculate the prevailing concentration at the base of the chimney 100 h after the start of the emissions, assuming a diffusivity in still air of 10^{-5} m²/s. (*Hint*: Use coordinates $z - z_o$, where z_o = chimney height.)

Answer:

4 g/m³ ≈ 4000 ppm

4.5 Twin Sources: The Superposition Principle

When a second-order differential equation is *linear* (i.e., does not contain nonlinear terms such as u^2, u^1, $(\partial u/\partial x)^2$, etc.), the following important *superposition principle* applies: Any two independent solutions that satisfy the equation and its boundary conditions can be added to obtain the *general* solution of the differential equation.

Fick's equation falls in this category. Use this fact to obtain the concentration response to *two* instantaneous point sources of equal strength Mp, a distance d apart, and emitting into infinite space (Figure 4.13).

(*Hint*: Draw a midplane between the two sources, and sketch the concentration profiles in its vicinity. Then use Practice Problem 4.4 as a guide. The superposition used here is referred to as the Method of Images.)

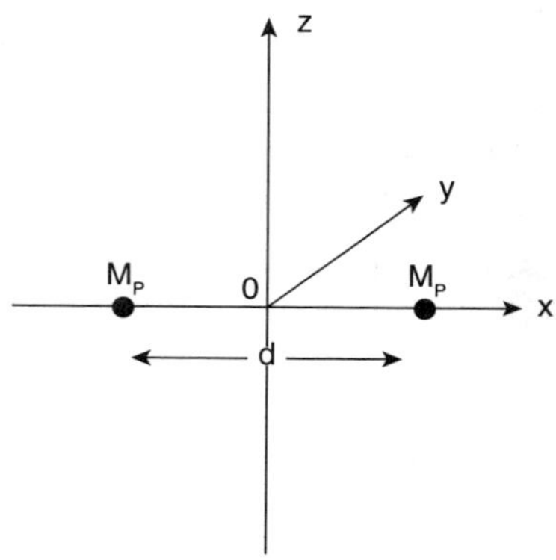

FIGURE 4.13
Emissions from twin-point sources.

4.6 Evaporation of a Solvent Spill

A load of solvent is spilled over a large area, resulting in a layer of liquid with an area density of 1 kg/m^2. The solvent has a vapor pressure of 10 kPa and an average molar mass of 100. Diffusivity is estimated at 10^{-5} m^2/s. Calculate the time of evaporation in the absence of any air currents. This is the *maximum* to be expected.

Answer:

136 h

4.7 Doping of a Silicon Chip with Boron

A junction in silicon is made by doping it with boron. This is done by first depositing a layer of boron on the chip (predeposition), followed by what is termed *drive-in diffusion*. If the deposition step requires 5 min, at what distance from the surface is the concentration of boron raised by 3×10^{18} atoms/cm^3 during this interval? The density of pure boron is 5.1×10^{20} atoms/cm^3, its diffusivity 5.8×10^2 μm^2/h.

Answer:

0.271 μm

4.8 Leaching of an Ore

The leaching of ores to recover valuable mineral components is a commonly applied operation in the field of hydrometallurgy. The extraction of gold with cyanide solution is a familiar example. Suppose the finely ground ore can be thought of as plane flakes, and that it is desired to carry the process to the point where no more than 2% of the original material remains at the midplane of the flake, which has a thickness of 2 mm. Diffusivity is estimated at 10^{-9} m^2/s. Estimate the time of leaching, assuming the external solution concentration to be very small.

Answer:

0.417 h

4.9 Batch Adsorption of a Trace Substance

When a diffusing solute partitions or adsorbs onto a solid matrix, one can often use standard solutions for nonsorbing solids to follow the course of adsorption by suitably modifying one of the solution parameters.

For the case of adsorption by spherical particles from a well-stirred solution of limited volume, for example, the parameter $V_{Sol'n}/V_{Spheres}$ in Figure 4.6 is replaced by $V_{Sol'n}/KV_{Spheres}$, where K is the partition coefficient or Henry's constant. Assume the following parameter values: $K = 10$, $V_{Sol'n}/V_{Spheres} = 10$, $D = 10^{-5}$ cm^2/s, and $R = 0.5$ cm. What is the fractional saturation of the adsorbent after 1 h?

Answer:

0.74

4.10 Response of a Water Basin to Changes in Atmospheric CO_2 Levels

The earth's water basins, such as oceans and lakes, are important "sinks" for greenhouse emissions and are often cited in dismissive reports on global warming. To obtain a sense of their potential, consider the following example:

A large water basin with an irregular shoreline but a fairly constant depth of 20 m is exposed to a 2 ppm increase in the carbon dioxide content of air. The task is to calculate the time required to attain a 10% saturation level of the lake waters assuming (1) a normal molecular diffusivity of $D_{CO_2} = 2 \times 10^{-5}$ cm^2/s and (2) a 100-fold higher value due to convection currents.

(*Hint*: Show that even though uptake here is one sided, the slab solution of Figure 4.4 for two-sided uptake still applies.)

Answer:

(1) 127 years

4.11 Electroplating by Intermittent Voltage Cycling

Conventional DC electroplating often results in uneven deposits and clustering. This is particularly undesirable and costly when precious metals (Pt, Pd) are to be deposited to act as catalysts on electrodes used in fuel cells.

One way of reducing the effect is to plate briefly and intermittently. The voltage is ramped up to a set value and maintained there for periods of the order of a millisecond. It is then shut off for about the same period to allow the depleted boundary layer to be replenished by diffusion from the bulk electrolyte.

Develop an equation to determine the total amount deposited (kg/m^2) in n half-cycles, each of duration Δt. Does the result depend on the *form* of the voltage half-wave (i.e., whether it is a half-square, half-sine, or half-triangular wave?) Discharge at the electrode is assumed to be instantaneous, and distance ℓ is set at 1 cm.

4.12 The Long-Time Solution in Environmental Work

In environmental studies, the release of pollutants from water to the atmosphere is often reported in terms of the "half-life" τ of the substance (i.e., the time required for the concentration in the water to drop to one half its initial value — that is, $M_t/M_\infty = 0.5$). For benzene the reported half-life is $\tau = 4.81$ h, for a water phase with characteristic length = Volume/Area = 1 m. Derive a value for the liquid-phase mass transfer coefficient k_L. If the water phase is a 2 mm diameter droplet ($\ell = 1$ mm), what would the half-life be?

Answer:

$k_L = 0.144$ m/h

4.13 Blood Coagulation and the Higbie Equation

Blood coagulation is triggered by the enzyme thrombin that, upon reaching an injured site or other anomaly, forms strands of fibrin that incorporate platelets to form the clot (see Practice Problem 1.). Assuming instantaneous conversion of thrombin (i.e., a diffusion-controlled process), give a persuasive argument that the short-time solution applies during the first few seconds of the coagulation process and that the mass transfer coefficient during this period is given by the Higbie equation: $D = 10^{10}$ m^2/s.

4.14 Nanoparticles for Drug Delivery

Nanoparticles in the range 10 to 100 nm containing a dissolved drug are the latest device being investigated for direct drug delivery to malignant tumors. In many instances, they are drawn into the cancerous cells where small changes in the pH may trigger drug release. The shape of these particles, the internal and external diffusivities, or whether release proceeds in part at a constant rate are all unknown parameters. Given these uncertainties, is there a way of, at least, setting a *lower limit* to the release time (say to 95% of contents)?

4.15 A Conduction Analogue: Heat Sealing of Plastic Sheets

Plastics sheets and other materials are often bonded together by applying heated elements (platens) to the surface of the sheets. In order to establish the proper heating cycle, it is necessary to know the time t required for the interface to reach a specified sealing temperature. Calculate t using the following data:

Initial temperature: T_o = 25°C

Platen temperature: T_p = 200°C

Sealing temperature: T_s = 100°C

Thermal diffusivity: $\alpha = 10^8$ m²/s

Sheet thickness: 2 mm each

4.16 Catalyst Pellet Performance

Critically examine the following statements pertaining to a fixed-bed catalytic reactor:

(1) Conversion doubles if the pellet porosity is doubled.

(2) A fourfold increase in the (first-order) rate constant results in a fourfold increase in product yield.

(3) An increase in pellet size from 1.5 to 2 cm is always undesirable.

4.17 Raschig Rings as Catalyst Pellets

Hollowed out cylindrical pellets resembling Raschig rings are sometimes used to increase the exposed area available for diffusion and hence increase the effectiveness factor. An additional advantage is the resulting reduction in pressure drop. These gains are to varying degrees offset by the attendant *decrease* in catalyst mass per unit volume of reactor.

Consider two cylindrical pellets, one a regular solid with radius R_o (= ∫ length L), the other hollowed out with an inner radius R_i and outer radius R_o. Show that the hollow cylinder will at most *double* the conversion. (*Hint*: Examine the product $E \times$ Pellet Volume. For the solid cylinder, R_o takes the place of L in Figure 4.7, for the hollow cylinder, $R_o - R_i$.)

4.18 The Catalyst Pellet under Nonisothermal Conditions

Derive the energy balance for a flat-plate catalyst pellet operating under nonisothermal conditions (first-order exothermic reaction). Give a plausible argument why the effectiveness factor can in this case exceed unity.

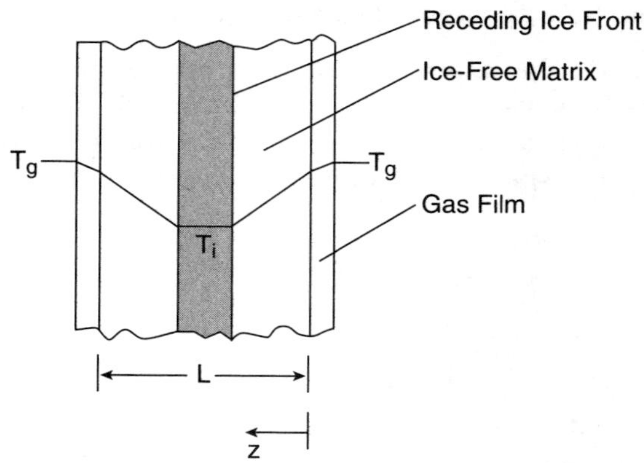

FIGURE 4.14
Freeze-drying of meat.

4.19 A Model for a Burning Fuel Droplet

Decompose the combustion process of a liquid fuel spray into individual steps (evaporation, reaction, heat transfer, mass transfer) that, taken together, might yield a satisfactory model for the process. Describe in words only.

4.20 A Heat Transfer Counterpart to the Shrinking Core Model: Freeze-Drying of Food

In the process considered here, it is desired to derive a model that would allow us to obtain relevant transport coefficients from freeze-drying rate data. The food to be dried (e.g., a slab of frozen poultry meat) has an initial (frozen) water content of m_o kg. It is heated with an electric heater and, in the experiment in question, provided with thermocouples to measure surface temperature T_g (see Figure 4.14). Sublimation of the ice takes place in a vacuum chamber, and water loss is monitored by means of a spring balance. As sublimation progresses, the core ice front, assumed to be at the constant temperature T_i, recedes into the interior, exposing an ice-free matrix that increases in thickness with time. Heat conduction through this matrix is assumed to be at a quasi-steady state so that a linear temperature gradient prevails at any given instant. Start in the usual fashion by first making mass and energy balance about the core, followed by an energy balance on the ice-free matrix. Use the fraction of ice removed, f, as the dependent variable.

Answer:

$t/f = af + b$

4.21 Gas Absorption with a Reactive Solvent

State the advantages and disadvantages of using a reactive solvent and some of the requirements that need to be met.

4.22 The Enhancement Factor E_h

Give a derivation of Equation 4.25c for the enhancement factor, using the procedure used to derive the catalyst effectiveness factor.

4.23 Diffusion into a Biofilm

Derive Equation 4.26h. (*Hint*: Use the third boundary condition to eliminate X.)

5

More about Mass Transfer Coefficients

In Chapter 1 the notion of a mass transfer coefficient was introduced and the connection to what was termed *film theory* was shown. In essence, this approach assumes the resistance to mass transfer to be confined to a thin film in the vicinity of an interface in which the actual concentration gradient is replaced by a linear approximation. The result is that the rate of mass transport can be represented as the product of a mass transfer coefficient and a linear concentration difference, or concentration driving force. Thus,

$$N_A/A = k_C \Delta C$$

It was further shown that individual transport coefficients could be combined into overall mass transfer coefficients to represent transport across adjacent interfacial layers. The underlying concept was referred to as two-film theory. Chapter 1 was confined to simple applications of the mass transfer coefficient which is either assumed to be known or is otherwise evaluated numerically in simple fashion.

This chapter seeks to enlarge our knowledge of mass transfer coefficients by compiling quantitative relations and data for the use in actual calculations applied to practical systems. In Chapter 1, the range over which these coefficients vary was explored and an estimate of the upper and lower bounds to which they are confined was given. This was done by considering the extremes of mass transfer into highly turbulent flow and the much slower transport by diffusion from a flat surface to an unbounded medium. The results, shown in Table 1.5, indicate that the two extremes are separated by at least two orders of magnitude after 1 h of diffusion, and that these limiting coefficients steadily diverge with the passage of time. It was also shown that for gases, k_C values exceed those for liquids of normal viscosity by three orders of magnitude, irrespective of the transport mechanisms or the nature of the solute. These values will serve as useful guidelines in what follows.

When the system under consideration is in laminar flow, it is often possible to give precise analytical expressions of the transport coefficients. In most other cases, including the important case of turbulent flow, the analytical approach generally fails, and we must resort to semiempirical correlations, arrived at by the device known as dimensional analysis, which involves the use of dimensionless groups.

To represent these facts in an organized fashion, we start our deliberations with a brief survey of the dimensionless groups pertinent to mass transfer operations. One of these, the Biot number Bi_m, is singled out for a more detailed examination because it serves as an important criterion in mass transfer to and from particles. We next turn to transport coefficients that apply to systems in laminar flow and show how these coefficients are extracted from the solutions of the pertinent partial differential equation (PDE) models. This is followed by an analysis of systems in turbulent flow where the approach of dimensional analysis is used. We describe the method and present the results obtained for some simple geometries, including flow in a pipe and around spheres and cylinders.

More complex geometries involving commercial tower packings are discussed next. Such packings, used in separation and purification methods such as gas absorption and distillation, have their own peculiar characteristics, and our main source of information here is the data given in manufacturers' catalogs. Even more complex conditions apply to mass transfer operations carried out in stirred vessels. Configuration of the stirring mechanism and the speed of stirring enter the picture here, leading to rather complex expressions.

Finally, we turn our attention to transport in an environmental context. The methodology used by environmentalists in determining transport coefficients has its own peculiarities, which are discussed, and are related to the standard concepts used here.

5.1 Dimensionless Groups

In the dimensional analysis of convective mass transport, three new and important dimensionless groups make their appearance, all of which have well-known heat transfer counterparts, and all of which represent ratios of transport rates or transport resistances. They are, respectively, the Sherwood number Sh, the Schmidt number Sc, and the Biot number (for mass transfer) Bi_m, and they are defined as follows:

$$\text{Sh} = \frac{k_c \ell}{D} = \frac{\text{convective diffusivity}}{\text{molecular diffusivity}} \tag{5.1a}$$

$$\text{Sc} = \frac{\mu}{\rho D} = \frac{\nu}{D} = \frac{\text{momentum diffusivity}}{\text{molecular diffusivity}} \tag{5.1b}$$

$$\text{Bi}_\text{m} = \frac{(k_c \ell)_\text{ext}}{(D)_\text{int}} = \frac{\text{external diffusivity}}{\text{internal diffusivity}} \tag{5.1c}$$

where ℓ is some pertinent dimension of the system, such as the diameter of a pipe or a sphere, and k_c is the mass transfer coefficient in units of meter per second (m/s).

The *Sherwood number* can be viewed as describing the ratio of convective to diffusive transport and finds its counterpart in heat transfer in the form of the Nusselt number. It is high (»1) when flow is turbulent or the boundary layer "film" is very thin. The *Biot* number has the same form as the Sherwood number but refers to two adjacent phases or media. In one of these, which we term *internal*, transport is usually by diffusion. This can be in a gas bubble, a liquid drop, a porous solid particle, or some other entity. The adjacent ("external") phase is a liquid or a gas in relative motion to the particle and has an attendant "film resistance." Like the Sherwood number, Bi_m is high (»1) when the external phase is in turbulent motion (caused, for example, by stirring) or the boundary layer is very thin. We will have more to say about it in Illustration 5.1.

Two additional dimensionless groups, the Peclet number and the Stanton number, are also used, although with lesser frequency. Both of these numbers are composites of other dimensionless groups, which frequently occur in unison. Thus, the Reynolds and Schmidt numbers often crop up combined as a product, which leads to the Peclet number:

$$\mathrm{Pe} = \mathrm{Re\,Sc} = \frac{\ell v \rho}{\mu}\frac{\mu}{\rho D} = \frac{\ell v}{D} \tag{5.1d}$$

The Stanton number is a combination of Sherwood, Reynolds, and Schmidt numbers, which likewise often appear in unison. It is defined as

$$\mathrm{St} = \mathrm{Sh} / \mathrm{Re\,Sc} = \frac{k_c}{v} \tag{5.1e}$$

The result here is particularly simple and in essence represents the ratio of two velocities: the "velocity" of mass transfer k_c (in units of m/s) and the velocity of flow (likewise in units of m/s). These and other groups for both mass and heat transfer processes are summarized in Table 5.1. The last entry, the Grashof number, describes the special case of free convection triggered by density differences.

Illustration 5.1: The Biot Number in Mass Transfer

The most fruitful way of utilizing the Biot number is to express it in terms of *mass transfer resistances* and use it as a criterion to determine which of the two adjacent phases controls the mass transfer process. We write

$$(\mathrm{Bi})_m = \frac{(k_C \ell)_{ext}}{(D)_{int}} = \frac{\ell / D}{1 / k_C} = \frac{\text{Internal resistance}}{\text{External resistance}} \tag{5.2a}$$

TABLE 5.1

Summary of Dimensionless Groups Used in Mass and Heat Transfer Processes

Mass Transfer		Heat Transfer	
Sherwood number	$\text{Sh} = \frac{k_c \ell}{D}$	Nusselt number	$\text{Nu} = \frac{h\ell}{k}$
Schmidt number	$\text{Sc} = \frac{\mu}{\rho D}$	Prandtl number	$\text{Pr} = \frac{C_p \mu}{k}$
Reynolds number	$\text{Re} = \frac{\ell v \rho}{\mu}$	Reynolds number	$\text{Re} = \frac{\ell v \rho}{\mu}$
Peclet number	$\text{Pe}_\text{m} = \frac{\ell v}{D}$	Peclet number	$\text{Pe}_\text{h} = \frac{\ell v \rho C_p}{k}$
Stanton number	$\text{St}_\text{m} = \frac{k_c}{v}$	Stanton number	$\text{St}_\text{h} = \frac{h}{C_p v \rho}$
Fourier number	$\text{Fo}_\text{m} = \frac{Dt}{\ell^2}$	Fourier number	$\text{Fo}_\text{h} = \frac{\alpha t}{\ell^2}$
Biot number	$\text{Bi}_\text{m} = \frac{(k_c \ell)_\text{ext}}{(D)_\text{int}}$	Biot number	$\text{Bi}_\text{h} = \frac{(h\ell)_\text{ext}}{(k)_\text{int}}$
Grashof number	$\text{Gr}_\text{m} = \frac{\ell^3 \rho g \Delta\rho}{\mu^2}$	Grashof number	$\text{Gr}_\text{h} = \frac{\ell^3 \rho^2 g \beta \Delta T}{\mu^2}$

from which it follows that large Biot numbers are associated with a high internal or particle resistance to mass transfer, and vice versa.

It is common to divide the range of $(\text{Bi})_\text{m}$ values into three regions:

1. For $(\text{Bi})_\text{m} > 10$, the preponderant part of the resistance, over 90%, resides in the *internal phase*. Mass transfer to and from that phase is described by Fick's equation (a PDE), and the uptake or release of solute can be read from Figure 4.6, Chapter 4, or calculated from the "long-time" solution (an ordinary differential equation [ODE]) listed in Table 4.4. Internally, the concentration profile varies with time; externally, it is flat and constant, with interface concentration C_g equal to the bulk fluid concentration C_p (see Figure 5.1a).
2. For $(\text{Bi})_\text{m} < 0.1$, the reverse holds true. Over 90% of the resistance is now in the *external phase*. The internal profile is flat but varies with time as solute is released or taken up (see Figure 5.1c). It is described by the following ODE mass balance over the internal phase:

   Rate of solute in or out = Rate of change of solute content

$$\pm k_C A(C_\text{int} - C_b) = V \frac{dC_\text{int}}{dt} \tag{5.2b}$$

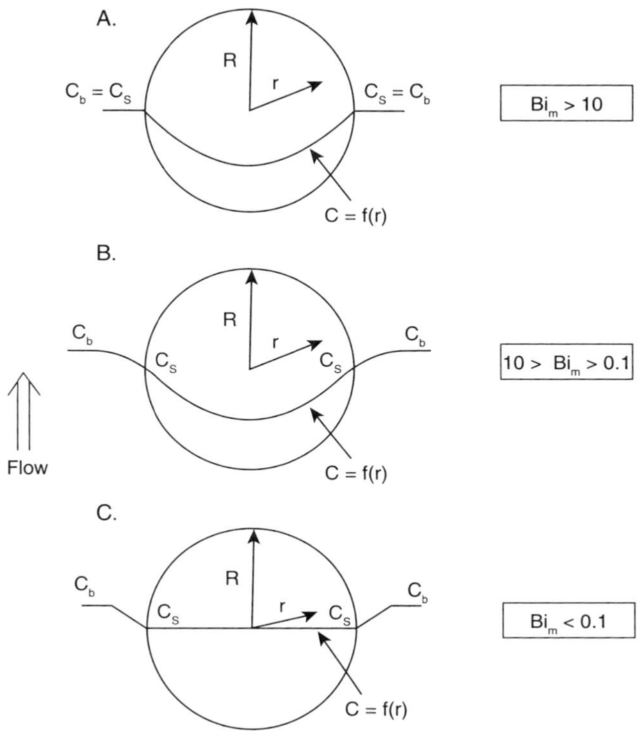

FIGURE 5.1
Transient mass transfer between a fluid in motion and a permeable particle: (a) internal resistance only (most common case); (b) internal and external resistance; and (c) external resistance only.

3. For $10 > (\mathrm{Bi})_\mathrm{m} > 0.1$, both resistances are significant (i.e., >10%, see Figure 5.1b), but one is always greater than the other except for $(\mathrm{Bi})_\mathrm{m} = 1$. Even in this region, it is common to identify a "controlling" resistance, if it is, say, over 70%.

In practice, the internal phase tends to be the controlling resistance in the majority of applications. Consider, for example, an adsorbent particle of 5 mm diameter, a size normally used in packed bed adsorbers for the separation and purification of gases. Even assuming only mildly turbulent flow ($k_\mathrm{C} \approx 0.1$ cm/s) and an internal "effective" diffusivity D_e of 10^{-3} cm^2/s, we still have

$$(\mathrm{Bi})_\mathrm{m} = \frac{k_\mathrm{C} d}{D_e} = \frac{0.1 \times 0.5}{10^{-3}} = 50 \tag{5.2c}$$

that is, particle resistance controls. In liquid-phase applications, including ion exchange, both k_C and D_e change by the same factor (three orders of magnitude, see Table 1.5) so that the result is still the same.

Much the same situation holds if the porous solid is replaced by a liquid droplet, say a raindrop of 1 mm diameter, exposed to a pollutant during its fall (see Practice Problem 5.3). Using $k_C = 0.1$ cm/s as before and a liquid-phase diffusivity of 10^5 cm^2/s, we obtain

$$(\mathrm{Bi})_\mathrm{m} = \frac{k_C d}{D_e} = \frac{0.1 \times 0.1}{10^{-5}} = 10^3 \tag{5.2d}$$

that is, particle resistance is again the controlling factor.

A notable exception occurs in the case of mass transfer from rising gas bubbles to a surrounding liquid. Here one sees a shift of the resistance in its entirety to the external medium, even if it is in a state of turbulence. For *single* bubbles, the Higbie equation (Equation 4.11g) provides the appropriate k_C value. Mass transfer from *swarms* of bubbles is taken up in Illustration 5.5.

Because k_C in external flow is only weakly dependent on internal dimension (or d), it follows that particle resistance will be even more pronounced for larger geometries. Any system exceeding the millimeter or lower range of particle size will therefore be invariably controlled by the mass transfer resistance within that particle.

Comments

The Biot number was originally established as a criterion for gauging relative resistances for the analogous *heat transfer* case. Here the span of Biot numbers extends into both regions of controlling resistances. This is in large part due to the wide range of thermal conductivities of solids seen in practice. Thus, heat transfer from a turbulent stream of air to a sheet of copper (k = 390 W/mK) would have external (air) resistance control, while in heat transfer to a sheet of asbestos (k = 0.15 W/mK) under the same conditions the principal resistance would reside in the solid. The reach of the Biot number in mass transfer is more limited but still of some importance.

5.2 Mass Transfer Coefficients in Laminar Flow: Extraction from the PDE Model

Mass transport in laminar flow in a tubular geometry or around simple submerged shapes is generally modeled by PDEs because there is more than one direction of diffusional flow involved. These PDEs have been solved analytically for a number of cases and generally lead to fairly formidable expressions representing the concentration and velocity profiles in the geometries in question. As indicated on a number of previous occasions, concentration profiles, which represent the primary information obtained from the PDE model, are often not directly useful for engineering purposes. In the present case, the useful quantity we wish to extract from the primary information is an equivalent mass transfer coefficient k_C.

Let us demonstrate its derivation using transport in a tube as an example. The situation here is one in which solute diffuses in the radial direction, either as a result of release from the wall, or in consequence of transport to and ultimately through a permeable wall. The model for this case was presented in Section 2.8 and was referred to as the Graetz problem in mass transfer.

The concentration profiles that arise in this case are distributed in both the radial and axial directions, as diffusion in one direction is superposed on convective transport in the other direction. To obtain a mass transfer coefficient from this information, we perform a mass balance at the tubular wall, equating the diffusional transport rate to an equivalent "convective" rate expressed by means of a mass transfer coefficient. Thus,

$$-D\left.\frac{\partial C}{\partial r}\right|_{r=R} = k_c(C_m - C_{r=R}) \tag{5.3a}$$

To evaluate k_c, two quantities need to be obtained from the primary concentration profile. One is the derivative at the tubular wall given above, which is obtained by differentiating the solution $C(r,z)$; the second is the mean integral concentration C_m in the flowing fluid. This latter quantity is obtained from the following expression:

$$C_m = \frac{\int_0^R C(r,z)2\pi r dr}{\pi R^2} \tag{5.3b}$$

Note that both of these items are part of the "information package" contained in a model that we alluded to in Table 2.2.

With these two quantities in place, Equation 5.3a can be solved for k_c and the latter tabulated. Similar calculations can be carried out for systems involving flow around simple geometries.

5.2.1 Mass Transfer Coefficients in Laminar Tubular Flow

In the case of tubular mass transfer coefficients, we distinguish between mass transfer in the so-called entry or Lévêque region, in which concentration changes are confined to a thin boundary layer $\delta(x)$ adjacent to the wall, and the so-called fully developed region, in which the concentration changes penetrated into the fluid core. The situation is depicted in Figure 5.2 and represents a tubular wall coated with a soluble material of solubility C_s dissolving into pure solvent.

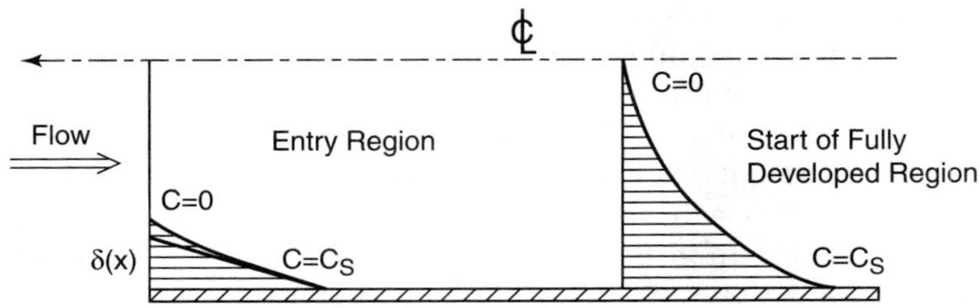

FIGURE 5.2
Mass transfer from a coated tubular wall into a flow of pure solvent.

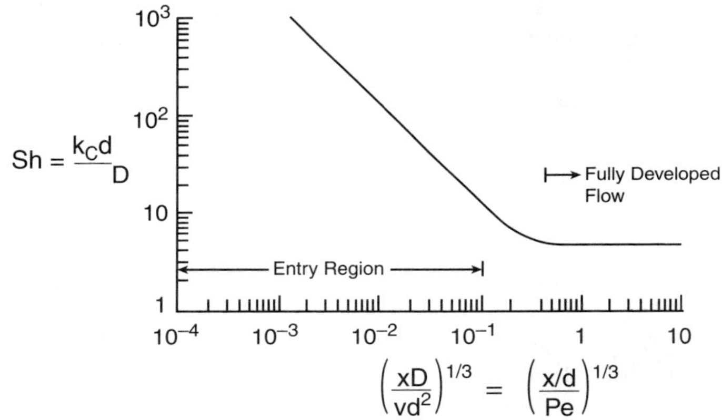

FIGURE 5.3
Sherwood numbers for laminar flow in a cylindrical duct.

Because of the thinness of the boundary layer, mass transfer in the entry region is very rapid, with Sherwood numbers in excess of 1000 attained near the tubular entrance (Figure 5.2). As we move away from the entrance in the downstream direction, the boundary layer gradually thickens, and the Sherwood number diminishes with the one-third power of axial distance x. Eventually it levels off and attains a constant value as the fully developed region is reached (Figure 5.2 and 5.3). Table 5.2 lists some of the relevant Sherwood numbers obtained in ducts of various geometries and constant wall concentration. The quantity x/d is referred to as the "entry length."

For engineering calculations, it is often more convenient to deal with mass transfer coefficients that were averaged over the entire length of the entry region. The result is expressed in terms of the dimensionless Reynolds and Schmidt numbers and takes the following form for laminar flow in cylindrical tubes:

$$\mathrm{Sh} = 1.86\ (\mathrm{Re\ Sc}\ d\ /\ L)^{1/3} = 1.86\left(\frac{vd^2}{DL}\right)^{1/3} \tag{5.6}$$

TABLE 5.2

Mass Transfer Coefficients in Ducts of Various Geometries for Laminar Flow

Duct Geometry	Entry Region	Fully Developed Region
Cylinder	$\mathrm{Sh} = 1.08\left(\frac{xD}{vd^2}\right)^{-1/3} = 1.08\left(\frac{x/d}{Pe}\right)^{-1/3}$	Sh = 3.66
Parallel planes	$\mathrm{Sh} = 1.23\left(\frac{xD}{vd^2}\right)^{-1/3} = 1.23\left(\frac{x/d}{Pe}\right)^{-1/3}$	Sh = 7.54
Square	—	Sh = 2.98
Triangular	—	Sh = 2.47

TABLE 5.3

Mass Transfer Coefficients in Laminar Flow around Simple Geometries

Geometry	Correlation	Re
Flat plate	$\mathrm{St} = 0.66(\mathrm{Re})^{1/2}(\mathrm{Sc})^{2/3}$	$<10^5$
Sphere	$\mathrm{Sh} = 2.0 + 0.60(\mathrm{Re})^{1/2}(\mathrm{Sc})^{1/3}$	$<10^1$
Cylinder	$\mathrm{Sh} = 0.43 + 0.53(\mathrm{Re})^{1/2}(\mathrm{Sc})^{0.31}$	$1 < \mathrm{Re} < 10$

where L is the length of pipe in question. The mass transfer coefficient here is a constant, rather than a function of distance, and can be directly incorporated in the usual tubular mass balances. This will be shown in Illustration 5.2.

Mass transfer in the entry region is far from being a rare event. Transport in the larger blood vessels lies entirely in the entry region, and so does mass transfer in commercial reverse osmosis desalination plants and other membrane processes (see Chapter 8). Entry lengths of several meters are not uncommon, particularly for low solute diffusivities. The only requirement is that flow must be in the laminar regime (Re < 2000).

5.2.2 Mass Transfer Coefficients in Laminar Flow around Simple Geometries

In the case of mass transfer involving laminar flow about simple geometries, no distinction is made between regions near the leading front of the particle and portions farther downstream. Instead, we report transfer coefficients averaged over the entire particle. The results are expressed as correlations of the Sherwood or Stanton numbers as a function of Reynolds and Schmidt numbers.

The correlation for three simple geometries, the flat plate, the sphere, and the cylinder, are tabulated in Table 5.3. The flat plate here is taken to be positioned parallel to the direction of flow, while the cylinder has its axis normal to the flowing medium.

Illustration 5.2: Release of a Solute into Tubular Laminar Flow: Transport in the Entry Region

Consider the transport from a tubular wall coated with a soluble substance into a solvent in laminar flow. We set the diameter at $d = 1$ cm, velocity v at 1 cm/s, and the length x of the tube at 1 m. The solubility C^* of the solute is 10 g/L, and its diffusivity $D = 10^{-9}$ m²/s. The problems we wish to address are the following:

1. What is the boundary layer thickness at the exit of the tube?
2. What is the mean concentration at that position?

To answer question 1, we draw on the definition of a mass transfer coefficient, which was set out in the film theory given in Chapter 1, Equation 1.11b. The rearranged version of this expression has the following form:

$$k_c = \frac{D}{z_{FM}} = \frac{D}{\delta} \tag{5.5a}$$

from which the boundary layer thickness δ can be extracted.

To obtain δ, we start by calculating the mass transfer coefficient k_c for which we draw on Table 5.3. Note that because $(xD/vd^2)^{1/3} = (1 \times 10^{-9}/10^2 \times 10^4)^{1/3} = 0.1$, the flow does in fact fall in the Lévêque region (Figure 5.2). We have

$$\frac{k_c d}{D} = 1.08\left(\frac{xD}{vd^2}\right)^{-1/3} \tag{5.5b}$$

from which we obtain

$$k_c = \frac{1.08 \times D}{d}\left(\frac{vd^2}{xD}\right)^{1/3} = \frac{1.08 \times 10^{-9}}{10^{-2}}\left(\frac{10^{-2} \times 10^{-4}}{1 \times 10^{-9}}\right)^{1/3} \tag{5.5c}$$

and, consequently,

$$k_c = 1.08 \times 10^{-6} \text{ m/s} \tag{5.5d}$$

Substitution of this value into Equation 5.5a and solving for δ yields the boundary layer thickness:

$$\delta = \frac{D}{k_c} = \frac{10^{-9}}{1.08 \times 10^{-6}} = 0.93 \times 10^{-3}\,\text{m} = 0.93\,\text{mm} \tag{5.5e}$$

Thus, the boundary layer thickness is approximately 10% of the tubular diameter at the exit of a 1 m tube.

We next turn to the calculation of the concentration at the tubular exit. This requires setting up a differential solute balance over a tube segment and its subsequent integration over the entire length of the tube. For a finite tubular increment Δx and volumetric flow rate Q, we can write

$$\text{Rate of solute in} - \text{rate of solute out} = 0$$

$$\begin{bmatrix} QC_{1x} \\ +k_c \pi d \Delta x (C^* - C) \end{bmatrix} - QC_{1x+\Delta x} = 0 \tag{5.6a}$$

where C is the tubular concentration averaged over the entire cross section.

Dividing by Δx and letting $\Delta x \to 0$ yields

$$Q\frac{dC}{dx} - k_c \pi d(C^* - C) = 0 \tag{5.6b}$$

or

$$v(d/4)\frac{dC}{dx} - k_C(C^* - C) = 0$$

To keep the integration simple, we use the mean integral Sherwood number given by Equation 5.4, rather than the distance-dependent local coefficient listed in Table 5.2. We have

$$k_C = 1.86\frac{D}{d}\left(\frac{vd^2}{DL}\right)^{1/3} \tag{5.6c}$$

$$k_C = 1.86\left(\frac{vd^2}{DL}\right)^{1/3} \tag{5.6d}$$

$$k_C = 1.86\left(\frac{10^{-2} \times 10^{-18}}{10^{-2} \times 1}\right)^{1/3} = 1.86 \times 10^{-6} \text{ m/s} \tag{5.6e}$$

This value is used in the solution of the ODE (Equation 5.6b) obtained by the separation of variables:

$$\int_0^C \frac{dC}{C^* - C} = (4/vd)1.86 \times 10^{-6} \int_0^L dx \tag{5.6f}$$

$$\frac{C^* - C}{C^*} = \exp\left(-\frac{4 \times 1.86 \times 10^{-6}}{10^{-2} \times 10^{-2}} \times 1\right) \tag{5.6g}$$

so that

$$1 - C/C^* = 0.93 \tag{5.6h}$$

and $C = 0.7$ g/L (i.e., the well-mixed solution is approximately 7% saturated at the exit). Note that the *core* concentration remains constant and equal to zero over the entire tube length, and average concentration changes very slowly. We use this fact to simplify the model for reverse osmosis taken up in Illustration 8.6.

5.3 Mass Transfer in Turbulent Flow: Dimensional Analysis and the Buckingham π Theorem

Both heat and mass transfer in turbulent flow are generally not amenable to analytical treatment. It is customary in these cases to resort to what is termed *dimensional analysis*. This device consists of grouping the pertinent physical parameters of the system into a number of dimensionless groups, thus reducing the number of variables that have to be dealt with and evaluating the undetermined coefficients experimentally. It is a powerful tool for arriving at a first qualitative description of complex systems and eases enormously the experimental work required to quantify the relationship.

5.3.1 Dimensional Analysis

The process of dimensional analysis can be carried out in three deceptively simple steps. Some comments are necessary to ensure the proper application of this scheme:

Step 1. *List all variables that affect system behavior.* This is by far the most important and difficult step and confronts the user with the task of deciding which independent variables to include in the analysis. There is no easy recipe for carrying out this step, but the following suggestions may be found useful:

- Apply physical reasoning to identify the pertinent variables.

- Use experimental observations to amplify this list and to verify the final result.
- If no functional relationship of dimensional consistency results, reexamine the situation by adding or omitting parameters.
- Make sure that only truly independent variables are included. Thus, if mass and acceleration are chosen as parameters, force cannot be added to the list because it is related to the former through Newton's law.

Step 2. *Write down the dimensions of these variables.* The basic dimensions that are commonly chosen in dimensional analysis are those of mass (M), length (L), time (θ), and temperature (T). All other quantities are expressed in terms of these fundamental dimensions. Thus, force has the units of $ML\theta^2$ by virtue of Newton's law. Joule (J) is not a fundamental dimension but is instead expressed as $ML^2\theta^2$.

Step 3. *Combine the variables into a functional relationship involving dimensionless groups or some other dimensionally consistent form.* The ultimate aim of this step, and of the analysis as a whole, is to express system behavior in terms of the functional relationship:

$$F(\pi_1, \pi_2 \ldots \pi_p) = 0 \tag{5.7a}$$

where π is the symbol commonly used for a dimensionless group. If $x_2 \ldots x_p$ are independent variables, and x_1 is the dependent variable, the dimensionless groups can be represented in the form

$$\pi_1 = \frac{x_1}{x_2^{a_1} x_3^{a_2} \ldots} \qquad \pi_2 = \frac{x_i}{x_2^{\beta_1} x_3^{\beta_2} \ldots} \quad \text{etc.} \tag{5.7b}$$

Often, it is convenient to solve this relationship for the dependent variable. We then obtain

$$x_1 = x_2^{a_1} x_3^{a_3} \ldots F(\pi_2, \pi_3 \ldots \pi_n) \tag{5.7c}$$

To aid the reader in implementing this analysis, a short list of the dimensions of important variables is presented in Table 5.4.

5.3.2 The Buckingham π Theorem

The application of dimensional analysis is relatively simple when the number of independent variables equals the number of fundamental dimensions. Straightforward application of Steps 1 through 3 then yields a result expressed in terms of a *single* dimensionless group. This case, however, is

TABLE 5.4

Variables and Their Dimensions

	Variable	Symbol	Dimension
1. Fundamental	Mass	m	M
	Length	d	L
	Time	t	q
	Temperature	T	T
2. Mechanical	Velocity	v	$L\theta^{1}$
	Acceleration	a	$L\theta^{2}$
	Density	r	ML^{3}
	Viscosity	m	$ML^{3}\theta^{1}$
	Force	F	$ML^{1}\theta^{2}$
	Pressure	p	$ML^{3}\theta^{2}$
3. Thermal	Thermal conductivity	k	$ML\theta^{3}T^{1}$
	Specific heat	C_p	$L^{2}\theta^{2}T^{1}$
	Heat transfer coefficient	h	$M\theta^{3}T^{1}$
4. Diffusional	Concentration	C	ML^{3}
	Rate of mass flow	N	$M\theta^{1}$
	Diffusivity	D	$L^{2}\theta^{1}$
	Mass transfer coefficient	k_c	$L\theta^{1}$

relatively rare. It is more common for the number of variables to exceed the number of dimensions, which means that more than one dimensionless group will be involved. The question then arises as to how many such groups are to be sought out, and how they are to be composed. These problems were first addressed by Buckingham and resolved in his famous π theorem. Simply stated, that theorem reads as follows:

Given p variables, $x_1, x_2, \dots x_p$ are related to a physical phenomenon that can be expressed in terms of r fundamental dimensions. Then these variables, which include the dependent variable, can be gathered into p-r dimensionless groups π and cast in the functional form

$$F(\pi_1, \pi_2, \dots \pi_{p-r}) = 0 \tag{5.8}$$

In other words, we managed to replace a functional relation, which involves p variables, with one that involves only p-r variables. This constitutes a considerable saving.

To implement Buckingham's π theorem, we retain the three steps we had formulated earlier but amplify Step 3 in the following fashion:

> Step 3a. Select a set of variables that equals the number of dimensions r and does not include the dependent variable. Raise each variable to some unknown power $\alpha, \beta, \dots$ and form a product of the result. This product is placed in the *denominator* of each dimensionless group π.
>
> Step 3b. Place the remaining (p-r) variables, which now include the dependent variable, in the *numerators* of the (p-r) dimensionless groups. This results in the set of π terms shown in Equation 5.8.

Illustration 5.3: Derivation of a Correlation for Turbulent Flow Mass Transfer Coefficients Using Dimensional Analysis

The problem addressed here is that of expressing mass transfer coefficients that apply to turbulent flow conditions in a tube in terms of appropriate dimensionless groups. To implement Step 1, both fluid mechanical and transport properties must be taken into account. The former determine the degree of turbulence or ability to form eddies, and hence the rate at which mass is transported to or from the tubular wall, while the transport parameter determines the rate at which mass is conveyed through the film adjacent to the interface. It is proposed to use velocity v, density ρ, and viscosity μ as the fluid mechanical properties, as each of these parameters either promotes or resists the formation of eddies. Transport through the film is determined by only one parameter, the diffusivity of the conveyed species. In addition to these factors, we expect pipe diameter to play a role, because it determines the distance over which the mass is to be transported and also plays a role in the degree of turbulence generated in the system.

Note that in choosing these parameters, we had to proceed in a somewhat intuitive fashion. This is the difficult and challenging part of dimensional analysis that we alluded to in Step 1 of our outline. It requires physical insight as well as some inspired guesswork.

We can now proceed to Step 2 of the procedure and list the dimensions of these variables, making use of Table 5.4 for this purpose.

Variable	Symbol	Dimension
Velocity	v	$L\theta^{1}$
Density	r	ML^{3}
Viscosity	m	$ML^{3}\theta^{1}$
Mass transfer coefficient	k_c	$L\theta^{1}$
Diffusivity	D	$L^{2}\theta^{1}$
Diameter	d	L

There are six variables, including the dependent variable k_c. Hence, by Buckingham's theorem, there are 6 − 3 = 3 dimensionless groups to be established.

As recommended in Step 3a and Step 3b, we place ρ, v, d in the denominator, and k_c, μ, D in each of the numerators. Thus,

$$\pi_1 = \frac{k_c}{\rho^{\alpha_1} v^{\alpha_2} d^{\alpha_3}} \Rightarrow \frac{L\theta^{-1}}{(ML^{-3})^{\alpha_1}(L\theta^{-1})^{\alpha_2}(L)^{\alpha_3}} \tag{5.9a}$$

$$\pi_2 = \frac{\mu}{\rho^{\beta_1} v^{\beta_2} d^{\beta_3}} \Rightarrow \frac{ML^{-3}\theta^{-1}}{(ML^{-3})^{\beta_1}(L\theta^{-1})^{\beta_2}(L)^{\beta_3}} \tag{5.9b}$$

$$\pi_3 = \frac{D}{\rho^{\gamma_1} v^{\gamma_2} d^{\gamma_3}} \Rightarrow \frac{L^2\theta^{-1}}{(ML^{-3})^{\gamma_1}(L\theta^{-1})^{\gamma_2}(L)^{\gamma_3}} \tag{5.9c}$$

We find by inspection, or by formally equating coefficients in the numerator and denominator,

$$\alpha_1 = \alpha_3 = 0 \qquad \alpha_2 = 1 \tag{5.9d}$$

$$\beta_1 = \beta_2 = \beta_3 = 1 \tag{5.9e}$$

$$\gamma_1 = 0 \qquad \gamma_2 = \gamma_3 = 1 \tag{5.9f}$$

Hence, we obtain

$$\pi_1 = f(\pi_2, \pi_3) \tag{5.9g}$$

or

$$\frac{k_c}{v} = f\left(\frac{\mu}{\rho v d}, \frac{D}{vd}\right) \tag{5.9h}$$

In Table 5.1, we can see that there is a first grouping in terms of the dimensionless Stanton, Reynolds, and Peclet numbers — that is,

$$\text{St} = F(\text{Re}, \text{Pe}) \tag{5.9i}$$

or, because Pe = ReSc,

$$\text{St} = F(\text{Re}, \text{Sc})$$

A different set of dimensionless groups may be arrived at as follows: We divide π_3 by π_2, divide π_1 by π_3, and invert π_2, resulting in

$$\frac{k_c d}{D} = G\left(\frac{\rho v d}{\mu}, \frac{\mu}{\rho D}\right) \tag{5.9j}$$

or, equivalently,

$$\text{Sh} = G(\text{Re}, \text{Sc}) \tag{5.9k}$$

This is the more common form of dimensionless grouping seen in the literature and states that the Sherwood number Sh, which contains the mass

TABLE 5.5

Correlations for Mass Transfer Coefficients in Turbulent Flow

Range	Correlation
1. Flat plate Re > 10^6	$St = 0.036\ (Re)^{0.2}\ (Sc)^{0.67}$
2. Sphere Unlimited	$Sh = 2.0 + 0.60\ (Re)^{0.5}\ (Sc)^{0.33}$
3. Inside tubes Re > 20,000	$Sh = 0.026\ (Re)^{0.8}\ (Sc)^{0.33}$
4. Packed bed of spheres Re > 50	$St = 0.61\ (Re)^{0.41}\ (Sc)^{0.67}$
5. Fluidized bed	$St = 1/[0.010 + 0.863/(Re^{0.58} - 0.483)]Sc^{0.67}$
6. Bubble swarms (d > 2.5 mm)	$Sh = 0.42\ Gr^{0.33}\ Sc^{0.5}$

transfer coefficient as the dependent variable, is a function of both Reynolds and Schmidt numbers.

Evidently, for Equation 5.9k to be of practical use, it must be rendered quantitative. This is done by assuming that the functional relation is of a power form — that is, we set

$$Sh = a\ Re^b\ Sc^c \tag{5.9l}$$

and evaluate the coefficients experimentally. This approach has been proven successful in most undertakings of this kind. Note that the amount of experimentation is considerably reduced by dimensional analysis, as the original six variables are replaced by three dimensionless groups.

We compiled a list of the most frequently used correlations and tabulate them in Table 5.5. They include mass transfer correlations for turbulent flow about simple geometries as well as in tubes and in packed and fluidized beds.

Illustration 5.4: An Intermezzo: Terminal Velocity of Rising or Falling Spheres

Frequently, mass transfer takes place to and from rising or falling particles, or particles kept in suspension by the upward thrust of a flowing fluid. Practical applications are numerous and varied, ranging from mass transfer in spray towers, liquid extraction columns, and fluidized beds to particles falling through the atmosphere or gas bubbles rising through a liquid.

In order to extract the pertinent mass transfer coefficients from the correlations of Table 5.3 and Table 5.5, one needs to know the terminal velocity v_t of the particle in order to specify the Reynolds number Re. v_t can be obtained from a simple steady-state force balance, which for a falling particle reads

$$V_p\rho_p g - C_D\rho_f A_C(v_t^2/2) - V_p\rho_f g = 0 \tag{5.10a}$$

Gravity Drag Buoyancy

where subscripts p and f refer to the particle and fluid, respectively, and A_C is the cross-sectional area normal to the line of motion. For spherical particles, the drag coefficient C_D equals 24/Re in viscous (Stokes) flow, in the turbulent region, $C_D = 0.44$; and in between those two extremes, the suggested value is $C_D = 18.5/\text{Re}^{3/5}$. Solving Equation 5.10a for the terminal velocity in meter per second (m/s), one obtains the following:

VISCOUS REGION

$$\text{Re} < 0.1 \qquad v_t = (0.545/\nu)d^2(\rho_p/\rho_f - 1) \tag{5.10b}$$

TURBULENT REGION

$$500 < \text{Re} < 2\times10^5 \quad v_t = 5.45\left[d\left(\rho_p/\rho_f - 1\right)\right]^{1/2} \tag{5.10c}$$

INTERMEDIATE REGION

$$2 < \text{Re} < 500 \quad v_t = (0.781/\nu^{0.60})d^{1.14}(\rho_p/\rho_f - 1)^{0.714} \tag{5.10d}$$

where $\nu = \mu/\rho$ = kinematic viscosity. For rising particles, the density factor becomes $(1 - \rho_p/\rho_f)$. Because both v_t and d need to be known to specify Re, a trial-and-error procedure is usually called for. This can be avoided by using the Re inequalities to set upper and lower bounds on particle diameter d. For the range of densities and viscosities of relevance in mass transfer, the following guideline applies: *Particles with* d > *2 mm operate entirely in the turbulent region; those with* d < *0.02 mm operate entirely in the viscous region. For intermediate values of* d*, the region closest should be tested first.*

Consider mass transfer to a falling raindrop of d = 1.5 mm, close enough to d = 2 mm for the turbulent region to be tested first. Setting air density at 1.2 kg/m^3, one obtains, from Equation 5.10c,

$$v_t = 5.45\left[1.5\times10^{-3}(1000/1.2 - 1)\right]^{1/2} = 6.1\,\text{m/s} \tag{5.10e}$$

and the corresponding Reynolds number,

$$\mathrm{Re} = \frac{dv_t}{\nu} = \frac{1.5 \times 10^{-3} \times 6.1}{1.6 \times 10^{-3}} = 570 \qquad (5.10f)$$

that is, we are within the specified turbulent region.

Consider next a fine slurry containing valuable material that is to be washed in a cascade of settling tanks (see Chapter 7 for a more detailed account of this operation). Although particle size in these operations is typically in the range d = 0.5 to 1 mm, it is specified that no particles over 0.05 mm diameter should be lost to the overflow, and sufficient residence time should be allowed for the limiting size to settle a distance of z = 5 m. Because we are close to the Stokes region, d (0.02 mm), a first trial is made using Equation 5.10b. With ρ_p set at 2000 kg/m³, and kinematic viscosity of water $\nu = 9 \infty 10^7$ m²/s, we obtain

$$v_t = (0.545 / 9 \times 10^{-7})(5 \times 10^{-5})^2 \left(\frac{2000}{1000} - 1 \right) = 1.5 \times 10^{-3} \text{ m/s} \qquad (5.10g)$$

and

$$\mathrm{Re} = \frac{v_t d}{\nu} = \frac{1.5 \times 10^{-3} \times 5 \times 10^{-5}}{9 \times 10^{-7}} = 0.083 \qquad (5.10h)$$

that is, we are within the viscous region. The residence time works out to

$$t = \frac{z}{v_t} = \frac{5}{1.5 \times 10^{-3}} \frac{1}{3600} = 0.9 \text{ hrs} \qquad (5.10i)$$

This is a *minimum* time requirement since neighboring particles in close proximity will tend to slow down the process.

Illustration 5.5: Aeration of a Wastewater Treatment Trench

Effluents from paper mills and other industrial plants are often treated in large aerated lagoons or trenches in which a slow biodegradation of the pollutants takes place (see Figure 5.4). Pollutant concentration C_L is usually expressed as an equivalent biochemical oxygen demand (*BOD*, mg O_2/L), and the biodegradation process is expressed in terms of a first-order rate constant k_r (day s¹).

The task to be accomplished is twofold: (1) dimensioning of the trench, which is usually of a long rectangular configuration and (2) determination of the air requirements.

FIGURE 5.4
Aerial view of paper mill in Northern Ontario, Canada. Aerated Lagoons for effluent treatment appear on lower left. (Courtesy H. Tran, University of Toronto.)

Trench dimensions can be easily established from a conventional tubular plug flow reactor model that yields

$$C_L/C_L^0 = BOD/BOD_0 = \exp(-k_r A_C z/Q) \quad (5.11a)$$

where A_C is the cross-sectional area of the trench, and z is its length.

Suppose we set volumetric flow rate Q at 0.2 m^3/s, k_r at 1.0 day^1, BOD_0 at 500 mg/L, and A_C at 200 m^2, typical for such processes. BOD is to be reduced to 10% of its inlet value. The length of the trench is then

$$z = Q / k_r A_C \ln(BOD / BOD_0) \quad (5.11b)$$

$$= -\frac{0.2 \times 24 \times 3600}{1.0 \times 200} \ln 0.1 \quad (5.11c)$$

$$z = 199 \text{ m} \quad (5.11d)$$

Velocity within the trench is $v = Q/A_C = 0.2/200 = 1$ mm/s, and the residence time $t = A_C z/Q = 200 \times 199/0.2 = 1.99 \times 105$ s = 2.3 days.

The calculation of air requirements is somewhat less straightforward, because the oxygen driving force $\Delta C = (C_{O_2}{}^* - C_{O_2})$ varies in both horizontal and vertical directions. We can, however, calculate a *minimum* air requirement by assuming that oxygen content in the bubbles *remains constant* at 21 mol% and C_{O_2} is small compared to the oxygen solubility $C_{O_2}{}^*$. The value for the latter is 0.266 mol/m³ (Henry's constant H = 70 kPa m³ mol¹).

We are now in a position to calculate the *minimum* total bubble area A_b required from which one can calculate the airflow required using the bubble residence time. The calculations are somewhat circuitous but involve well-known principles:

OXYGEN SUPPLY REQUIRED

$$N = BOD_0 \times Q/32 = 500 \text{ g/m}^3 \times 0.2/32 = 3.14 \text{ moles } O_2/\text{s} \quad (5.11e)$$

OXYGEN TO BE SUPPLIED BY BUBBLES

$$N = k_C A_b C^* = 3.14 \text{ moles } O_2/\text{s} \quad (5.11f)$$

K_L FOR BUBBLES

From the correlation for bubble swarms (Table 5.5), using d = 3 mm, $\mu = 10^3$ Pa s, $\rho = \Delta\rho$ = 1000 kg/m³, $D = 10^{-9}$ m²/s, one obtains

$$k_L = 1.1 \times 10^{-4} \text{ m/s} \quad (5.11g)$$

TOTAL MINIMUM BUBBLE SURFACE AREA A_N

From Equation 5.11f and Equation 5.11g,

$$A_b = 3.14/1.1 \times 10^4 \times 0.266 = 1.07 \times 10^5 \text{ m}^2 \quad (5.11h)$$

TERMINAL VELOCITY OF BUBBLES V_T

From Equation 5.10c,

$$v_t = 5.45\ [10^2(1 - 1/1000)^{1/2}] = 0.545 \text{ m/s}] \quad (5.11i)$$

RESIDENCE TIME OF BUBBLES t_b

Assuming a liquid depth of L = 10 m,

$$t_b = L/v_t = 10/0.545 = 18.3 \text{ s} \quad (5.11j)$$

VOLUME OF BUBBLES IN TRENCH, V_B

$$V_b = A_b(d/6) = 1.07 \times 10^5\ (3 \times 10^3/6) = 52.8 \text{ m}^3 \quad (5.11k)$$

MINIMUM AIRFLOW RATE REQUIRED $(Q_A)_{MIN}$

$$(Q_a)_{\text{Min}} = V_b/t_b = 52.8/18.3 = 2.88 \text{ m}^3/\text{s} \tag{5.11l}$$

We can make an estimate of how much higher the *actual* air requirements would be by drawing on the Streeter–Phelps oxygen sag curve referred to in Practice Problem 2.4. The equation for that curve is given by

$$C_{O_2}{}^* - C_{O_2} = \frac{k_r (BOD)_0}{k_L a - k_r} \exp(-k_r z / v)$$

$$+\left[\left(C_{O_2}{}^* - C_{O_2}\right)_0 - \frac{k_r (BOD)_0}{k_C a - k_r}\right] \exp(-k_L a z / v) \tag{5.11m}$$

where a = bubble surface area/trench volume.

The reader can verify that $k_L a$ » k_r (2.96×10^4 » 1.16×10^5 s^1) so that the second term is negligibly small. One obtains the following:

$$C_{O_2}{}^* - C_{O_2} \cong \frac{k_r (BOD)_0}{k_L a} \exp(-k_r z / v) \tag{5.11n}$$

At a distance from the inlet of z = 100 m, we have

$$C_{O_2}{}^* - C_{O_2} = \frac{1.16 \times 10^{-5} \times 15.7}{2.98 \times 10^{-4}} \exp(-1.16 \times 10^{-5} \times 100 / 10^{-3}) \tag{5.11o}$$

$$C_{O_2}{}^* - C_{O_2}\Big|_{z=100\,\text{m}} = 0.10 \text{ mol/m}^3 \tag{5.11p}$$

that is, the driving force dropped by only 28% from the assumed value of 0.266 mol/m^3. At the outlet this has risen to 78%. This suggests that the actual air requirement will be about 1.5 times the minimum of 2.88 m^3/s we previously calculated. This actual flow would supply 38 mol O_2/s (25°, 1 atm) compared to the 3.14 mol/s transferred, so that our second assumption, that of a nearly constant 0_2 partial pressure in the bubbles, is satisfied.

5.4 Mass Transfer Coefficients for Tower Packings

To this point in our narrative, we confined ourselves to mass transfer in and around simple geometries such as channels of various types and exterior flow about flat plates, cylinders, and spheres.

In this section we turn our attention to more complex shapes represented by tower packings used in operations such as gas absorption, stripping, and distillation. The operation of such columns is addressed in more detail in Chapter 8.

Tower packings are used to fill the interior of large upright cylindrical shells in which two phases, usually a gas and a liquid or two liquids, are brought into intimate contact for the purpose of transferring or exchanging certain components contained in these phases. In gas absorption, for example, the aim is to remove an objectionable component from the gas stream or to recover a valuable component contained in it by contacting it with an appropriate liquid solvent. Both phases flow through the tower, usually in countercurrent fashion.

Packed towers and tower packings have been in use for more than 100 years. Some early examples of their application involve the production of sulfuric acid and the purification of coke oven gas. Prior to 1915, these towers were filled with coke, random-sized and -shaped quartz, broken glass, or broken crockery. Tower performance was unpredictable, and no two towers would perform alike. The development in 1915 of Raschig Ring made it possible for the first time to impart a degree of predictability and dependability to tower performance. These first uniformly shaped packings not only improved tower performance but also enabled engineers to translate the performance of one tower to others. Today, modern tower packings greatly exceed the capacity and the efficiency of the early Raschig Rings.

Figure 5.5a displays the older types of packings, some of which are still in current use. The more modern shapes shown in Figure 5.5b are the result of decades of development work by packing manufacturers. Although the general purpose of this work was to enhance overall tower performance, it is worth noting the specific areas targeted for improvement. They include the following:

Pressure drop. This item, which is often overlooked in preliminary design work, dictates the pumping costs and is the principal contributor to the operating cost of a tower. Even a 10 to 20% reduction in ΔP is considered a major victory.

Promotion of lateral distribution and surface renewal of the liquid solvent. Both of these items improve the mass transfer rate and result in a reduction in tower height.

Resistance to corrosion and high temperatures. Packings are now available in a wide variety of materials, including plastic and high-performance metals and alloys.

High strength, low weight. Both of these are desirable characteristics to resist crushing when used in deep beds and to reduce the load on the support structure.

High wettability. Here again, the aim is to increase the mass transfer rate by maximizing the liquid surface area in contact with the gas or vapor.

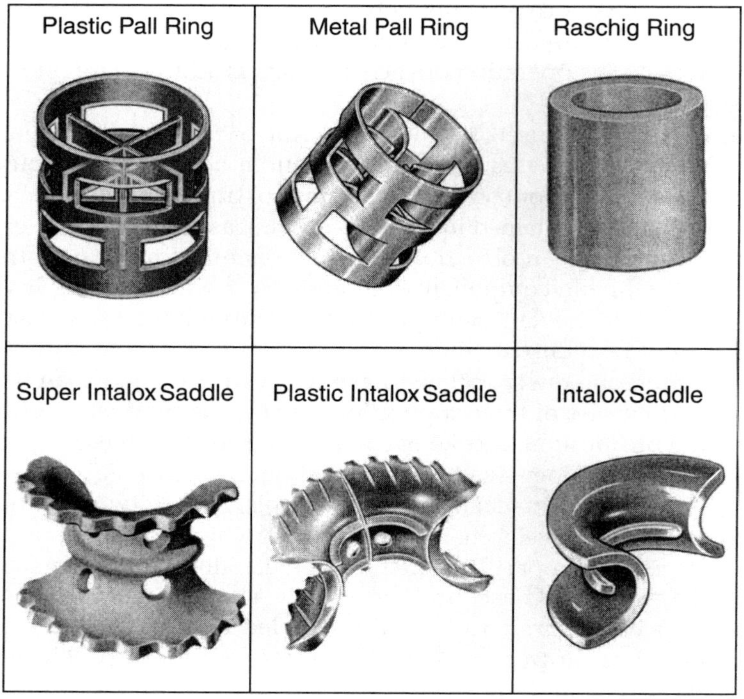

FIGURE 5.5
(a) Classical packings.

The mass transfer characteristics obtained on packings are reported in several different ways. At the more fundamental level, we extract volumetric mass transfer coefficients from the experimental performance data. These coefficients, which we encountered in Illustrations 2.2 and 2.3, consist of the product of a film coefficient and the nominal specific surface area a (in m^2/m^3) of the packing, expressed in units of square meter (m^2) per cubic meter (m^3) of packing. If we use the molar concentration-based coefficient k_c or k_L that has units of meter per second (m/s), then the volumetric coefficient given by the product $k_c a$ or $k_L a$ will have units of reciprocal seconds, which is the same as that of a first-order reaction rate constant. Specific surface areas of packings span the range of 100 to 1000 m^2/m^3.

Listed in Table 5.6 are values of $k_c a$ and $k_L a$ extracted from the literature for four systems and a range of conventional packings and superficial flow rates per square meter column cross section. These are all based on experimental performance data obtained over a range of gas and liquid flow rates. The liquid flow rates listed in Table 5.6 represent the range commonly used in gas–liquid operations. They fall between the extremes of too low a flow, which fails to thoroughly wet the packing, and an excessively high flow, which causes the tower to flood. Neither of these conditions is desirable, and the range listed, typically $L = 1 - 10$ kg/m^{2}s, provides a guideline for

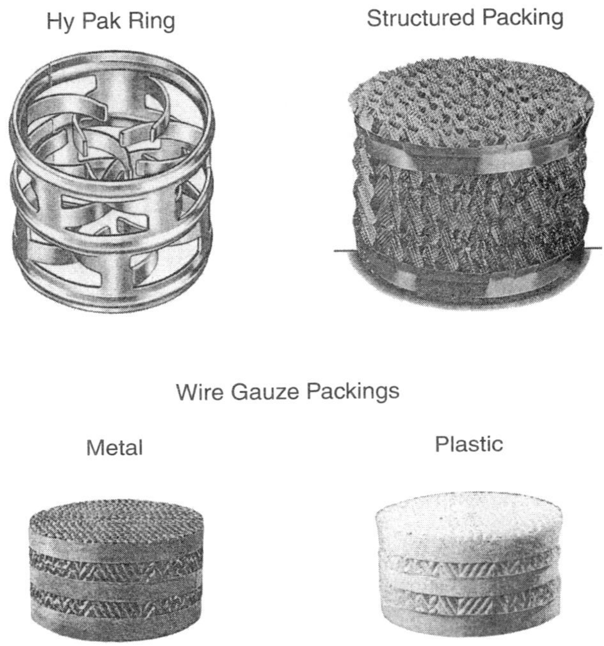

FIGURE 5.5 (continued)
(b) Samples of modern packings.

avoiding these extremes. The gas flow rates, which lie an order of magnitude lower at $G = 0.1 - 1$ kg/m^{2}s, fall between two different extremes, that of insufficient flow to force the gas through the column, and an excessively high flow, which would tend to entrain liquid out of the column and lead to excessive pressure drops. These two undesirable conditions can be avoided by operating within the range indicated in Table 5.6. The mass transfer coefficients listed can be used as a guide for sizing packed columns.

Quantitative prediction methods for volumetric mass transfer coefficients that rely on empirical coefficients for each particular packing and packing size have been developed and can be found in the relevant literature. Complete predictions of this type are not often resorted to, and it is more common to use relations that will extend known coefficients, such as those listed in Table 5.6, to a different set of conditions. This can be done in an approximate fashion using the following proportionalities:

For the gas-film coefficient:

$$k_c a \ (s^1) \propto D_G^{0.67}(G\rho_G)^{0.8}(L/\rho_L)^{0.5} \tag{5.12a}$$

For the liquid-film coefficient:

$$k_L a(s^1) \propto D_L^{1/2}(L/\rho_L)^{0.75} \tag{5.12b}$$

TABLE 5.6

Mass Transfer Coefficients in Various Commercial Packings

Packing	G, kg/m²s	L, kg/m²s	$k_c a$, s¹
System: CO_2-aqueous NaOH ($k_C a$)			
1-in. Raschig ceramic	0.61–0.68	1.4–14	0.14–0.33
2-in. Raschig ceramic	0.61–0.68	1.4–14	0.09–0.31
1-in. Raschig metal	0.61–0.68	1.4–14	0.16–0.32
1-in. Pall plastic	0.61–0.68	1.4–27	0.16–0.33
2-in. Pall plastic	0.68	1.4–54	0.13–0.33
1-in. Pall metal	0.61–0.68	1.4–27	0.19–0.44
1-in. Intalox plastic	1.2	2.7–27	0.28–0.43
2-in. Intalox plastic	1.2	4.1–41	0.20–0.30
∫-in. Intalox ceramic	0.61–0.68	1.4–14	0.30–0.51
1-in. Intalox ceramic	0.61–0.68	1.4–14	0.17–0.36
3-in. Intalox ceramic	1.2	1.4–54	0.04–0.22
2-in. Hyac rings	1.2	3.0–40	0.21–0.63
System: NH_3-water ($k_C a$)			
1-in. Raschig ceramic	0.54	0.68–6.1	1.3–5.2
2-in. Raschig ceramic	0.54	0.68–6.1	0.87–2.6
1-in. Berl ceramic	0.54	0.68–6.1	1.7–4.3
2-in. Berl ceramic	0.54	0.68–6.1	1.3–4.0
50-mm Pall plastic	0.45–2.5	4.2	2.0–7.0
System: CO_2-water ($k_L a$)			
50-mm Pall plastic	0.4–2.0	4.2	11
System: O_2-water ($k_L a$)			
1.5-in. Raschig ceramic	0.054–0.51	2.7	0.14–1.4

A second and less fundamental way of expressing packing performance is through the concept of the height equivalent to a theoretical plate, or HETP. The theoretical plate, which is a concept we will encounter in Chapter 7, is a hypothetical gas–liquid contacting device in which the two phases are brought into intimate contact and exit under equilibrium conditions. The number N_p of such theoretical plates required for a particular separation performance is easily derived by means of algebraic or graphical procedures to be described in Chapter 7. The total height H of packed column needed is then established by the product of N_p and the equivalent height of the theoretical plate — that is, we have

$$H = N_p \times \text{HETP} \tag{5.13}$$

For rough estimates of the HETP, the following recommendations, given in English units and nominal packing size d_p, hold:

1. Pall Rings and similar high-efficiency random packings with low-viscosity liquids:

$$\text{HETP, ft} = 1.5\, d_p, \text{ in.} \tag{5.14a}$$

2. Absorption with viscous liquids in general:

$$\text{HETP} = 5 \text{ to } 6 \text{ ft} \tag{5.14b}$$

3. Vacuum service:

$$\text{HETP, ft} = 1.5\, d_p, \text{ in.} + 0.5 \tag{5.14c}$$

4. Small-diameter columns, $d_T < 2$ ft:

$$\text{HETP, ft} = d_T, \text{ ft, but not less than 1 ft} \tag{5.14d}$$

5. Wire gauze packing:

$$\text{HETP} = 4 \text{ in. to } 1 \text{ ft} \tag{5.14e}$$

In spite of its empirical nature, the HETP concept remains in universal use for the description of packing performance by manufacturers and users.

Illustration 5.6: A First Estimation of Packed Tower Diameter

The height of a packed tower required for a given service is solely determined by the mass transfer characteristics of the system and is usually the first item to be addressed in design procedures. The tower diameter, which can be equally important, often appears as an afterthought toward the end of the proceedings. What we wish to do here is provide some guidelines for its early determination.

A basic requirement for the proper functioning of a packed column is that its diameter should be at least ten times the nominal size of the packing (see Table 5.6) to ensure adequate distribution of the gas and, particularly, the liquid. Redistribution sections, in which the liquid down-flow is collected and redistributed over entire tower cross sections, are often inserted to ensure good performance. On the other hand, the tower cannot be too large for structural and economic reasons. A normal diameter range is 0.5 to 5 m, but diameters as high as 14 m have been reported.

An approximate way to determine it is to use manufacturer's data such as those shown in Table 5.6. Although the diameter is not directly listed, it can be extracted by the simple device of dividing the feed rate by the superficial velocities L and G given in Table 5.6. Suppose, for example, that the gas flow to a tower designed to remove ammonia from air is set at

$G = 10$ kg/s (approximately 10 m^3/s). Then using a G_s value of 0.5 kg/m^{2}s from Table 5.6, we obtain for the cross-sectional area A_C for the tower,

$$A_C = \frac{G}{G_s} = \frac{10}{0.54} = 18.5\,\text{m}^2 \tag{5.15a}$$

and, hence, for the tower diameter,

$$d = \left(\frac{4}{\pi}18.5\right)^{1/2} = 4.9\,\text{m} \tag{5.15b}$$

Comments

We used this section, as well as the last illustration, to present an introduction to tower packings and packed towers performance and to prepare the ground for a more thorough discussion of these topics, which appears in Chapter 8. It will also have become apparent that much of the information required in the design and use of these devices will have to come from the vendors. This notion will be further reinforced in the chapters to follow.

5.5 Mass Transfer Coefficients in Agitated Vessels

Mass transfer involving tower packings, which was considered in the previous section, is our first introduction to systems with complex and highly irregular geometries. Agitated vessels represent yet another example of an unusual and not easily quantifiable geometry. The prominent irregularity here is the shape and size of the impeller and the geometry of its blades. Internal baffles, which are frequently used to enhance transport rates, are an additional unusual feature.

Agitated vessels find their use in a considerable number of mass transfer operations. At the simplest level, they are employed to dissolve granular or powdered solids into a liquid solvent in preparation for a reaction or other subsequent operations. The reverse process of precipitation or crystallization is likewise carried out in stirred vessels. Agitation is also used in leaching operations, or its reverse counterpart, adsorption, which is used to remove objectionable materials from a liquid solution or to recover valuable substances. Liquid extraction processes are often carried out in a batch mode using agitated tanks, as are a host of heterogeneous "stirred tank" chemical reactions. In all of these operations, we are concerned with establishing mass transfer coefficients that determine the rate of transport to the continuous phase.

Fundamental work in this area dates back to the 1940s and 1950s and has been refined in subsequent decades. These studies revealed that mass transfer

TABLE 5.7

Mass Transfer Coefficients in Agitated Vessels

System	Correlation
Solid–liquid baffled vessel Re = 10^4–10^6	$\frac{k_c d_v}{D} = 1.46\left(\frac{d_i^2 N\rho}{\mu}\right)^{0.65}(\text{Sc})^{0.33}$
Solid–liquid unbaffled vessel Re = 10^2–10^5	$\frac{k_c d_v}{D} = 0.402\left(\frac{d_i^2 N\rho}{\mu}\right)^{0.65}(\text{Sc})^{0.33}$
Liquid–liquid	$\frac{k_c d_v}{D} = 0.052\left(\frac{d_i^2 N\rho}{\mu}\right)^{0.833}(\text{Sc})^{0.5}$

coefficients in these systems can be correlated by the same combination of Sherwood, Reynolds, and Schmidt numbers we encountered in simpler geometries, provided the former two are suitably modified to account for the altered system geometry and operation. These modifications are implemented as follows:

- For the Sherwood number, the dimensional length to be used is the vessel diameter d_v.
- For the Reynolds number, the dimensional length is represented by the impeller diameter d_i and the dimensional velocity by the product $d_i N$, where N represented the number of revolutions per unit time. All other parameters are used in the same fashion as before (i.e., μ, ρ, and D are the viscosity, density, and diffusivity, respectively, of the continuous phase).

Using these modified dimensionless groups, it was found possible to correlate a host of experimental data for a wide range of operations (Table 5.7).

Illustration 5.7: Dissolution of Granular Solids in an Agitated Vessel

The assumption made at the outset is that the concentration at the surface of the particles equals the saturation concentration C_s of the solid material, and that the mass transfer is driven by the linear potential $(C_s - C)$, where C is the prevailing concentration in the liquid at any particular instant.

An initial unsteady mass balance over the solid leads to the following expression:

$$\text{Rate of solid in} - \text{Rate of solid out} = \begin{matrix}\text{Rate of change}\\\text{of solid contents}\end{matrix}$$

$$0 - k_c A_s(C_s - C) = \frac{d}{dt}m \tag{5.16a}$$

Note that both the surface area A_s and concentration in the liquid C vary with time or indirectly with the remaining mass m. For the area, which can be quite irregular, we stipulate that it varies with the two-thirds power of volume, so that for a single particle,

$$A_p = \alpha V_p^{2/3} = \frac{\alpha}{\rho_p^{2/3}} m_p^{2/3} \tag{5.16b}$$

where α is some shape factor and equals 4.83 for sphere. For the total surface area A_s of the solids, Equation 5.16b becomes

$$A_s = N_p A_p = \alpha N_p^{1/3}(N_p V_p)^{2/3} \tag{5.16c}$$

or

$$A_s = \alpha N_p^{1/3}(m/\rho_p)^{2/3} = \beta m_p^{2/3} \tag{5.16d}$$

where N_p = number of particles, m is the *total* mass of solids at any time, and $\beta = \alpha N_p^{1/3} \rho_p^{2/3}$.

To obtain an expression for the external concentration C, we apply a simple cumulative mass balance that reads

Initial solid = Solid left over + Solid in solution

$$m_o = m + CV \tag{5.16e}$$

and, consequently,

$$C = \frac{m_o - m}{V} \tag{5.16f}$$

Substituting Equation 5.16d and Equation 5.16f into the original mass balance (Equation 5.16a), we obtain

$$-k_c \beta m^{2/3}\left(C_s - \frac{m_o - m}{V}\right) = \frac{dm}{dt} \tag{5.16g}$$

which yields, after integration by separation of variables,

$$k_C \beta t = \int_{m_o}^{m} \frac{dm}{m^{2/3}(m / V + C_s - m_o / V)} \tag{5.16h}$$

Evaluation of the integral is by numerical or graphical means, which we do not address here. We consider instead the special and frequent case when the final solution concentration is considerably below the saturation value — that is,

$$C_s >> \frac{m_o - m}{V} \tag{5.16i}$$

Evaluation of the integral then yields

$$t = \frac{3\rho_p^{\ 2/3}}{\alpha k_C C_s}\left(\frac{m_o}{Np}\right)^{1/3} = \frac{3\rho_p^{\ 2/3}}{\alpha k_C C_s} m_p^{\ 1/3} \tag{5.16j}$$

where t is the total dissolution time and the charge is given either by the total initial mass m_o divided by the number of particles or its equivalent, the particle mass m_p. More commonly one specifies *particle size or diameter* d_p rather than its mass, which is introduced by setting $m_p = \rho_p V_p$ (= $\rho_p \pi d_p 3/6$ for spherical particles). Equation 5.16j is then transformed into the equivalent expression,

$$t = \frac{3\rho_p(\pi / 6)^{1/3}}{\alpha k_C C_s} d_p \tag{5.16k}$$

or

$$\boxed{t = 0.50\, \rho_p d_p / k_C C_s} \tag{5.16l}$$

This is the working equation used to determine dissolution time. Its application is demonstrated in Practice Problems 5.13 and 5.14.

Comments

There are several points of note in these expressions. First, time of dissolution varies inversely with the mass transfer coefficient k_C and solubility C_s. This is in line with physical reasoning. High values of these parameters imply a high mass transfer rate, resulting in shorter dissolution times. More startling is the one-third power dependence on initial mass (at constant N_p) or on particle mass m_p. This one-third power relation translates into a *linear* dependence of t on particle diameter d_p. Dissolution time, in other words, does not vary linearly with particle volume, or even particle area, as one might have expected, but varies instead with its linear dimension.

A similar result is obtained in the reverse process of *crystallization*, which is taken up in Practice Problem 5.14. The time required to form a given quantity of crystals from a saturated solution is proportional to the linear dimension of the crystals, not their volume or mass. We see here yet another example of the power of modeling to reveal the unexpected.

5.6 Mass Transfer Coefficients in the Environment: Uptake and Clearance of Toxic Substances in Animals — The Bioconcentration Factor

In the past three decades, a dramatic increase in awareness of the effect of various toxic substances on animal life, in particular on aquatic species, has been seen. Among the identified culprits, the chlorinated hydrocarbons used in pesticides, the polychlorinated biphenyls (PCBs), and various organometallic compounds stand out.

Research in this area has focused, on the one hand, on the physiological consequences of exposure to these substances, and, on the other, on tracking their fate both in the environment and in the affected wildlife and their specific organs. It is the latter aspect that is considered here.

The fate of toxic substances within an animal is monitored in two ways: first, by measuring the concentration changes during uptake from a controlled environment, and second, by following the decline in concentration after exposure has ceased. The latter process is termed clearance, elimination, or depuration.

Interpretation of the experimental data is usually carried out by means of compartmental models. The simplest of these, the one-compartment model, yields the following results.

During uptake, the relevant mass balance over the animal body takes the form

$$\text{Rate of toxin in} - \text{Rate of toxin out} = \begin{array}{c}\text{Rate of change}\\ \text{of toxic contents}\end{array}$$

$$k_{cu}A_uC_w - k_{cd}A_dC_a = V\frac{dC_a}{dt} \tag{5.17a}$$

where k_{cu} and k_{cd} are the mass transfer coefficients in units of meter per second (m/s) for uptake and depuration; A_u and A_d are the associated transfer areas; C_w denotes concentration in the water or other medium, taken to be constant; and C_a is the time-varying concentration within the animal or one of its organs.

Before integrating, it is customary to divide Equation 5.17a by volume, with the results that k_{cu} and k_{cd} are converted into volumetric mass transfer coefficients $k_u = k_{cu}A_u/V$ and $k_d = k_{cd}A_d/V$. Equation 5.17a consequently becomes

$$k_u C_w - k_d C_a = \frac{dC_a}{dt} \tag{5.17b}$$

The coefficients now have units of reciprocal time (s[1]) and are precisely of the same type that were encountered in Section 5.4 in connection with mass transfer in packed towers. They were denoted as $k_c a$ and $k_L a$, where a is the specific surface area of the packing in units of square meter per cubic meter packing. The corresponding term here is the ratio of transfer areas to body or organ volume A/V. The advantage of this procedure is that both of these factors, which are either unknown or not well known with precision, are lumped into a single empirical transfer coefficient that is determined experimentally.

The fact that these volumetric coefficients have units of reciprocal time has led to the erroneous impression, and even statements, that the process is one of *chemical reaction*, with k_u and k_e playing the roles of first-order rate constants. The mechanism of uptake is clearly one of mass transfer, and that of elimination is probably a combination of reaction and transport. Mass transfer is therefore the key phenomenon that dominates these processes.

We now turn to Equation 5.17a and obtain, by separating variables and integrating,

$$C_a/C_w = (k_u/k_d)[1 - \exp(-k_d t)] \tag{5.17c}$$

This result expresses the toxin concentration in the animal, C_a, as a function of time. We note that as $t \to \infty$, a balance between uptake and elimination is obtained, and a steady-state concentration ratio $(C_a/C_w)_{ss}$ results. This ratio is termed the bioconcentration factor (BCF) and expresses the magnification of toxic concentration in the animal over that prevailing in the surrounding water. It equals the ratio of the two mass transfer coefficients and is given by

$$\text{BCF} = (C_a/C_w)_{ss} = k_u/k_d \tag{5.17d}$$

Turning next to the elimination process, a mass balance similar to that performed for the uptake step leads to the following result:

$$\text{Rate of toxin in} - \text{Rate of toxin out} = \begin{matrix}\text{Rate of change}\\ \text{in toxin contents}\end{matrix}$$

$$0 - k_{Cd}A_dC_a = V\frac{dC_a}{dt} \quad (5.18a)$$

or, equivalently,

$$-k_dC_a = \frac{dC_a}{dt} \quad (5.18b)$$

which upon integration by separation of variables yields

$$C_a/(C_a)_{ss} = \exp(-k_dt) \quad (5.18c)$$

Here $(C_a)_{ss}$ is taken to be the steady-state concentration attained during uptake.

The uptake and elimination coefficients k_u and k_e can in principle be calculated by first extracting k_d from Equation 5.18c using measured-clearance histories, and then substituting it into Equation 5.17c and performing a similar analysis of uptake-concentration histories. In practice, it is more common to perform an independent evaluation of k_u by using the initial portion of the uptake process, which is unaffected by the relatively slow elimination process. Equation 5.18b then assumes the reduced form:

$$k_uC_w = \frac{dC_a}{dt} \quad (5.19a)$$

which is integrated to yield

$$C_a = k_uC_wt \quad (5.19b)$$

This expression is used to calculate k_u from a linear plot of the initial uptake date.

Table 5.8 presents uptake and elimination constants obtained on bluegill fish exposed to an aqueous environment containing anthracene and benzopyrene and on ring doves that were fed pellets containing various PCBs. This is supplemented, in Table 5.9, by the BCF found in various organs of

TABLE 5.8

Uptake and Depuration of Some Toxic Substances

	Toxin	k_u (hr^1)	k_d (hr^1)
Bluegill fish	Anthracene	36	0.04
	Benzopyrene	49	0.01
Ring doves	PCBs	(0.3–4.4)10^3	0.094–0.24

TABLE 5.9

Bioconcentration Factor (BCF) in Various Organs of Bluegill Fish

Tissue	BCF	
	Anthracene	Benzopyrene
Gallbladder	1,800	14,000
Liver	561	1,600
Viscera	640	770
Brain	555	90
Carcass	42	30

the bluegill fish. Of note in the latter is the extraordinary magnification that takes place particularly in the gallbladder of this species, which attests to the dangerous effect of these toxins.

Illustration 5.8: Uptake and Depuration of Toxins: Approach to Steady State and Clearance Half-Lives

Two topics are addressed here. The first involves the calculation of the time required to attain 95% of the ultimate steady state during toxin uptake. This conveys a sense of the speed with which this process takes place. In the second calculation, we seek to quantify the depuration process by calculating its half-life (i.e., the time required for the toxin concentration to drop to one half its original steady-state concentration). This again serves as an indicator of the speed with which depuration proceeds.

1. Approach to 95% of steady state

Here we make use of Equation 5.17c and Equation 5.17d, which upon division by each other yield

$$C_a/(C_a)_{ss} = 1 - \exp(-k_d t) \tag{5.20a}$$

and, consequently,

$$t = -\frac{1}{k_d}\ln[1 - C_a / (C_a)_{ss}] \tag{5.20b}$$

Applying this to the uptake of benzopyrene by bluegill fish listed in Table 5.8, we obtain

$$t = (-1 / 0.01)\ln[1 - 0.95] \tag{5.20c}$$

$$t = 300\ \text{h} \tag{5.20d}$$

2. Half-life $t_{1/2}$ of depuration

The pertinent expression here is obtained from Equation 5.19c by setting $C_a/(C_a)_{ss} = 0.5$. Hence,

$$0.5 = \exp(-k_d t) \tag{5.21a}$$

and, consequently,

$$t_{1/2} = \frac{\ln 2}{k_d} \tag{5.21b}$$

Applying this to the same data as before, we have

$$t_{1/2} = \frac{\ln 2}{0.01} = 69\ \text{h} \tag{5.21c}$$

To obtain a more direct comparison with the uptake case, we allow $C_a/(C_a)_{ss}$ to drop to 5% of its original value. This time, there results

$$t = \frac{\ln 20}{0.01} = 300\ \text{h} \tag{5.21d}$$

Comments

The surprising fact that emerges from these calculations is that the uptake and depuration processes proceed at the same speed. This is a direct consequence of the linearity of the rate equations. Both processes are long, drawn-out events taking place over a period of days. This is encouraging news for the uptake step, because it implies no harmful effects for brief exposure times of, say, 10 min. The long depuration period, on the other hand, is disturbing, because the animal will require many days in a clean environment to recover from its exposure. We have here a mix of good and bad news, which is often the norm in environmental events.

The BCF mentioned above has its benign counterpart in the effective therapeutic concentration (ETC), which was encountered in Practice Problem 2.3. Both result when the rate of inflow of the material is exactly balanced by the rate of elimination. Thus, although the two substances are at opposite poles, one toxic and the other therapeutic, the mechanisms by which they reach their plateau values are identical.

Practice Problems

5.1 The Nusselt and Prandtl Numbers

Express these dimensionless groups in terms of fundamental heat transport coefficients.

5.2 The Wall Sherwood Number Sh_w

A dimensionless group not listed in Table 5.1 is the so-called Wall Sherwood Number that represents the resistance to mass transfer through a fluid in laminar flow divided by the resistance within the tubular wall. It is used to gauge the relative importance of these resistances in industrial membrane processes as well as those occurring in living organisms (see Chapter 8).

Derive an expression for this parameter in terms of the pertinent transport coefficients and tubular dimensions.

5.3 Acid Rain and the Biot Number

Illustration 5.1 had shown that mass transfer between particles and a fluid in relative motion was almost invariably controlled by diffusional resistance within the particle for a Biot number greater than ten. This applied even to small liquid or solid particles in relative motion to a surrounding fluid.

Consider a raindrop of diameter $d = 2$ mm that falls from a height of 500 m with a terminal velocity of 7.7 m/s. The surrounding air contains an average of 5 ppm by volume of SO_2 ($p = 0.5$ Pa). What will be the final concentration of SO_2 when the drop reaches the ground? What will be the pH? Neglect evaporation from the drop.

Data:

Diffusivity $D = 1.7 \times 10^{-5}$ cm²/s

Henry's constant $H = 0.015$ kPa $m^3 mol^1$

Dissociation constant $K_a = 1.2 \times 10^2$ mol L^1

5.4 The Biot Number in Heat Transfer: Design of a Thermocouple

Design a thermocouple (i.e., determine the diameter of the bead) that would respond to a 10°C temperature change in flowing air to within 0.5°C in less than 2 sec.

Data:

$C_p = 0.42$ kJ/kgK

$\rho = 9000$ kg/m³

$h = 0.46$ kJ/m²sK

5.5 The Entry Region

a. Does the entry length increase or decrease with increasing velocity? With increasing tubular diameter?
b. Determine the entry length for a 20-cm-diameter tube carrying (1) a typical gas and (2) a typical liquid at a Reynolds number of 1000.
c. Derive the analogous heat transfer relations from the mass transfer entries of Table 5.1.

5.6 The Drag Force in Flow around a Particle

Derive, by dimensional analysis, the standard equation for the drag force on a particle in relative motion to a fluid:

$$F_D = C_D A_C \frac{1}{2} \rho v^2$$

There are at least two additional factors that have to be taken into account even if the geometry is simple (flat plate, sphere, etc.). What are they?

5.7 Mass Transfer Coefficients in Terms of Shear Rate $\dot{\gamma}$

In physiological flows, such as that of blood, it is customary to replace velocity by the shear rate $\dot{\gamma}$, which equals the velocity gradient at the vessel wall. Show that the k_C in the entry region is then given by

$$k_C = 0.54 \left(\frac{\dot{\gamma} D^2}{x} \right)^{1/3}$$

(*Hint*: Use the parabolic velocity distribution for laminar flow as a starting point, and derive the relation $\dot{\gamma} = 8\, v/d$.)

5.8 Mass Transfer Regimes in Blood Flow: The Critical Blood Vessel Diameter

Shear rates $\dot{\gamma}$ in physiological blood flow typically lie in the range 100 to 1000 s^1. Show that for proteins ($D = 10^{-10}$ m^2/s), mass transfer in the "larger" blood vessels, $d > 1$ mm, falls entirely in the developing (Lévêque) region, while for $d < 10^{-2}$ mm, the concentration profile is fully developed. (*Hint*: Use the relation $\dot{\gamma} = 8\, v/d$.)

5.9 Mass Transfer in the Kidney: The Loop of Henle

The Loop of Henle is part of an intricate system of permeable channels that carry raw urine through the kidney and to the bladder. During its passage, the urine exchanges water and solutes with the surrounding tissue to ensure that not too much or too little of either is withdrawn from the body. A more detailed account of the process is given in Chapter 8.

The *total amount* of salt reabsorbed into the tissue depends in a complex way on flow velocity, and early workers attributed this to a flow-sensitive boundary layer within the tube. Show that this is not the case, and show also that one can still have flow-sensitive reabsorption.

Data: $d = 2 \times 10^{-3}$ cm, $D = 2 \times 10^{-5}$ cm²/s, $v = 10^{-1}$ cm/s, Wall Sherwood number ≤ 2.

(*Hint*: Show that transport is in the fully developed region.)

5.10 Controlled Release of Anticoagulants from Artificial Blood Vessels

The use of artificial grafts to replace diseased blood vessels has by now become a commonplace operation. The danger of blood clot formation, which attends these replacements, can be largely eliminated by incorporating an anticoagulant in the graft, which is slowly released to the bloodstream. One such anticoagulant is the protein heparin, which effectively prevents the onset of coagulation. It has been estimated that to achieve this, a micro-environment with a heparin concentration of $C_s = 0.5$ μg/cm³ must be provided at the blood vessel surface. The problem to be addressed here is to calculate the release rate N (μg/cm²s) required to maintain this concentration within 0.5 cm from the tubular entrance. Flow is laminar, and mass transfer resides entirely within the entry region. Shear rate is set at 1000 s^{1}, and the heparin diffusivity is 7.5×10^{7} cm²/s.

Answer:

5.2×10^{4} μg/cm²s

5.11 Solution Mining of Potash

One proposed method for harvesting underground deposits of potash (KCl or sylvite) is the technique called *solution mining*. The procedure consists of cracking the deposit open by pumping high-pressure water through an "injection well" drilled at one end of the deposit ("hydrofracturing"). A second well, the so-called production well, is drilled at the far end of the fracture. Water is then continuously pumped into the deposit through the injection well, dissolving potash as it passes through the open fracture and returning to the surface through the production well as a KCl solution. It is desirable in these operations to have an *a priori* estimate of the KCl concentration in the effluent, as this will determine the production rate of the mine. Consider in this connection the following data:

Fracture height d = 1 m
Fracture length L = 500 m
Fracture perimeter P = 500 m
Water velocity v = 0.1 m/s

As a result of the geothermal gradient, temperature near the fracture is estimated to be 50°C. The following physical parameters were calculated with this factor in mind:

Density ρ: 985 kg/m^3
Viscosity μ: 0.6×10^{-3} Pas
Diffusivity D: 1.8×10^{-9} m^2/s

The problem we set ourselves is to calculate the degree of saturation C/C^* attained at the far end of the fracture.

Simplifying assumptions: We assume isothermal operation even though the endothermic nature of the dissolution process will cause a drop in temperature and, consequently, of the solubility C^*. The variation in fracture height that occurs with time is neglected, as are the variations in physical properties due to the changing KCl concentration. The correlation for tubular flow listed in Table 5.5 is assumed to hold, with fracture height taking the place of tubular diameter.

Answer:

$C/C^* = 0.047$

5.12 Packed Column Diameter and Packing Size

List pros and cons of large/small column diameters and packing size.

For more problems on this topic, see Chapter 8.

5.13 Coffee with Sugar

Confirm what you know about the dissolution time of sugar in a stirred cup of coffee by calculating t for (a) granulated sugar with d_P = 150 μm (100 mesh) and (b) a sugar cube with equivalent diameter of d_P = 0.8 cm. ρ_p = 1.59 g/cm^3, C_s = 179 and 487 g/100 mL water at 0° and 100°C for sucrose. Use Table 1.4 as a guide for k_C. Does the dissolution time increase if one uses two lumps instead of one?

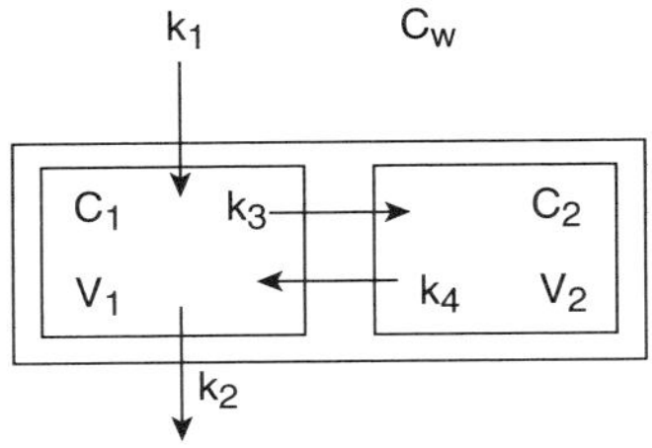

FIGURE 5.6
Two-compartment model for bioconcentration.

5.14 Neutralization of Spent H_2SO_4 with Limestone

Solutions of sulfuric acid too dilute for economic recovery are often rendered benign before disposal by neutralizing them with limestone to calcium sulfate (gypsum):

$$CaCO_3 + H_2SO_4 = CaSO_4 + H_2O + CO_2$$

The rate-determining step is normally taken to be the dissolution of $CaCO_3$, which has a low solubility, at least in water, and hence dissolves very slowly. Once in solution, it dissociates and reacts quickly. The problem is to set an appropriate value for the solubility C_s which is pH dependent. If we assume the boundary layer to be saturated with CO_2, $C_s = 1.2$ g/L. At pH = 7, it is only 0.015 g/L H_2O.

Consider a continuous feed of spent acid and limestone to a stirred baffled vessel with diameter $d_v = 4$ m ($V \approx 50$ m^3), impeller diameter $d_i = 1.5$ m, and $N = 30$ rpm. If particle diameter is set at $d_p = 150$ μm (100 mesh), what is the required minimum residence time? Use your best judgment in choosing a value for C_s. Other parameters can also be adjusted.

Data: $\rho_p = 2.7$ g/cm^3, $\rho_f = 1000$ kg/m^3, $D = 10^{-9}$ m^2/s, $\mu = 10^{-3}$ Pas.

5.15 Batch Crystallization

Crystallization can be considered as the reverse of the dissolution process described in Illustration 5.7 and can be modeled in nearly identical fashion, differing in only one or two minor details.

In the classical process, the well-stirred solution containing the solute to be crystallized is subcooled some 1 to 5° below the saturation level and "seeded" with crystals of mass m_o. Solute diffuses through a film and is assumed to be incorporated instantaneously into the crystal structure (see Figure 5.6).

a. Show that an integral unsteady balance about the growing mass of crystals m yields, upon integration,

$$t = \frac{\rho_p^{2/3}}{\alpha k_C N_p^{1/3}} \int_{m_o}^{m} \frac{dm}{m^{2/3}\left(C_o - \dfrac{m - m_o}{V} - C_s\right)}$$

where C_o = initial concentration of the solution, and t = elapsed time from start of crystallization solution.

b. Show that during the initial stages of the process, or if the solution volume is large, the above expression reduces to the form

$$t = \frac{\rho_p^{2/3}(m_p^{1/3} - m_{po}^{1/3})}{\alpha k_C (C_o - C_s)} = \frac{\rho_p}{a\rho k_C \left(C_o - C_s\right)} \Delta L$$

where L is some linear dimension of the crystals, and $\Delta L = L - L_o$. The expression, known as the *McCabe ΔL Law,* was first proposed in 1929 and has been confirmed since by a considerable body of experimental findings. By converting L to mass, we obtain the same one-third power relation we saw in Illustration 5.7.

5.16 Clearance of Toxins from Aquatic Life

An aquatic species was exposed to a toxin for an unspecified period of time. It is known from independent measurements that after 1000 h of exposure, the toxin level in the animal is approximately twice that observed after 100 h of exposure. How long would it take the animal to attain 95% recovery after emissions have ceased and the water is clear of toxin?

Answer:

430 h

5.17 Bioconcentration in a Two-Compartment Model

Consider the two-compartment model shown in Figure 5.6. Toxins enter the first compartment from the water phase with first-order rate constant k_1 and k_2. Simultaneously, a reversible exchange of toxins takes place with an adjacent second compartment with first-order rate constants of k_3 and k_4, respectively. The first compartment may be viewed as the circulatory system; the second compartment represents the tissue, which ultimately, at steady state, equilibriates with the first unit. We define a relative mass for the first compartment, given by

$$f_1 = \frac{m_1}{m_1 + m_2}$$

Show that with this definition in place, the total BCF for the two compartments is given by

$$\text{BCF}_{\text{Tot}} = C_{\text{Tot}}/C_{\text{w}} = f_1(k_1/k_2) + (1 - f_1)(k_1/k_2)(k_3/k_4)$$

(*Hint*: Set up the ultimate steady-state balances for the two compartments.)

6

Phase Equilibria

Throughout the preceding chapters, it is evident that, along with transport coefficients, phase equilibria play a crucial role in determining overall mass transfer rates. In processes involving single-film resistances, equilibrium compositions or pressures constitute the anchor of the driving force responsible for the transport of mass. This is shown most vividly in Figure 1.8a. The driving forces here are all of the form $C^* - C$ or $p^* - p$, where the asterisked quantities represent equilibrium compositions and partial pressure, and the plain symbols denote the same quantities in the bulk fluid. Evidently, transport will continue only as long as the two quantities are unequal and will come to a halt when $C^* = C$ and $p^* = p$. The two phases are then said to be in equilibrium.

Phase equilibria also appear in processes involving two-film resistances in series, but their role here is somewhat more complex. They still appear as one of the two partners constituting the driving forces in the individual films. This comes about because, in two-film theory, the interface separating the two phases is postulated to be at equilibrium (see Figure 1.7). The use of individual coefficients and driving forces, however, is awkward because conditions at the interface are generally unknown. It then becomes convenient to express transport in terms of overall driving forces that bridge the interface and extend from one bulk concentration to the other. Driving forces still retain their previous form, $C^* - C$ and $p^* - p$, but the asterisked quantities are now *hypothetical* compositions and pressures denoting equilibrium with the second phase at a particular point of the system. This is described in detail in Section 1.5.

Yet another important aspect of phase equilibria rests on the fact that they determine certain maximum or minimum quantities associated with the process. Suppose, for example, that a liquid evaporates into an enclosure. By allowing the process to proceed to equilibrium (i.e., to full saturation), we are able to determine the *maximum amount of liquid* that will have evaporated, or, conversely, the *minimum mass of air* that can accommodate that amount of vapor. Suppose next that the same liquid (e.g., water) adheres to a solid that is to be dried by passage of air over it. Then by using very low flow rates we can ensure that the air leaving the drying chamber is fully saturated and that, consequently, the air consumption is at a minimum.

The picture that emerges from this brief discussion is that the role of equilibrium is considerable. Equilibrium, or the departure from it, determines the driving potential of the process and sets upper and lower limits to the enrichment or depletion attainable and the material inventory involved. Equilibrium can be viewed as one of the two key players in mass transport — the other being the transport coefficients. Because each of these factors has a distinct role to play, it has become customary to examine the component factors separately and, subsequently, combine the results for a complete description of the event. In this procedure one starts by first ignoring the effect of transport resistance and allowing the two phases to come to equilibrium. The device in which this step is carried out is termed an *equilibrium* or *ideal stage*. Concentration or pressure changes that result from the procedure are noted and set aside. The effect of transport resistance is examined next. In the case of staged operations, this involves as assessment of *stage efficiency*, which is a measure of the effect of transport resistance on the amount transferred. When resistance is negligible, the efficiency is 100% and the process proceeds to complete equilibrium. When efficiency is 50%, only half of the attainable enrichment is obtained. In the final step, the results of this dual scrutiny are combined to arrive at an overall description of the process. These steps, as well as the underlying concept of an equilibrium stage, are taken up in the next chapter.

While some of the topics in this chapter will be new to the reader, others may be known from previous courses in thermodynamics. They are repeated to provide a refresher and a link to subsequent chapters.

6.1 Single-Component Systems: Vapor Pressure

Most pure substances that the reader will be familiar with can exist in the solid, liquid, and vapor phases. The principal exceptions are high-molecular-weight solid compounds, such as proteins, carbohydrates, and polymers, which decompose before they can pass into the liquid or vapor phase. These will not be of concern here. The wider class of substances, which are addressed in this chapter, is capable of existing in all three phases. Their behavior is best illustrated by means of a pressure-temperature phase diagram; a representative example is shown in Figure 6.1a.

The diagram is divided into three regions representing the solid phase (S), the liquid phase (L), and the vapor phase (V). The dividing boundaries between these regions are the melting point or freezing point curve A; the sublimation curve B, which separates the solid and vapor phases; and the vapor pressure or boiling point curve C, which is the dividing line between liquid and vapor phases. It is principally the latter curve and the relation between vapor pressure and temperature that we will be concerned with here. Also marked on the diagram are four specific pressure–temperature

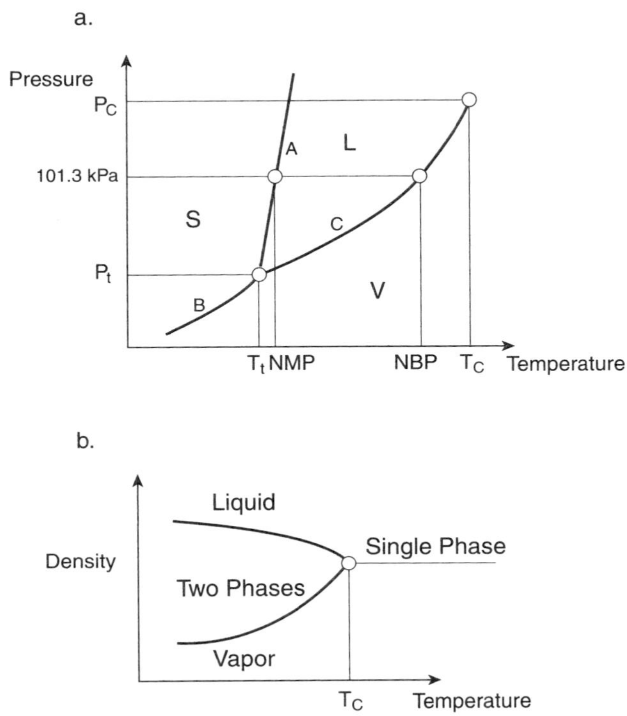

FIGURE 6.1
(a) Phase diagram of a pure substance; (b) approach to the critical point.

pairs that characterize the system. NMP and NBP are the "normal" melting and boiling points — those that prevail at a pressure of 1 atm or 101.3 kPa. The prefix *normal* is often dropped and the shorter version is used to denote conditions of 1 atm. The pair $P_t - T_t$ marks the *triple point* and represents the conditions under which all three phases can coexist. For water, the pressure and temperature values are 0.006 atm and 0.01°C, respectively.

Suppose, now, that a liquid sealed in a container and in equilibrium with its vapor is heated up starting from room temperature. As the temperature is raised, liquid passes into the vapor phase, thus increasing the density of the latter, while the liquid undergoes expansion with an attendant decrease in density (Figure 6.1b). A point is eventually reached where the densities of the two phases become identical and the two phases merge into a single entity. The specific condition ($P_c - T_c$) at which this occurs is termed the critical point of the system. Beyond it (i.e., above the critical temperature T_c), a substance can exist only as a vapor and no liquefaction can occur, no matter how high the pressure is raised. For water, this point is reached at a temperature of 374.4°C and a corresponding pressure of 219.5 atm.

Note that while a substance can exist only as a gas above the critical temperature, its density is nevertheless high enough that it can also qualify as a quasi-liquid. This dual behavior has certain advantages that are exploited in a process termed *supercritical fluid extraction*, which is taken up in Section 6.2.5.

We now turn to the consideration of the equilibrium between liquid and vapor. The aim is to establish a quantitative relation for the curve C in Figure 6.1a (i.e., for the vapor pressure of the liquid in the interval between the triple point T_t and the critical temperature T_c). The starting point is given by the Clapeyron equation, which is the thermodynamic expression of phase equilibrium for a pure substance:

$$\frac{dP^o}{dT} = \frac{\Delta H}{T\Delta V} \tag{6.1a}$$

Here ΔH and ΔV denote the molar enthalpy and volume changes that occur during the passage from one phase to another. Equation 6.1a is confined to pure substances but is otherwise quite general and capable of expressing the transition between any two of the phases shown in Figure 6.1a. Thus, for a passage from the solid to the liquid phase, ΔH will represent the latent heat of fusion, while ΔV denotes the difference between liquid and solid molar volumes. The derivative dP^o/dT, which appears on the left side of Equation 6.1a, gives the slope at any point of the phase boundary curve B or C shown in Figure 6.1a.

Recall from thermodynamics that Equation 6.1a integrates to the celebrated Clausius–Clapeyron equation, of which we present two versions below:

$$\ln P^o = A - \frac{\Delta H_v}{RT} \tag{6.1b}$$

$$\ln P_2^o / P_1^o = \frac{\Delta H_v}{R}\left(\frac{1}{T_1} - \frac{1}{T_2}\right) \tag{6.1c}$$

Although these equations are used routinely in the plotting, computation, and extrapolation of vapor pressure, it should be kept in mind that they are based on the following three assumptions:

1. Constant latent heat ΔH_v
2. Validity of ideal gas law
3. Negligible condensed phase molar volume — $\Delta V \approx V_{vapor}$

Variations of ΔH_v with temperature are often considered negligible over a wide range of T but are, in fact, a thermodynamic requirement (see Practice Problem 6.2). All three assumptions tend to break down with increasing

TABLE 6.1

Antoine Constants for Various Liquids

Substance	A	B	C
Methyl chloride	7.09349	948.582	249.336
Methylene chloride	7.40916	1325.938	252.616
Chloroform	6.95465	1170.966	226.252
Carbon tetrachloride	6.87926	1242.021	226.409
Acetone	6.11714	1210.595	229.664
Diethyl ether	6.92032	1064.066	228.799
Methanol	8.08097	1582.271	239.726
Ethanol	8.11220	1592.864	226.184
n-Hexane	6.88555	1175.817	224.867
n-Octane	6.91874	1351.756	209.100
Benzene	6.89272	1203.531	219.888
Toluene	6.95805	1346.773	219.693
o-Xylene	7.00154	1476.393	213.872
Water	7.96681	1668.210	228.000

pressure, with the deviations escalating as the critical point is approached. At the critical temperature, both ΔV and ΔH vanish, and Equation 6.1a through Equation 6.1c become completely invalid. Various semiempirical modifications to extend their range were proposed, of which the Antoine equation (1888) has proved to be particularly successful. It takes the following form:

$$\log P^{o} = A - B/(T + C)$$

where P^{o} is the liquid vapor pressure (in mmHg); T the temperature (in degrees Celsius); and the values of A, B, and C for some common liquids appear in Table 6.1.

The log P-inverse temperature relation is not unique to vapor–liquid and vapor–solid equilibria. It applies equally well to *gas–liquid* and *gas–solid* equilibria, provided assumptions 1. to 3. hold, and the concentration in the condensed phase is assumed to be *constant*. We have, in other words, the following:

For gas *absorption* (at constant liquid-phase concentration C)

$$\ln\left(\frac{p_2}{p_1}\right)_C = \frac{\Delta H_{\text{abs}}}{R}\left(\frac{1}{T_1} - \frac{1}{T_2}\right) \tag{6.1d}$$

For gas *adsorption* (at constant amount adsorbed X)

$$\ln\left(\frac{p_2}{p_1}\right)_X = \frac{\Delta H_{\text{ads}}}{R}\left(\frac{1}{T_1} - \frac{1}{T_2}\right) \tag{6.1e}$$

Because the heat values of solution in water and of adsorption (ΔH_{abs} and ΔH_{ads}) cluster about a value of 25 kJ/mol for most common gases, except very low-boiling species, the following guideline emerges from Equation 6.1d and Equation 6.1e:

> *For each 20°C increase in temperature, the partial pressure of the dissolved or adsorbed gas approximately doubles.*

This is a convenient rule of thumb to use in the absence of experimental data, and we briefly demonstrate its application with the following example.

Illustration 6.1: Henry's Constant for Sulfur Dioxide in Water

At low concentrations of a dissolved gas, the ratio of partial pressure to concentration, p/C, assumes a constant value known as Henry's constant H, to be discussed more fully in Section 6.2.1. Here we use it to recast Equation 6.1d in the following form:

$$\ln(p_2 / p_1)_C = \ln\frac{p_2 / C}{p_1 / C} = \ln H_2 / H_1 = \frac{\Delta H}{R}\left(\frac{1}{T_1} - \frac{1}{T_2}\right) \tag{6.1f}$$

Consider now the following example: The Henry's constant for SO_2 dissolved in water at 25°C has been measured at 0.015 kPa m^3 mol^1 and its heat of solution ΔH_{abs} at 25.6 kJ/mol. Hence, by Equation 6.1f, its Henry's constant H_2 at 5°C is given by

$$H_2 / H_1 = \exp\left[-\frac{25.6\times10^3}{8.31}\left(\frac{1}{298} - \frac{1}{278}\right)\right] = 0.48 \tag{6.1g}$$

Our rule of thumb, on the other hand, predicts that the drop in temperature of 20°C should give a ratio of partial pressures, and hence of Henry's constants of 0.5, in reasonable agreement with the experimental value.

Illustration 6.2: Maximum Breathing Losses from a Storage Tank

During a rise in ambient temperature, solvent-laden air in the headspace of storage tanks expands and is partially expelled into the atmosphere through a vent pipe. When the temperature drops, the process reverses itself, and fresh solvent-free air enters the headspace. Subsequent cycles of rising and falling temperature cause a cumulative loss in solvent. An accurate calculation for this loss would require a knowledge of the time- and space-dependent concentrations and temperatures in the tank, and would thus call for the solution of partial differential equations (PDEs; mass and energy balances) along with the appropriate equilibrium relation. The somewhat irregular geometry (tank and vent pipe) and the possibility of both conductive and free convective transport, plus uncertainties in the external heat transfer coefficient, make this a formidable problem to solve.

In this first elementary treatment of this problem, these complications are avoided by confining ourselves to the calculation of the *maximum* loss that can occur in the course of a single temperature cycle. This is achieved by assuming that the tank contents are well mixed and in thermal and phase equilibrium at the maximum temperature attained in a cycle. This would be the maximum daily loss. The assumption is not outlandish, because the inevitable thermal convection currents will bring about a considerable degree of mixing. The result is an enormous simplification of the problem, because we are now dealing merely with algebraic expressions representing the vapor pressure of the system and the appropriate gas laws.

Suppose that the stored liquid in question is benzene and that the headspace of the storage tank is 100 m^3. We assume that the temperature rises from 15 to 30°C in the course of a day and that, as a result, some 5% of the headspace air is ejected. The task is to calculate the maximum amount of solvent lost.

We start by computing the vapor pressure at the maximum temperature of 30°C using the Antoine constants for benzene listed in Table 6.1. Thus,

$$\log P^o = 6.89272 - \frac{1203.531}{30 + 219.888}$$

and, consequently,

$$P^o = 119.3 \text{ mmHg}$$

Substituting this value into the ideal gas law, we obtain

$$m = \frac{M P^o V}{RT}$$

or

$$m = \frac{78(119.3 / 760)1.013 \times 10^5 \times 5}{8.314 \times 303}$$

and, therefore,

$$m = 2.46 \text{ kg}$$

Although the amounts calculated represent a theoretical maximum, through the use of more-refined calculations and tests, these losses were shown to be substantial and would in most cases be considered unacceptable. Provision is therefore often made to recover the escaping vapors by compression condensation or by adsorption, or to cover the tank with a floating top. An attendant benefit of this procedure is the avoidance of the adverse effect of such emissions on the environment.

6.2 Multicomponent Systems: Distribution of a Single Component

We begin our discussion of multicomponent equilibria by drawing a distinction between systems in which several or all components are present in the two equilibrated phases and those in which only one component plays a key role by distributing itself in significant amounts between the phases in question. Vapor–liquid equilibria of mixtures and other similar multicomponent systems involving the appearance of several solutes in each phase are the prime example of the former, while distributions of a single component occur in a number of different contexts, which we take up in turn below. They include the equilibrium of a single gas with a liquid solvent, or a solid (gas absorption and adsorption), and the distribution of solutes between a liquid solution and an immiscible solvent (liquid extraction) or solid (liquid-phase adsorption). We note that although more than one component may be present in both phases, the appearance of other components does not affect the distribution of the principal component under consideration. Thus, in gas absorption, solvent vapor is inevitably present in the gas phase but does not interfere in any way with the distribution of the main solute. Similarly, in liquid extraction, the two solvents may not be perfectly immiscible, but this does not significantly affect the distribution of the solute.

6.2.1 Gas–Liquid Equilibria

Examples of gas–liquid equilibria abound both in the physical world surrounding us and within an industrial context. Gases of both a benign and a toxic nature are taken up or released by bodies of water. The example of dissolved oxygen, which is essential to the sustenance of aquatic life, immediately comes to mind. On the industrial scene, both valuable and objectionable gases are often selectively removed or recovered by gas absorption or gas scrubbing. We alluded to this process in Illustration 2.3 and Section 5.4, and more on this process appears in Chapter 8.

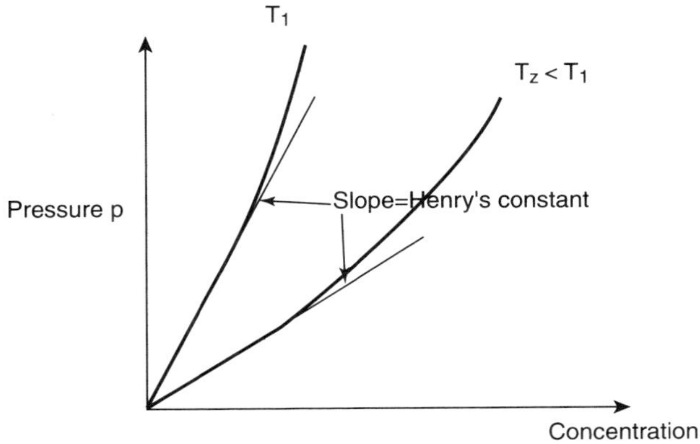

FIGURE 6.2
Gas–liquid equilibrium isotherm.

The phase equilibrium between a gas and a liquid solvent is usually expressed in terms of the amount absorbed or liquid-phase concentration as a function of gas pressure. A diagram of this relation appears in Figure 6.2. The concentration of the dissolved gas is seen to increase with pressure, and it does so indefinitely (i.e., no limiting saturation value is attained). This is in contrast to gas–solid and liquid–solid adsorption equilibria in which the solid surface ultimately becomes saturated with solute. An increase in temperature, on the other hand, diminishes the solubility of the gas and, hence, its concentration. One notes in addition that at the lower end of these diagrams the plot becomes linear. The slope of this linear portion is termed the Henry's constant H, and the phase equilibrium in this range is said to follow Henry's law, given by

$$p = HC \tag{6.2}$$

Here p is the gas pressure usually expressed in kilopascal (kPa), and C is the concentration of the dissolved gas (mol/L or mol/m^3). Henry's constants at 25°C for some common gases are displayed in Table 6.2. Note that large values of H denote *low solubility*, and vice versa.

The validity of Equation 6.2 for these gases extends to several atmospheres, and in the case of permanent gases, such as H_2, O_2, and N_2, to several tens of atmospheres. A good deal of useful information can therefore be gathered through the use of Henry's law and the associated Henry's constants. We demonstrate this with the following illustrations.

Illustration 6.3: Carbonation of a Soft Drink

It is common practice in the soft drink industry to carbonate drinks by dissolving a *fixed volume* of carbon dioxide in the liquid, rather than by

TABLE 6.2

Henry's Constants for Gases in Water at 25°C

Gas	H (kPa m^3 mol^1)
Hydrogen	130
Helium	260
Carbon monoxide	100
Nitrogen	150
Oxygen	79
Carbon dioxide	2.9
Hydrogen sulfide	1.0
Sulfur dioxide	0.015
Ammonia	0.0025
Methane	71
Ethane	53
Propane	66
n-Butane	80

applying a *prescribed pressure* to the contents. That volume is set at three to five times the volume of the liquid contents.

Consider a standard 1.5-L soft drink bottle with a headspace of 5%. The task is to calculate the pressure in the bottle after carbonation and the consumption of carbon dioxide in a plant bottling 10,000 containers per day. We assume a CO_2 charge equal to 5 L.

Taking account of the 5% headspace, and assuming a bottling temperature of 298 K together with a molar volume of STP of 22.4 L, this leads to a carbon dioxide volume of

$$V_{CO_2} = 5000 - 0.05 \times 5000 = 4750\ \text{L}/\text{m}^3 \tag{6.3a}$$

or, equivalently,

$$C = (4750/22.4)(273/298)\ \text{moles}\ CO_2/\text{m}^3 \tag{6.3b}$$

We apply this value to Henry's law (Equation 6.2) and use a Henry's constant of 2.90 taken from Table 6.2 to obtain

$$p = H \times C = 2.9(4750/22.4)(273/298) \tag{6.3c}$$

or

$$p = 563\ \text{kPa}$$

To this value has to be added the initial air pressure of 100 kPa, which brings the total pressure in the container to slightly above 6.5 atm.

To obtain this result in 10,000 bottles of 1.5 L each, one requires a CO_2 volume of

$$V_{CO_2} = 10,000 \times 1.435 \times 5\,\text{L} \tag{6.3d}$$

where the headspace of 0.075 L was subtracted from the total bottle volume of 1.5 L. The corresponding mass of CO_2 in the carbonated drink $(m_{CO_2})_d$ is given by

$$(m_{CO_2})_d = \frac{10,000 \times 1.425 \times 5}{22.4}(273 / 298)M \tag{6.3e}$$

where M = molar mass of carbon dioxide = 44. Consequently,

$$(m_{CO_2})_d = 128\,\text{kg} \tag{6.3f}$$

The amount of carbon dioxide in the headspace is a small fraction of this value and is given by

$$(m_{CO_2})_h = 10,000\frac{PV}{RT}M = 10,000\frac{5.63 \times 10^5 \times 0.075 \times 10^{-3}}{8.314 \times 298} \times 44 \tag{6.3g}$$

$$(m_{CO_2})_h = 7499\,\text{g} = 7.5\,\text{kg} \tag{6.3h}$$

Thus, a total of 135.5 kg of CO_2 will be required in the daily operation of the plant.

The question we now wish to address is whether Henry's law can be extended to *mixtures* of sparsely soluble gases. We note from Table 6.2 that most of the gases listed have solubilities at 1 atm pressure, of the order of 0.01 mol%. Even a moderately soluble gas such as carbon dioxide falls below the concentration level of 0.1 mol%. These values are so low, and the distance between dissolved molecules so large, that their solubilities will not be significantly altered by the presence of other, sparely soluble components. In other words, we feel justified in applying Henry's law for a *pure* gas to each individual component in a *mixture*, with the understanding that p is now the *partial pressure* of the gas in question. We apply this principle in the following illustration.

Illustration 6.4: Absorption of Gas Mixtures: Recovery of Methane from Landfill Gases

Gases emanating from landfill sites are typically composed of approximately 45% each of methane and carbon dioxide, a few percent of nitrogen, and lesser amounts of H_2S and other objectionable trace components. The volumes in question are such that it is now deemed profitable to recover the methane for use as a fuel. Technologies to achieve this are either already in place or in the process of development.

One proposed method consists of absorbing CO_2 in a packed tower using water at elevated pressures after the removal of H_2S and other minor components has been accomplished. The question to be addressed here is whether the process can achieve substantially complete removal of CO_2 without undue losses in methane.

We consider a feed consisting of 1000 mol each of CH_4 and CO_2 at 20 atm pressure, with a stipulated CO_2 removal of 99%. The tools available for modeling the process are the usual mass balances, together with the relevant equilibrium relations (Henry's law). Using superscripts L and G to denote the two phases, we have the following:

CO_2 BALANCE

$$\left(n_{CO_2}^G\right)_{in} = \left(n_{CO_2}^G\right)_{out} + \left(n_{CO_2}^L\right)_{out} \tag{6.4a}$$

CH_4 BALANCE

$$\left(n_{CH_4}^G\right)_{in} = \left(n_{CH_4}^G\right)_{out} + \left(n_{CH_4}^L\right)_{out} \tag{6.4b}$$

PHASE EQUILIBRIA

$$p_{CO_2} = H_{CO_2} C_{CO_2} \tag{6.4c}$$

$$p_{CH_4} = H_{CH_4} C_{CH_4} \tag{6.4d}$$

Dividing the latter two equations yields

$$\left(n_{CH_4}^L\right)_{out} = \frac{2.9}{71}\left(n_{CO_2}^L\right)_{out} \tag{6.4e}$$

while the CO_2 mass balance now becomes

$$1000 = 0.01 \times 1000 + \left(n_{CO_2}^L\right)_{out} \tag{6.4f}$$

Substitution of Equation 6.4f into Equation 6.4e leads directly to the amount of methane lost in the effluent:

$$\left(n_{CH_4}^L\right)_{out} = 40 \text{ moles}$$

This represents 4% of the total incoming feed.

Comments

A noteworthy feature of these results is the absence of any effect of total pressure P_T unless one operates outside the Henry's law region. This will likely be the case when P_T exceeds 20 atm by a considerable margin. At those levels, however, one can also expect to see the beginnings of interference by CO_2 with the equilibrium uptake of methane. This is welcome news, because it will tend to reduce the methane losses.

6.2.2 Liquid and Solid Solubilities

A second example of a binary system in which one component is considered to be confined to one phase only involves the solubility of liquids and solids in a solvent. The confined species here is the solvent, which is in contact with a second phase containing a pure liquid or solid. At equilibrium the solvent phase has become saturated with the dissolved species or solute and no further dissolution takes place. The concentration corresponding to this state is termed the *solubility* of the liquid or solid in question.

The number of possible solute–solvent combinations is evidently quite large, and the corresponding number of required measurements infinite if temperature is considered an additional variable. It is therefore customary to deal with only a small number of solvents, principally water, and to perform measurements at a standard temperature of 25°C. Most reported data were obtained within this framework.

A listing of the solubility in water of a number of solutes, mostly organic in nature, appears in Table 6.3. Sodium nitrate and glucose show high values of close to 50%, as expected, while solubilities of DDT and mercury are measured in parts per billion. The latter are nevertheless sufficiently significant to be of environmental concern. The reader will also note the relatively high solubility of diethyl ether (~7.5%), which is generally thought of as being insoluble. Unusually high values are also shown by chloroform and benzene. Some caution should therefore be exercised before declaring a solvent "insoluble" in water. The amounts dissolved may be considerable, and their presence in discharged process water must be duly accounted for. This is illustrated in the following example.

Illustration 6.5: Discharge of Plant Effluent into a River

We consider here the case of process water saturated with benzene being discharged into a river. The question to be addressed is whether the diluting

TABLE 6.3

Solubilities of Liquids and Solids in Water at 25°C

Substance	Solubility (g/L water)
Chloroform	11
Carbon tetrachloride	0.8
Diethyl ether	75
n-Hexane	0.15
n-Octane	6.6×10^{-4}
Benzene	1.8
Toluene	0.52
o-Xylene	0.18
Naphthalene	3.3×10^{-2}
DDT	1.2×10^{-6}
Glucose	820
Calcium carbonate	1.5×10^{2}[a]
Calcium sulfate	2.0
Sodium chloride	360
Mercury	3.0×10^{5}

[a] The value rises to 1.2 g/L in water saturated with CO_2.

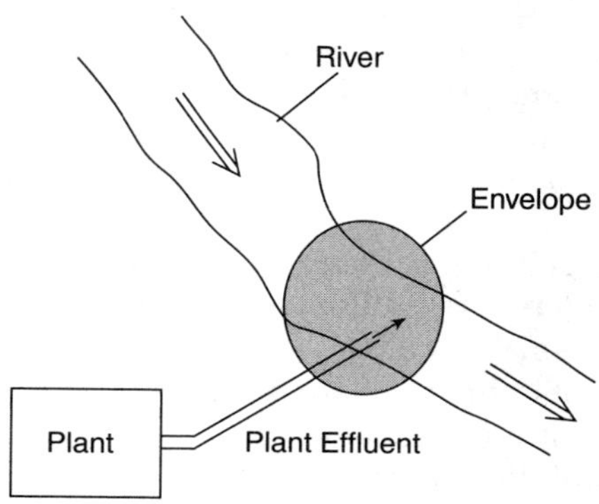

FIGURE 6.3
Discharge of a plant effluent into a river.

effect of the river flow is sufficient to reduce the effluent concentration to within permissible limits and, if not, how much of the offending substance has to be removed to meet environmental standards. A depiction of this effluent and river flow is shown in Figure 6.3.

Consider an effluent discharge of 150 L/min and a river flow that varies seasonably from 23,000 L/s to 50,000 L/s. Note that because the regulatory limit has to be met at all times during the year, the lower summer flow rate of 23,000 L/s must be used. The standard used here is that of the U.S. Environmental Protection Agency, which set the maximum permissible level of benzene in drinking water at 0.05 mg/L or 5 ppb. The effluent is assumed to be saturated with benzene at the solubility level of 1.8 g/L (1.8×10^6 ppb) given in Table 6.3.

We commence with a mass balance around the juncture of effluent and river flow shown in Figure 6.3 and assume the contents of the envelope to be well mixed and to have attained a steady state. Thus,

$$\text{Rate of benzene in} - \text{Rate of benzene out} = 0$$

$$Q_e C_e - (Q_e + Q_r)C_r = 0 \tag{6.5a}$$

where the subscripts *e* and *r* refer to effluent and river, respectively, and *Q* denotes volumetric flow rates. Substituting the given data into this equation, we obtain

$$(150/60)1.8 \times 10^6 - [(150/60) + 23{,}000]C_r = 0 \tag{6.5b}$$

Solving for C_r yields a downstream river concentration of

$$C_r = \frac{4.5 \times 10^6}{23002.5} = 195\,\text{ppb} \tag{6.5c}$$

With the allowable concentration set at 5 ppb, the required fractional removal *R* is given by

$$R = 1 - \frac{5}{196} = 0.974 \tag{6.5d}$$

In other words, slightly more than 97% of the benzene in the plant effluent will have to be removed to meet the aforementioned standard. It is likely that an adsorption purification process using activated carbon can be used to achieve this goal. Such a process will be taken up in Illustration 6.5.

6.2.3 Fluid–Solid Equilibria: The Langmuir Isotherm

In Section 6.2.1 we considered the case in which a gas is absorbed into and comes into equilibrium with a liquid solvent. This process of absorption, in which the solute gas permeates the entire body of the liquid, differs from adsorption, which is essentially a surface phenomenon. Here the solute also penetrates the porous structure of the solid, and the process is therefore initially at least akin to absorption. Ultimately, however, the solute molecules come to rest on the walls of the porous structure and remain confined there in dynamic equilibrium with the surrounding pore space. Thus, while permeation of the solid structure takes place, the solute molecules are not uniformly dispersed but are rather localized on the internal surface of the solid matrix. We show this, as well as the differences between adsorption and absorption, in Figure 6.4a and Figure 6.4b.

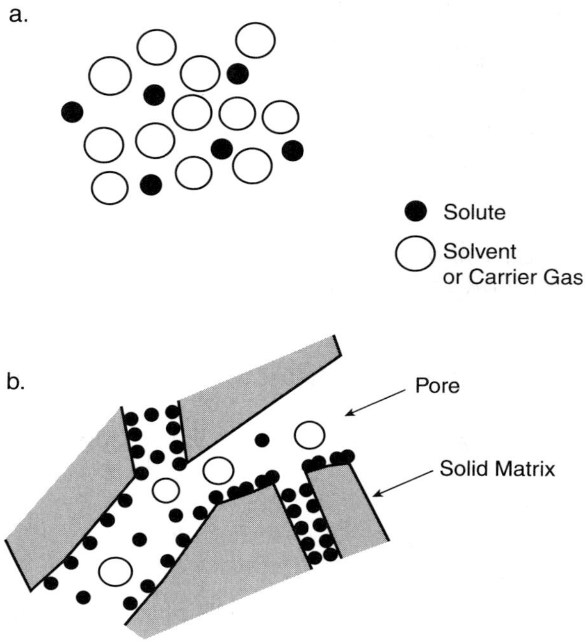

FIGURE 6.4
(a) Absorption: dispersion of solute molecules throughout liquid solvent. (b) Adsorption: localization of solute molecules on pore walls.

There is a further distinction to be made between the two processes. A liquid has in principle an unlimited capacity for dissolving a gas, although that capacity diminishes asymptotically as gas pressure is increased (see Figure 6.2). In adsorption, the surface area available for accommodating solute molecules is limited and finite. Here an increase in pressure or solute concentration will ultimately lead to complete coverage by a "monolayer" or saturation of the surface. On reaching this state, no further adsorption can take place. This is indicated by the asymptotic saturation capacity shown by the isotherm in the phase diagram of Figure 6.5. An exception occurs when the solute gas is within reach of a state of condensation. The solute may then form multiple adsorbed layer or multilayers and ultimately fill the entire pore space by condensation ("capillary condensation"). This less frequent case leads to inflecting isotherm curves and is shown in Figure 6.6. Most fluid–solid equilibria, however, lead to monolayer coverage and are well represented by the isotherms shown in Figure 6.5.

One of the earliest attempts to derive a theoretical expression for these isotherms, and to date also the most successful one, was that by Langmuir. In his derivation, Langmuir postulated that the adsorption equilibrium was the result of two rate processes, equal in magnitude but opposite in direction. They were the rate of adsorption, which was taken to be proportional to

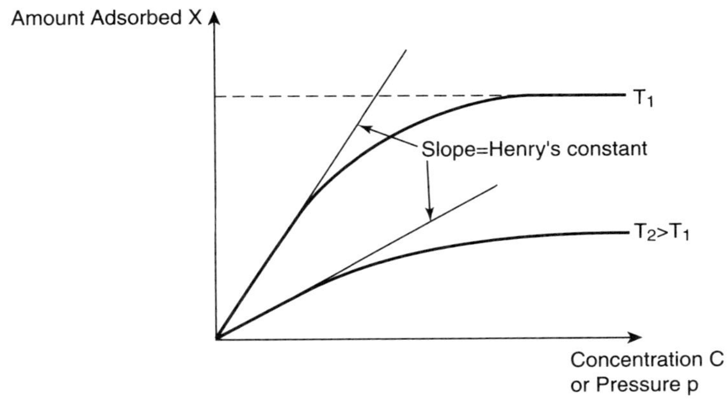

FIGURE 6.5
The Langmuir adsorption isotherm.

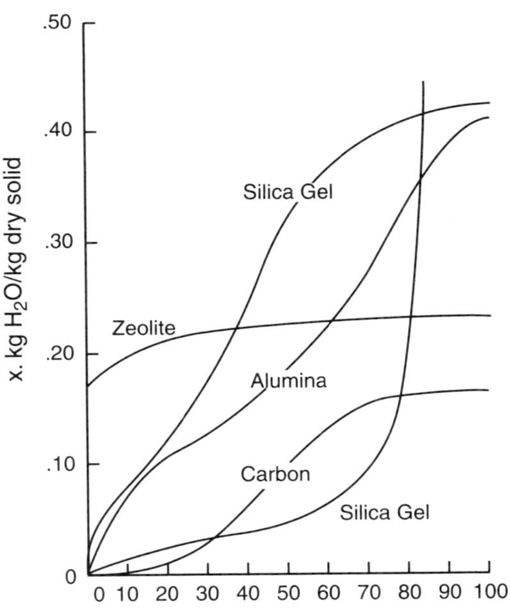

FIGURE 6.6
Moisture adsorption isotherms.

solute pressure or concentration and the available free surface area, and a desorption rate, which varied directly with the fractional surface coverage. The result of equating these two rate expressions can be expressed, after some manipulation, in the following form:

$$X = \frac{a'p}{1+b'p} \tag{6.6a}$$

or

$$X = \frac{a'C}{1+b'C} \tag{6.6b}$$

Here a' and b' are semiempirical constants; p and C are gas partial pressure and solute concentration, respectively; and X represents the amount adsorbed. For general engineering purposes, it is often more convenient to replace p and C by a single fluid-phase concentration Y (expressed in units of kg solute/kg inert gas, or kg solute/kg solvent). Equation 6.5a and Equation 6.5b can then be coalesced into a single expression of the form

$$X = \frac{HY}{1+bY} \tag{6.6c}$$

where a' and b' are replaced by the new empirical constants H and b. The amount adsorbed, X, is generally expressed in units of kilogram solute per kilogram solid or, less frequently, as moles solute per gram solid. Evidently, when dealing with pure solute gases, the fluid ratio Y can no longer be used, and we must revert to Equation 6.5a. The need to do this rarely arises in practice, because in most practical applications the gas phase contains an inert, nonsorbable component such as air.

Let us now examine the asymptotic behavior of the Langmuir isotherm. At low values of the fluid-phase concentration Y, the term bY becomes small compared to 1, and the Langmuir isotherm approaches the limiting linear form:

$$X = HY \tag{6.6d}$$

$$Y \to 0$$

The slope of this line is often referred to as Henry's constant, and Expression 6.5d as Henry's law for adsorption, analogous to the corresponding case of gas–liquid equilibrium (see Equation 6.2).

At the other extreme of high values of Y, bY becomes the dominant factor, and the Langmuir isotherm converges to the following form:

$$X_{Y \to 0} = H/b = X_{Sat} \tag{6.6e}$$

Both of these limiting cases nicely agree with the features shown in Figure 6.5.

TABLE 6.4

Henry's Constants for Aqueous Solutions on Carbon at 25°C

Solute	*H* (kg H_2O/kg C)
Methyl chloride	6.6
Methylene chloride	14
Chloroform	74
Carbon tetrachloride	360
n-Pentane	2,200
n-Hexane	10,400
Benzene	400
Styrene	600
Chlorobenzene	500
PCBs (upper limit)	1.5×10^8

We note that high values of H correspond to high adsorbent capacities, in contrast to gas absorption, where large values of Henry's constant are associated with *low* solvent capacities.

Adsorption Henry's constants are central to adsorptive purification processes of dilute streams and also reach, as will be seen in the next section, into areas of environmental concern. We compiled values of H on carbon for some important trace solutes in aqueous solution, which are displayed in Table 6.4. Of note here is the extremely high value for PCBs, which dominates the table. The reader should be reminded, however, that this is partly offset by the extremely low solubility of PCBs.

Although the great majority of adsorbed solutes show Langmuir-type behavior, a considerable number of substances exhibit inflecting isotherms. This is particularly the case with vapors in the vicinity of saturation. Figure 6.6, which shows moisture isotherms on a variety of adsorbents, illustrates this type of behavior. Zeolitic sorbents are the only ones among them which display Langmuir-type behavior. They are most effective at low humidities where they show a substantial uptake of water. Silica gel and activated alumina have high uptakes at higher humidities, where they exceed zeolite capacities by factors of two or more. Carbon, because of its hydrophobic nature, has only limited affinity for water.

In what follows, we use the tabulations of Table 6.4 to explore a particular and highly useful limiting case of the adsorptive purification of an aqueous solution.

Illustration 6.6: Adsorption of Benzene from Water in a Granular Carbon Bed

The purification of both potable and wastewater by adsorption is a widespread practice, which in the United States alone consumes over 100 million kilograms of activated carbon a year. Typically in such an operation, the water is passed through a fixed bed of the granular adsorbent which becomes

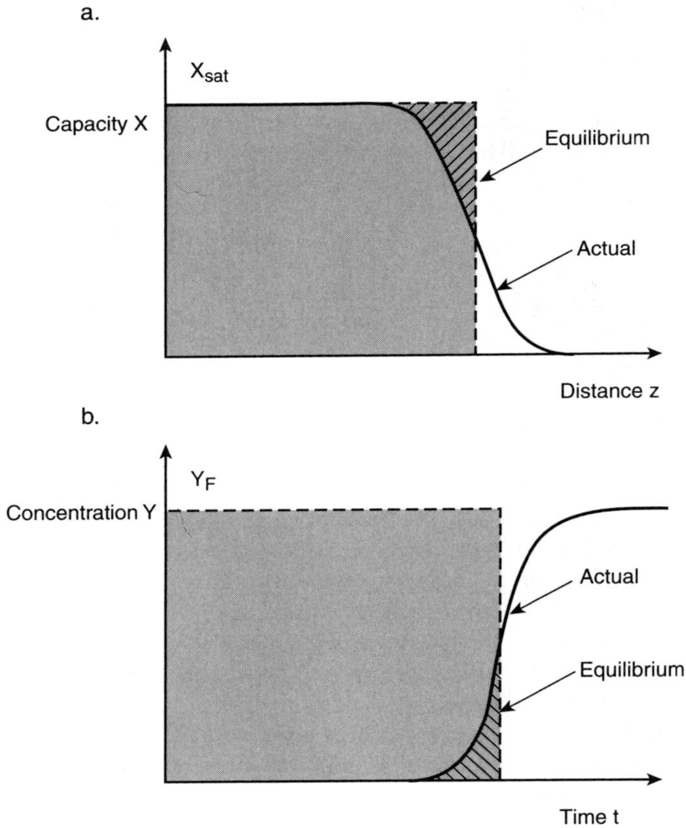

FIGURE 6.7
Adsorption in a fixed bed: (a) adsorbent concentration profiles and (b) fluid-phase concentration breakthrough curve.

progressively saturated with the impurities. In the course of this process, a concentration profile develops within the bed, which takes the form of an S-shaped curve ranging in level from full saturation q_s to essentially zero impurity content. This profile, which is shown in Figure 6.7a, progresses steadily through the adsorber until it reaches the end of the bed. An observer stationed at this position would at this point see the first traces of impurity break through. It would then rise in level, again in the form of an S-shaped curve, until the feed concentration Y_F is attained. This is, however, not allowed to happen, and the operation is instead terminated when the effluent concentration reaches a prescribed maximum permissible level, Y_P. This level and the associated concentration breakthrough curve are shown in Figure 6.7b. The operation is continued by switching the feed stream to a second clean bed, held on hand while the spent adsorbent is either regenerated or replaced by a fresh charge of the material.

Adsorption in a fixed bed is a complex system to model. Concentrations evidently vary with distance, and although they ultimately attain a steady form of distribution, they also vary with time. The model would consequently consist of two mass balances, one for the fluid phase and a second one for the solid phase, and both of these would be partial differential equations in time and distance, which generally have to be solved numerically. To avoid this complication, a procedure has come into use in which mass transfer resistance is neglected and the two phases are everywhere assumed to be in equilibrium. The concentration then propagates in the shape of a rectangular front, shown in Figures 6.7a and 6.7b and denoted "Equilibrium." The movement of this front and its dependence on flow rate and feed concentration can be analyzed by means of a simple cumulative mass balance, which takes the following form:

$$Y_F v \rho_f A_C t \quad = \quad X_F \rho_b A_C z \quad + \quad Y_F \rho_f A_C z \tag{6.7a}$$

Amount introduced to time t	Amount retained by adsorbent	Amount left in fluid

Here v and A_C are the fluid velocity and cross-sectional area of the bed, respectively, and ρ_f and ρ_b are the fluid and bed densities.

The last term in this equation is generally negligible, because the bulk of the impurity will reside in the adsorbent. It would otherwise not be a very efficient adsorbent. Equation 6.7a can then be recast in the following form:

$$z / t = \frac{\rho_f v}{\rho_b H} \tag{6.7b}$$

where we substituted Henry's constant for the concentration ratio X_F/Y_F.

Equation 6.7b can be used to calculate the time t it takes the front to reach a particular position or conversely, the position attained after a prescribed time interval. These quantities are, by necessity, *limiting lower values* because full saturation will be retarded by the mass transfer resistance. However, in many instances, the fluid flow is sufficiently slow that local equilibrium is attained, or will be nearly attained, during the interval of contact. We make use of this fact in Illustration 6.6 to analyze the contamination of soil that results from polluted groundwater.

This illustration addresses the problem of sizing a carbon bed to be used in the purification of a plant effluent. Our purpose here is best served by recasting Equation 6.7b in a form suitable for the calculation of bed requirements. This is done by first cross-multiplying the expression and then dividing and multiplying by the cross-sectional area A_C. We obtain

$$\frac{A_C z \rho_b}{A_C v \rho_f t} = \frac{(W_b)_{\text{Min}}}{G \times t} = \frac{(W_b)_{\text{Min}}}{v \rho_f t} = \frac{1}{H} \tag{6.7c}$$

or, equivalently,

$$\frac{\text{Minimum bed weight}}{\text{Weight of fluid treated}} = \frac{1}{H} \tag{6.7d}$$

Thus, the larger the Henry constant H, the smaller the required bed size. Note, however, that this is a *minimum* requirement, because we assumed the phases to be in equilibrium.

Let us see how this works out in practice. We refer to the preceding example of an aqueous plant effluent saturated with benzene, 97% of which has to be removed before being discharged. The rate of discharge was 150 L/min.

Suppose that we wish to size a bed of granular carbon, which will stay on-stream for 6 months before breakthrough occurs. The amount of effluent treated in this period comes to

$$G \times t = 150 \times 60 \times 24 \times 180 = 3.9 \times 10^7 \text{ kg} \tag{6.7e}$$

Using $H = 400$ for benzene, listed in Table 6.4, we obtain the following from Equation 6.7c:

$$(W_b)_{\text{Min}} = G \times t/H = 3.9 \times 10^7/400 = 9.8 \times 10^4 \text{ kg carbon} \tag{6.7f}$$

Now, the bed density ρ_b for granular activated carbon is, typically, of the order 500 kg/m³. If, therefore, a bed 3 m in diameter d is chosen, the packed height h of that bed is given by

$$h = \frac{(W_b)_{\text{Min}}}{(\pi d^2 / 4) \times \rho_b} = \frac{9.8 \times 10^4}{(\pi\, 3^2 / 4) \times 500} = 28 \text{ m} \tag{6.7g}$$

Because adsorbent beds typically range up to 10 m in height, this figure is excessive. We therefore reduce the time on-stream to 1 month and arrive at the more reasonable bed height of approximately 4.5 m. This allows some slack for *actual* bed requirements, which will be somewhat higher because of the neglected mass transfer resistance.

Illustration 6.7: Adsorption of a Pollutant from Groundwater onto Soil

Soils show a considerable sorptive affinity for pollutants which is principally brought about by the soil's carbon content (a result of decaying organic matter). That content typically ranges from 1 to 2% of the total mass. Adsorptive capacities per unit weight are consequently some 50 times lower than the carbon values shown in Table 6.4, and this figure was used to compose Table 6.5, listing Henry's constants for soils.

TABLE 6.5

Henry's Constants for Aqueous Solutions on Soil at 25°C

Solute	*H* (kg water/kg soil)
Methyl chloride	0.033
Methylene chloride	0.071
Chloroform	0.37
Carbon tetrachloride	1.8
n-Pentane	11
n-Hexane	52
Benzene	0.54
Styrene	3.0
Chlorobenzene	2.5
PCBs, upper limit	7.3×10^5

We consider here again an effluent saturated with benzene and assume that seepage into the groundwater has occurred. The task will be to calculate the stretch of soil that will have been contaminated after 10 days of exposure. Groundwater velocity is set at 1 mm/s and a soil density of 2500 kg/m³. Using a Henry's constant of 0.54 given in Table 6.5, we obtain from Equation 6.7b,

$$z / t = \frac{\rho_f}{\rho_b} \frac{v}{H} = \frac{1000}{2500} \frac{10^{-3}}{0.54} = 7.54 \times 10^{-4} \text{ m / s}$$

and, consequently,

$$z = 7.4 \times 10^4 \ t = 7.4 \times 10^4 \times 3600 \times 24 \times 10$$

$$z = 640 \text{ m}$$

Comments

The reader is reminded that the results obtained in both Illustration 6.6 and Illustration 6.7 represent *minimum* values (i.e., minimum carbon bed requirements and minimum penetration into the soil). Because of the mass transfer resistance, the concentrations on the percolating fluid run ahead of the rectangular front shown in Figure 6.7a and penetrate deeper into solid matrix than predicted. The corrections that must be applied often amount to no more than 20 to 30% of the ideal length, which is therefore a highly useful engineering estimate for the situation at hand. Higher corrections are needed in cases involving large particles or high fluid velocity.

Note that the rigorous PDE model would require, in addition to equilibrium data, the relevant transport parameter (i.e., the mass transfer coefficients both within the solid particle and the liquid). The solid-phase coefficient, in particular, requires fairly elaborate measurements and is usually unavailable to the general practitioner.

Illustration 6.8: Adsorption in Biology: The Oxygen Hemoglobin Isotherm — Tolerance of the Body to Altitude

The protein hemoglobin contained in red blood cells acts as the well-known carrier of oxygen in the human and animal body. Superficially unlike normal adsorbents, hemoglobin does share with them certain important features, which clearly places it in the same category.

1. Oxygen uptake varies directly with the partial pressure of the gas.
2. Retention of oxygen takes place at specific sites (Fe in the hemoglobin molecule). When all sites are occupied, the blood is saturated and no further uptake occurs. This is in line with the Langmuir model of adsorption.
3. Oxygen is taken up and released *reversibly*. At equilibrium, the rate of uptake equals the rate of release, again in line with the Langmuir model.
4. A plot of the amount retained versus oxygen pressure is of the same form as the Langmuir isotherm shown in Figure 6.5.

The physical details of the passage of oxygen to blood cells were laid out in Illustration 1.6. Recall from the description given there that the oxygen uptake by an adult human is at the rate of about 100 mL/min at ambient conditions or approximately 12 mol of O_2 every 24 h.

The oxygen inventory in the blood (i.e., the amount adsorbed by hemoglobin at any given moment) is minuscule in comparison. In the course of its passage from atmospheric air, the oxygen partial pressure drops from an initial value of 150 mmHg to 105 mm in the alveoli and ultimately to an equilibrium value in the blood of 95 mmHg. The corresponding amount adsorbed by the hemoglobin is shown on the isotherm plot of Figure 6.8. It amounts to 19 mL O_2 per 100 mL blood at 37°C and 95 mmHg. With the typical blood volume in an adult human set at 5 L, this corresponds to a minuscule 5×10^{-3} moles of oxygen or 160 mg. Still, without the adsorptive capacity of the blood, the inventory would be even lower by a factor of at least 100 (see Figure 6.9), incapable of sustaining the life of a human. This demonstrates the vital role played by the hemoglobin.

The equilibrium curve shown in Figure 6.8 has, as we noted, all the basic features of the Langmuir isotherm, the only departure being an inflection at low coverage. In this it resembles the water isotherm on silica gel (Figure 6.6).

The most popular analytical representation of oxygen uptake by hemoglobin is the semiempirical expression due to Hill:

$$X(\text{mL}\,O_2\,/\,100\,\text{mL}) = \frac{(p\,/\,p_o)^n}{1+(p\,/\,p_o)^n} \qquad (6.8)$$

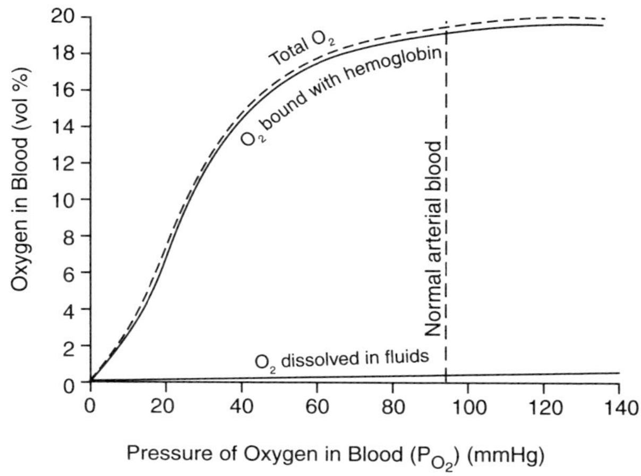

FIGURE 6.8
Equilibrium uptake of oxygen by blood. The difference between the dotted and full curves represents the portion dissolved in the water of the blood.

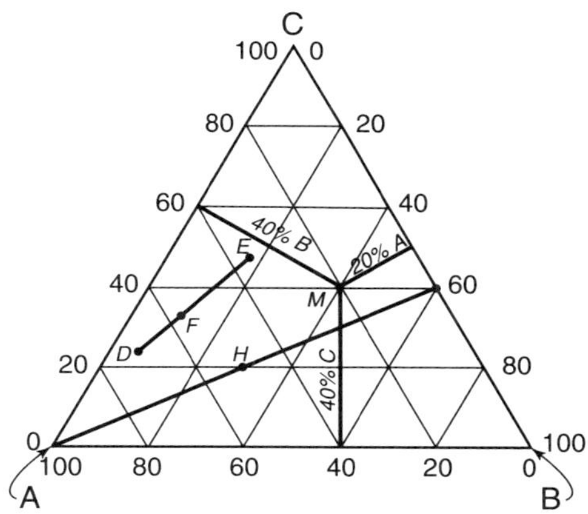

FIGURE 6.9
The triangular diagram.

n in this equation denotes the number of O_2 molecules adsorbed per hemoglobin molecule, which varies from one to four, and has an average value of 2.7. p_o is a normalizing factor equal to 27.2 mmHg, the oxygen equilibrium pressure at 50% saturation. Although Hill's equation is of the same form as the Langmuir isotherm, Equation 6.6, it lacks the required linearity in the Henry's law region. More precise theoretical expressions, which include this feature, have been developed but are too complex for practical use.

It is of some interest to calculate the oxygen content of blood at various altitudes. Atmospheric pressures can be calculated from the barometric formula, Equation 1.7b, and translated into oxygen uptake via Figure 6.9, assuming a proportional decrease of p_{O_2}. Some representative figures are given below:

Altitude (*m*)	Atmospheric Pressure (kPa)	O_2 in Blood (mL/100 mL)
0	100	19
3,000	70	17.5
5,000	49	16
8,850 (Mt. Everest)	30	5
10,000	16	3

Up to altitudes of 5 km, the decline may be considered tolerable, but thereafter it becomes precipitous, and at 10 km, life-sustaining breathing would be an impossibility even if breathing were at an extremely rapid rate. Note that the "Langmuir"-form of the isotherm is crucial. A linear or concave equilibrium curve would not be nearly as effective even at moderate altitudes.

Comments

The oxygen–hemoglobin system is but one of many examples of adsorption processes that occur in the human body. These are often irreversible in nature (i.e., involve chemisorption) and do not properly fall into the category of phase equilibria. All such events, however, do involve the transport of mass in one form or another, with both membranes and fluid barriers defining the mass transfer resistance. We will return to this topic in Section 8.2 dealing with membrane separation processes.

6.2.4 Liquid–Liquid Equilibria: The Triangular Phase Diagram

Liquid–liquid equilibria deserve our attention principally because of their widespread occurrence in industrial- and laboratory-scale extraction processes. They also play a role in assessing the effect of accidental spills of oil and organic solvents in lake and ocean waters.

On the industrial scene, the most prominent applications both in scale and number are seen in the petroleum industry. Liquid extraction is used here to separate petroleum fractions selectively and to purify or otherwise refine them. In the Edeleanu process, which is close to a century old, liquid sulfur dioxide is used to extract aromatics from various feedstocks. The removal of the ever-present sulfur compounds is accomplished by extraction with sodium hydroxide solutions. In addition, a wide range of organic solvents is used in the purification and refining of various lubricants.

Considerable use of liquid extraction is also made in the mineral processing industry for the separation and refining of metals, and in the food industry for the extraction of oils and fats and other edible products. The pharmaceutical industry uses liquid extraction to recover their end products, from the reaction mixture. In general, whenever one is dealing with heat-sensitive materials, extraction at low temperatures is often the process of choice.

A number of traditional solvents used in these processes, such as the chlorinated and fluorinated hydrocarbons, have in the past two decades come under increasing scrutiny and criticism because of health and environmental concerns. The effect of certain of these solvents on the ozone layer has received a good deal of publicity, as has their role as potential carcinogens. This has opened the door to the development of alternate processes, chief among them *supercritical fluid extraction*. We will address this topic in some detail in Section 6.2.5.

Liquid–liquid equilibria differ from the previous cases we considered in that the two phases will, with an appropriate shift in the relative concentrations of the three components, merge into a single homogeneous phase. Consider, for example, the two "immiscible" solvents benzene and water, to which the solute ethanol is added. Initially, at low solute concentrations, two distinct phases are maintained, with ethanol distributed between them in certain concentration ratios. As the ethanol content is increased, the two solvents begin to show greater mutual solubility. This trend continues until ultimately, at sufficiently high ethanol concentrations, the two phases coalesce into a single homogeneous phase.

To obtain a valid representation of the entire domain of liquid–liquid phase behavior, it is necessary to take account of the concentrations of all three components. This is done by using the trilinear coordinate system, which leads to a construction known as a *triangular diagram*, shown in Figure 6.9.

Each side of this diagram is scaled from 0 to 100%. The apexes of the triangle represent the pure components *A*, *B*, and *C*, respectively. *C* is usually taken to be the solute, and *A* and *B* represent the two mutually immiscible solvents. Any point on the side of the triangle denotes a two-component (binary) mixture, while the interior of the diagram represents the full complement of three-component (ternary) mixtures. Thus, the point *D* on the side *AB* represents 40% solute C and 60% solvent *B*, while the interior point *M* signifies a mixture containing 20% *A*, 40% *B*, and 40% *C*. Compositions are usually expressed as weight percent or, less frequently, as mole percent.

Liquid–liquid equilibria are determined experimentally by equilibrating the two phases having a total concentration *M* (Figure 6.10) and recording their compositions. These values are entered in the diagram and form the end points of what is termed a *tie-line*, shown in Figure 6.10. There is, in principle, an infinite set of such tie-lines, only a few of which appear in the diagram. They are ordinarily not parallel and change their slope slowly with changing composition. As more solute *C* is added to such a mixture, the mutual solubility of the two solvents *A* and *B* increases until the tie-line end points merge at the point *P*, known as the *plait point*. The line connecting all these points and passing through *P* is known as the *binodal* or *solubility* curve and encloses all mixtures showing two-phase behavior. Compositions lying outside this solubility curve represent homogeneous, single liquid-phase

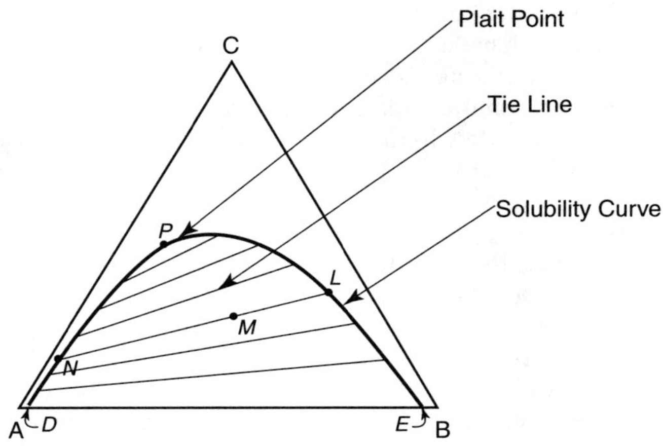

FIGURE 6.10
Representation of liquid–liquid equilibrium in trilinear coordinates.

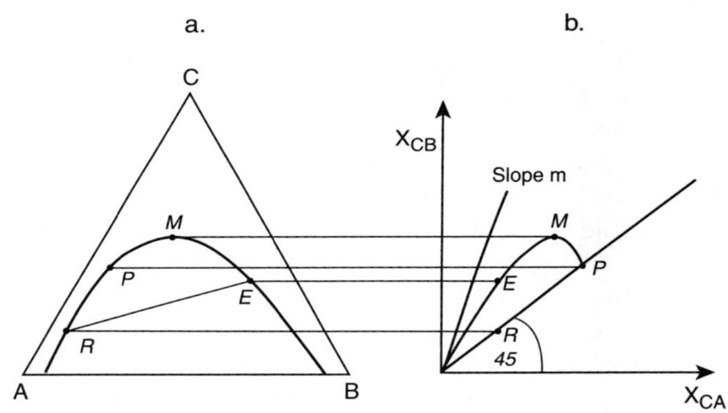

FIGURE 6.11
Translation from trilinear to rectangular coordinates: (a) triangular diagram and (b) distribution curve.

solutions, while the end points *D* and *E* denote the mutual solubility of the two solvents. When these are completely immiscible, *D* and *E* coincide with the two apexes *A* and *B*.

Representation of liquid–liquid equilibria in ternary diagrams is widely used in the graphical solution of extraction problems. This topic is discussed in considerable detail in the next chapter. Occasions arise, however, when it becomes convenient to use an alternate representation in rectangular coordinates, known as a *distribution curve*. This diagram, shown in Figure 6.11b, consists of a plot of the solute weight fraction in the two phases against each other. Its construction is accomplished by transferring the compositions represented by the tie-line end points to rectangular coordinate axes, as shown in Figure 6.11b.

TABLE 6.6

Low Concentration Solute Distributions in Water–Organic Solvent Systems (25°C)

Solute	Solvent	Mole %		Distribution Coefficient
		Aqueous Phase	Organic Phase	
Acetic acid	Carbon tetrachloride	5.088	0.916	5.56
	Hexane	14.810	1.614	9.18
	Toluene	7.850	2.440	3.22
Acetone	Chloroform	16.07	0.959	16.76
	Diethyl ether	1.519	5.446	0.279
Methanol	Benzene	4.067	0.798	5.10
	Toluene	5.945	0.286	20.79
Ethanol	Chloroform	4.187	1.784	2.35
	Diethyl ether	1.382	2.422	0.571
	Toluene	4.621	0.398	11.61
	Hexane	30.11	1.297	23.22
	Benzene	0.994	1.177	0.845

Clearly, this type of representation provides only a limited picture of the entire domain of compositions. Neither of the two solvent weight fractions, for example, can be deduced. Various other constructional features of the triangular diagram, which will be taken up below, likewise cannot be translated into rectangular coordinates. The distribution diagram does, however, convey several key features of two-phase behavior. It shows the maximum solute concentration $(X_{CB})_{Max}$ beyond which no phase separation can take place, and it locates the plait point on the 45° diagonal, which denotes equality of phase concentrations. Compositions that lie above the diagonal are richer in solute content than those of the companion phase, while points below it denote a depletion in solute. The greater the distance is from the diagonal, the greater the degree of enrichment or depletion. By noting these features, we gain an immediate sense of the potential of a particular system for extractive enrichment.

To quantify these properties, it is useful to define a quantity known as the distribution coefficient m, which equals the solute mole-fraction ratio in the two phases and is given by

$$m = \frac{x_{CB}}{x_{CA}} \tag{6.9}$$

Values of m greater than unity denote solute enrichment in the B layer, and those less than unity denote enrichment in the solvent A. In the limit of low concentration, or infinite dilution of the B layer, the distribution curve becomes linear and the distribution coefficient attains a maximum value. This limiting coefficient, shown in Figure 6.11b, is akin to the Henry's constants, which we had seen in gas–liquid and in fluid–solid equilibria, and is sometimes denoted as such. An indication of its magnitude may be obtained from the low-concentration distribution coefficients listed in Table 6.6 for a number of systems comprised of aqueous and organic solutions.

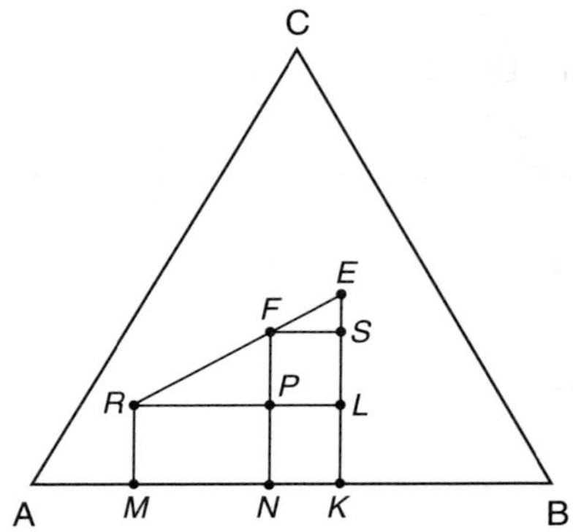

FIGURE 6.12
The mixture rule in a triangular diagram.

Illustration 6.9: The Mixture or Lever Rule in the Triangular Diagram

With this example, we wish to draw attention to certain simple geometric constructions that can be carried out in a triangular diagram to obtain important information in easy and rapid fashion. One such construction, sketched in Figure 6.12, leads to what is known as the *mixture rule,* or *lever rule,* for ternary liquid systems. Briefly stated, that rule asserts that the composition that results when two liquid solutions are mixed lies on a straight line connecting their compositions. Thus, if R kg of a mixture represented by point R is combined with E kg of a solution located at point E, the resulting composition F will lie on a straight line connecting points R and E. Furthermore, the location of F will be such that the line segments it defines stand in the ratio of the weights of the parent solutions — that is,

$$\frac{R}{E} = \frac{\overline{EF}}{\overline{RF}} \tag{6.10a}$$

We now proceed to present a proof of these statements. This is done by composing total and component mass balances for the mixtures and relating the resulting compositional changes to the line segments of Figure 6.12. Thus, for the total mass balance,

$$R + E = F \tag{6.10b}$$

where F = mass of final mixture, and for the component mass balance,

$$Rx_{CB} + Ex_{CE} = Fx_{CF} \tag{6.10c}$$

Eliminating F from these equations yields

$$\frac{E}{R} = \frac{x_{CF} - x_{CD}}{x_{CE} - x_{CF}} \tag{6.10d}$$

where x_{CF} will be located somewhere on the straight line connecting E and R. Its location is such that it subdivides the line in the ratio given by Equation 6.10a. From the diagram, it follows that

$$x_{CF} = \overline{FN}, \quad x_{CD} = \overline{RM}, \quad x_{CE} = \overline{EK} \tag{6.10e}$$

and, consequently,

$$\frac{E}{R} = \frac{x_{CF} - x_{CD}}{x_{CE} - x_{CF}} = \frac{\overline{SL}}{\overline{ES}} = \frac{\overline{FR}}{\overline{EF}} \tag{6.10f}$$

where the last equality follows from the similarity of the two triangles *FPR* and *ESF*. The proof is thus complete. Note that the closer the parent mixture F is to E, the more R is formed, and vice versa.

Comments

It will be seen from this discussion that the mixture or lever rule can be expressed both algebraically and in geometrical form. The algebraic version, given by Equation 6.10d, is the preferred tool for numerical calculation, while the geometrical construction of Figure 6.12 serves to provide a quick visual estimate of the quantities involved. Thus, if F lies midway between R and E, the two parent solutions will be equal in weight, and vice versa. It also follows that if an amount E with a composition located at the point E of Figure 6.13 is removed from the mixture, the point R representing the residue will lie on a straight line *EF* extended through F, and the relationship of Equation 6.10f will apply.

The lever rule finds its most notable application in the use of tie-lines to establish the compositions and relative proportions of the two solvent layers in an extraction process. Suppose, for example, that two liquid solutions with overall composition represented by M in Figure 6.10 are contacted and allowed to settle into a two-phase equilibrium. The compositions will then be given by the tie-line end points L and N, and the relative amounts of L and N will be in the ratio $\overline{MN}/\overline{ML}$. Here again the geometrical construction serves to convey a quick visual notion of the events, which can then be followed up by actual algebraic calculations.

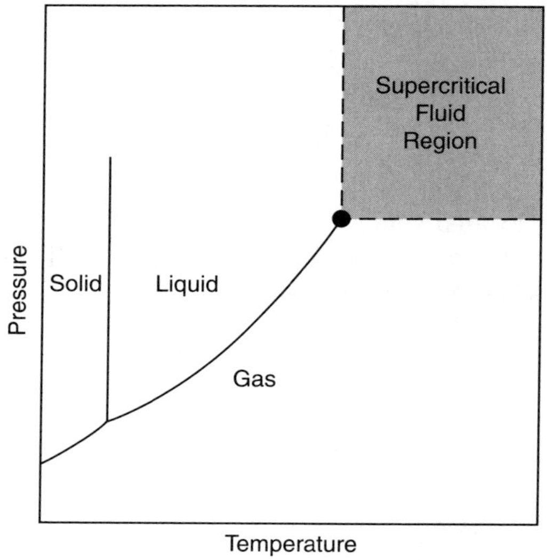

FIGURE 6.13
Temperature–pressure diagram for a pure substance and the region of supercritical fluid extraction.

6.2.5 Equilibria Involving a Supercritical Fluid

In Section 6.1 attention is drawn to the existence of a threshold temperature known as the critical temperature, above which a pure substance can exist only as a single phase. That phase possesses characteristics of a dual nature. It behaves on the one hand like a dense gas that still has a sufficiently open structure to allow rapid passage of solute molecules by diffusion. The relatively short distance between neighboring molecules, on the other hand, allows it to attract and hold substantial amounts of solute material, thus acting for practical purposes as an efficient solvent. These dual features have led to the use of supercritical fluids (SCFs) as solvents in a process known as supercritical fluid extraction (SCE). Its advantages over conventional liquid extraction are twofold: It shortens the required contact time because of the higher prevailing diffusivities, and it replaces costly and potentially harmful liquid solvents by inexpensive and benign gases such as carbon dioxide and the lower hydrocarbons. A number of commercial processes were developed that make use of this new technology, and there is an ongoing intensive quest for new applications.

To acquaint the reader with the key features of SCE, we display the supercritical region of interest (Figure 6.13) and a map of diffusivities in liquids and supercritical carbon dioxide (Figure 6.14), which clearly shows the superior transport properties of the latter. The increased extraction power of carbon dioxide with temperature and pressure becomes evident in Figure 6.15. We note in particular that doubling the pressure from 70 bar

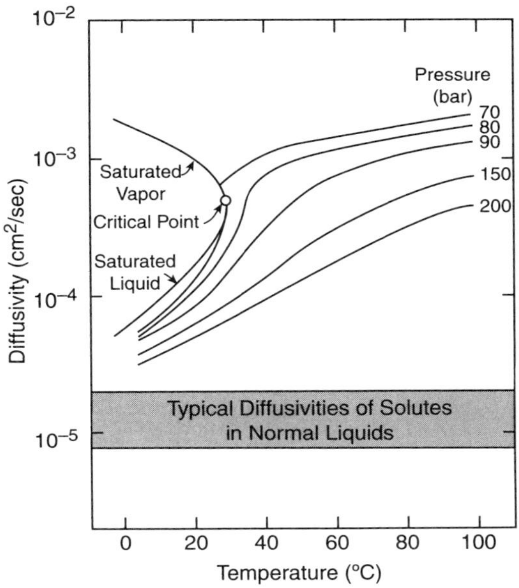

FIGURE 6.14
Comparison of diffusivities in supercritical carbon dioxide and normal liquids.

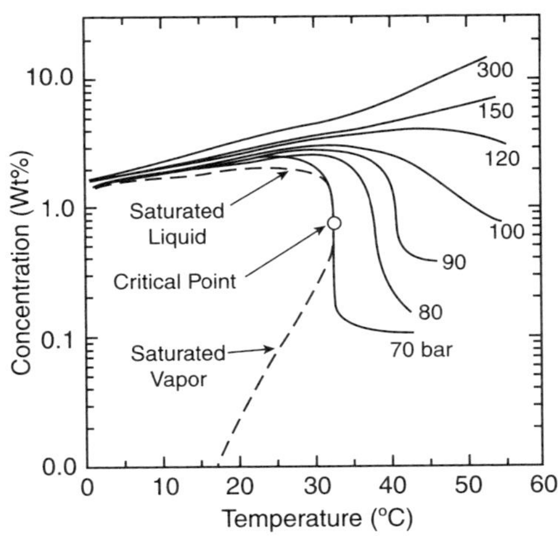

FIGURE 6.15
Solubility of naphthalene in supercritical carbon dioxide.

(~70 atm) to 150 bar at temperatures slightly above critical increases the solute concentration in the extracting medium by well over an order of magnitude.

The earliest indication of this property of SCF is to be found in the work of J.B. Hannay and J. Hogarth (1879), who reported that an increase in pressure of supercritical ethanol caused increased dissolution of certain

FIGURE 6.16
Extraction vessel used in Maxwell House® Coffee Company supercritical CO_2 decaffeination plant in Houston, Texas.

inorganic salts such as potassium iodide, while conversely, a relaxation in pressure resulted in the precipitation of the salts as "a white snow."

This early work was followed intermittently by a flurry of activity that intensified during the 1970s and 1980s, partly as a result of environmental concerns over the use of conventional solvents. Major commercial processes currently in use are the SCE of caffeine from coffee and tea, the SCE of spice aromas, and the fractionation and purification of polymers. Processes under investigation include the treatment of wastewaters, activated carbon regeneration, and the SCE edible oils and therapeutic agents from plant materials.

The scale of decaffeination processes is indicated by the extraction vessel shown in Figure 6.16, which measures about 2 m in diameter and 20 m in height. The operation is countercurrent, with CO_2 passing upward, and coffee being discharged intermittently at the rate of about 5000 kg/h. Operating pressure is typically in the vicinity of 300 atm.

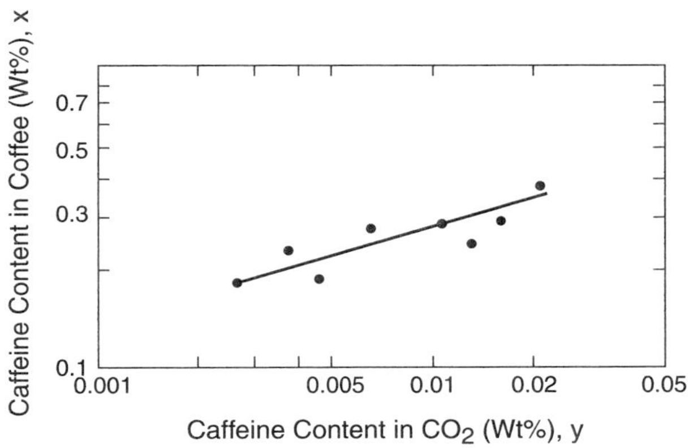

FIGURE 6.17
Caffeine equilibrium distribution at 60°C, 272 atm.

Illustration 6.10: Decaffeination in a Single-Equilibrium Stage

To gain an idea of the CO_2 requirements in decaffeination, we consider a hypothetical process in which the carbon dioxide is circulated through the extraction vessel until it is in equilibrium with its charge of coffee. We term this contact an *equilibrium stage*, a concept discussed more fully in Chapter 7.

The equilibrium distribution of caffeine under a particular set of conditions is shown in the log–log plot of Figure 6.17. The fitting of the data leads to the following expression:

$$x = 1.24\, y^{0.316} \tag{6.11a}$$

where x and y are the weight percentages of caffeine in the coffee and the carbon dioxide, respectively. This form of relation is known as the Freundlich isotherm.

Suppose the requirement is to reduce the caffeine content from 1 to 0.05%, which is typical of commercial decaffeination processes. The task is to calculate the ratio G/S kg CO_2/kg coffee required to achieve this reduction. The carbon dioxide is assumed to be devoid of caffeine initially. The equilibrium content in the gas at the end of the operation is given by

$$y^* = \frac{1}{1.24} 0.05^{1/0.36} \tag{6.11b}$$

$$y^* = 1.96 \times 10^{-4}\% \tag{6.11c}$$

A caffeine mass balance for the process leads to the expression

Rate of caffeine in – Rate of caffeine out = 0

$$S \times 0.01 - (Gy^* + S \times 0.0605) = 0 \tag{6.11d}$$

and, consequently,

$$\frac{G}{S} = \frac{0.0095}{1.96 \times 10^{-4}} \tag{6.11e}$$

$$\frac{G}{S} = 45.9\,\text{kg}\,\text{CO}_2 / \text{kg}\,\text{coffee} \tag{6.11f}$$

Comments

This is clearly an inordinate amount of CO_2 required, brought about by what appears to be a rather unfavorable distribution of caffeine between the extracting gas and the coffee. That seeming drawback can be overcome by "staging" the extraction process (i.e., providing a continuous contact between the two phases as is done in commercial countercurrent operations). This type of operation is taken up in Illustration 8.4. It is shown there that the gas requirements can be considerably reduced by adopting this type of contact, a fact that has been a major contributor to the commercial success of the process.

Supercritical extraction is a marvelous example of how "pure research," even of the most mundane kind, can lead, many years later, to a new technology of major importance. One hopes there are other such hidden gems waiting for an imaginative explorer to discover.

6.2.6 Equilibria in Biology and the Environment: Partitioning of a Solute between Compartments

In biology and the environment, the equilibria of interest are often those of a substance distributed or "partitioned" among the major compartments of the system. In biology we usually seek to establish the distribution of a substance between the circulating blood and the separate compartments of fat or lipids, muscle and bone. The substances in question can be toxic or benign (e.g., a drug).

In the environment, the major compartments of interest are air, water, and soil, and the solute involved is usually toxic or objectionable. An interesting confluence of biology and the environment occurs in the exposure of animals to toxic chemicals contained in the air or in water. The question then becomes one of establishing equilibrium concentrations between environmental and body compartments. In fish, for example, the usual aim is to determine the distribution of the solute between water and animal fat. A description of this case is given in Illustration 6.12.

The equilibrium concentrations between various compartments are usually taken to be in a constant ratio termed the *partition coefficient* K. Thus, for a substance distributed between blood and muscle, we have

$$\frac{C_B}{C_M} = K_{BM} \tag{6.12a}$$

or

$$C_B = K_{BM}\, C_M \tag{6.12b}$$

The relation is, in other words, *linear*, and the partition coefficient can be viewed as the equivalent of a Henry's constant, relating the two concentrations in question. These concentrations represent hypothetical maximum levels, which may appear briefly during the initial period of exposure. Thereafter, metabolic processes and excretion intervene to reduce the concentrations to new steady-state values. These can still be substantial, as was shown in Illustration 5.7.

Illustration 6.11: The Octanol–Water Partition Coefficient

In environmental work, extensive use is made of a special K value, the so-called *octanol–water partition coefficient*. It describes the distribution of a solute between octanol and water — that is, it is defined by

$$K_{OW} = \frac{C_o}{C_w} \tag{6.12c}$$

where K_{OW} is in units of cubic meter water per cubic meter octanol.

Octanol, or more properly 1-octanol, was chosen as a correlating substance because its carbon-to-oxygen ratio is similar to that of lipids and, in general, mimics the dissolution of solutes in organic matter. K_{OC} is also a direct measure of hydrophobicity (i.e., the tendency of a chemical to partition out of water) and is consequently an inverse measure of the solubility of a substance in water. The higher the K_{OC} is, the greater the effect of a chemical on an animal. A short list of K_{OC} values for various substances appears in Table 6.7.

TABLE 6.7

Octanol–Water Partition Coefficients for Various Substances

Solute	K_{OC} (m^3/m^3)
n-Hexane	13,000
Benzene	135
Styrene	760
Chloroform	93
Carbon tetrachloride	440
Chlorobenzene	630
DDT	1.6×10^6
Range of PCBs	$10^4 - 10^8$

The following serves as an example of the application of octanol–water partition coefficients. Suppose it is desired to estimate the effect on fish of the pesticide DDT dissolved in water. The lipid content of most fish clusters about a value of 4.8%. We can then define a fish–water partition coefficient K_{FW} and relate it to K_{OC} as follows:

$$K_{FW} = 0.048\ K_{OW} \tag{6.12d}$$

This relation expresses the assumption that a fish is composed of 4.8% by volume octanol. Using the value for DDT listed in Table 6.7, we obtain

$$K_{FW} = \frac{C_F}{C_W} = 0.0048 \times 1.6 \times 10^6 = 7.7 \times 10^4 \tag{6.12e}$$

that is, an almost *100,000-fold* increase in concentration over that in the surrounding water. Because DDT solubility in water is 1.2×10^{-3} mg/L, the concentration in the fish rises to $7.7 \times 10^4 \times 1.2 \times 10^{-3} = 92$ mg/L, or approximately 1/10 g in a fish of 1-L volume.

Remember that this is the maximum attainable in the absence of metabolic degradation. The steady-state bioconcentration factors (BCFs), some examples of which appeared in Table 5.9, are considerably lower but still sufficiently high to cause concern.

6.3 Multicomponent Equilibria: Distribution of Several Components

6.3.1 The Phase Rule

Note that in the equilibria considered so far, temperature was assumed to be constant or fixed. The wider question of how many such variables have to be prescribed to define a particular state of equilibrium was not addressed. It was tacitly assumed that once a temperature and the concentration or pressure in one phase were chosen, a unique composition in the second phase was automatically ensured. This approach worked without difficulty in the simple equilibria we considered so far, but it becomes somewhat tenuous when more complex systems are to be dealt with. Required here is a formalism that will tell us exactly how many variables have to be fixed to ensure a unique state of equilibrium. That formalism is provided by the Gibbsian phase rule, which states that the number of variables F to be prescribed equals the difference of the number of components and phases $C - P$, plus two. Thus,

$$F = C - P + 2 \tag{6.13a}$$

F, which is also referred to as the degrees of freedom of the system, is a measure of the latitude we have in arbitrarily assigning values to the independent variables of the system. In the case of a two-component liquid solution in equilibrium with its vapor, we have

$$F = C - P + 2 = 2 - 2 + 2 = 2 \tag{6.13b}$$

Thus, we can fix, for example, temperature and the vapor composition, and expect the system to set its own values of liquid composition as well as total pressure. If, on the other hand, we choose to prescribe pressure and vapor composition, the system will respond with a particular temperature (i.e., its boiling point), as well as a particular liquid composition. A third possibility is to fix both temperature and total pressure, in which case the system will set its own values of both liquid and vapor composition. All three cases are encountered in practice and are expressed in terms of appropriate phase diagrams, which are taken up below.

6.3.2 Binary Vapor–Liquid Equilibria

In the phase equilibria considered so far, the principal focus rested on the distribution of a single key component, usually referred to as the solute, between the constituent phases. Thus, in the gas–liquid and liquid–solid equilibria discussed in Sections 6.2.1 and 6.2.2, our concern was with only one of the components present, while the remaining bulk components, such as liquid solvent or solid adsorbent, were left out of consideration.

In vapor-liquid equilibria, all components participate in some measure in determining equilibrium behavior. No single substance dominates the picture, and none are relegated to the status of a passive constituent. Although these considerations apply to multicomponent vapor–liquid equilibria, our discussions here will be confined for illustrative purposes to binary systems. These systems possess two degrees of freedom, and this fact led to the development of several distinct phase diagrams. In these diagrams, discussed in detail below, either temperature or pressure have prescribed constant values, along with one of the phase compositions. Once these two variables are fixed, the system is uniquely defined, and all other variables fall into place. The graphical representations that result all have distinct features that are exploited in different ways to suit the needs of the user.

6.3.2.1 Phase Diagrams

There are three principal phase diagrams used to describe binary vapor-liquid equilibria. In the first, total pressure and one of the compositions are the prescribed variables. This leads to the boiling-point diagram, shown in Figure 6.18a, and the compositional x–y diagram that appears in Figure 6.18c. When total pressure is replaced by temperature as the prescribed variable, a third type of diagram results — the *vapor-pressure diagram*, shown in Figure 6.18b.

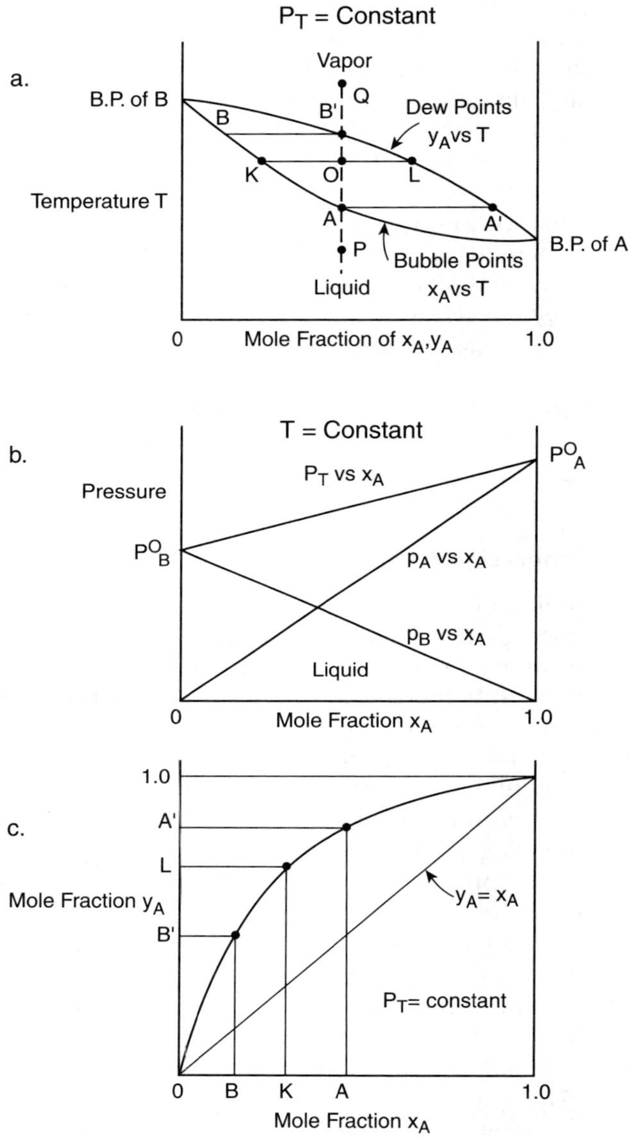

FIGURE 6.18
Binary vapor–liquid equilibria of ideal systems: (a) boiling-point diagram; (b) vapor-pressure diagram; and (c) x–y diagram.

The boiling-point diagram of Figure 6.18a provides the best global representation of binary vapor–liquid equilibria. It consists of a plot of temperature against both the vapor and liquid compositions, expressed as mole fractions. The lens-shaped domain encompasses the two-phase region, while the spaces above and below denote single-phase vapor and liquid behavior, respectively.

Consider first a point P below the lens. This point lies entirely in the liquid region. The system is below the boiling point corresponding to the prescribed total pressure P_T, and the vapor phase is completely absent. If we next move up the vertical axis PQ (i.e., raise the temperature), a point A is eventually reached where the first bubble of vapor is formed. That point, which lies on the *bubble–point curve*, is in equilibrium with a vapor of composition A', which is richer in the component A than the bulk composition of the parent liquid. As more of the mixture is vaporized, more vapor forms at the expense of liquid, giving rise, for example, to the liquid composition corresponding to the point K and a vapor composition denoted by L. The line connecting K and L, and similar horizontal lines connecting points on the two curves, are termed *tie-lines*. They play the same role as the tie-lines we have seen in liquid–liquid equilibria, and they obey the lever rule we derived in Illustration 6.10. Thus, the moles vapor V and the moles liquid L present in the two-phase mixture stand in the ratio of the corresponding line segments — that is, we have

$$\frac{V}{L} = \frac{\overline{KO}}{\overline{LO}} \tag{6.14}$$

Continuing our path along the PQ axis, a point B is eventually reached where the last drop of liquid is vaporized. Further increases in temperature result in superheated vapor. If the process is now reversed (i.e., the vapor mixture is cooled), all the phenomena reappear in reverse order. Condensation, for example, starts at point B, yielding the first drop of liquid "dew." The upper curve of the lens where this occurs is consequently termed the *dew-point curve*. Note the similarity of this diagram to the melting-point diagram for liquid–solid systems shown in Figure 2.9a.

When temperature replaces total pressure as one of the prescribed variables, the result is the vapor-pressure diagram, examples of which are shown in Figure 6.18b and Figure 6.19b (see Section 6.3.3). The plot here is one of vapor pressure at constant T against one of the liquid mole fractions. Both the component partial pressures P_A and P_B and the total pressure P_T are plotted and result in straight lines or curves, depending on whether the system is ideal or nonideal. Ideal behavior and deviations from it are taken up in greater detail in the next section.

A third type of plot, the *x–y diagram*, is shown in Figure 6.18c as well as in Figure 6.21c. The plotted quantities here are the vapor and liquid mole fractions of a particular component, while total pressure is held constant at a prescribed level. These diagrams play a *dominant* role in the analysis of distillation processes and will be encountered in considerable numbers in succeeding chapters.

6.3.2.2 Ideal Solutions and Raoult's Law: Deviation from Ideality

The prediction and, failing that, the correlation of vapor–liquid equilibria is a topic of considerable practical interest. In particular, we wish to address the following question: Given a prescribed liquid composition and certain standard physical properties of a system, is it possible to predict the corresponding vapor composition? Evidently, if this could be done, an immense amount of experimentation could be dispensed with. It turns out that this is accomplished most easily if the system shows what is termed *ideal behavior.* Such ideal systems, although relatively rare in practice, serve as a convenient reference for vapor–liquid equilibria in general. This is reminiscent of the concept of an ideal gas, which provides a similar yardstick against which the behavior of gases in general can be measured. These two cases, although similar in the approach used, differ in some important aspects.

For ideal gases, a total absence of intermolecular forces is assumed. Neither attractive nor repulsive forces are taken into account, and collisions between molecules are taken to be entirely elastic in nature. These assumptions hold well at low pressures because of the large distances between particles and the vanishingly short time spent in collision with each other. In liquids, the molecules are closely packed and in constant intimate contact with each other. To neglect intermolecular forces in this instance would be grossly unrealistic. What is done instead is to accept their existence and to assume that they are uniformly constant and independent of concentration. Thus, if in a binary solution a molecule A is replaced by a molecule B, the interaction between neighboring particles remains the same. This will clearly be the case only if the two species are similar in chemical structure and in size. The important consequence of this assumption is that the partial vapor pressure of a particular component is unaffected by the presence of other species and is subject only to its own molar concentration. Thus, if the mole fraction x_A is doubled, the partial pressure p_A of the component is similarly doubled. Continuing this process, a linear increase of partial pressure with mole fraction results until ultimately, at $x_A = 1$, the partial pressure equals the full pure component vapor pressure $P_A°$ of the species. We can consequently write

$$p_A = x_A P_A° \tag{6.15a}$$

or

$$y_A P_T = x_A P_A° \tag{6.15b}$$

where the product $x_A P_A°$ expresses both the linear increase in partial pressure with mole fraction and the ultimate attainment of the full vapor pressure $P_A°$. This expression is known as Raoult's law, and systems obeying it are said to be *ideal solutions*.

Several important subsidiary laws flow from this relation. In the first instance, we can extend Raoult's law to the second component of the binary mixture and obtain

$$p_B = x_B P_B^o = (1 - x_A)P_B^o \tag{6.15c}$$

Adding the two expressions then yields

$$p_A + p_B = P_T = x_A P_A^o + (1 - x_A)P_B^o \tag{6.15d}$$

where P_T is the total pressure.

All three of these expressions yield straight-line plots of vapor pressure versus liquid mole fraction, which are displayed in Figure 6.18b. Systems that obey these relations (i.e., ideal solutions) are relatively uncommon and are usually confined to neighboring substances taken from a homologous series and to isomer and isotope mixtures. We will address these in the context of the so-called separation factor of the system, which is taken up in a subsequent section.

Deviations from ideal behavior occur when there is a marked difference in the molecular structure of the participating species. Typical combinations that give rise to nonideal behavior are pairs of polar and nonpolar substances, in which the attractive and repulsive forces vary with compositional changes. When repulsive forces predominate, the vapor pressures rise above the values predicted by ideal solution theory, and one speaks of *positive deviations*. Most vapor–liquid equilibria fall in this category. A predominance of attractive forces, on the other hand, leads to a lowering of the vapor pressure, which is termed a *negative deviation*. The resulting nonideal vapor-pressure diagrams are displayed in Figure 6.19a and Figure 6.19b.

We note two interesting limiting cases that arise in systems with both positive and negative deviations. For low concentrations of a particular component, its partial pressure becomes linear in the mole fraction of the solution in question. This, in essence, is a Henry's-law-type relation — that is, one can write

$$\left(p_A\right)_{x_A \to 0} = H_A x_A \tag{6.16a}$$

and

$$\left(p_B\right)_{x_B \to 0} = H_B x_B \tag{6.16b}$$

where H_A and H_B are the Henry's law constants for the two components in question. If, on the other hand, the mole fractions are allowed to tend to unity, the partial pressure will, in the limit, approach the values given by Raoult's law. We have, in this case,

$$\left(p_A\right)_{x_A \to 1} = x_A P_A^o \tag{6.16c}$$

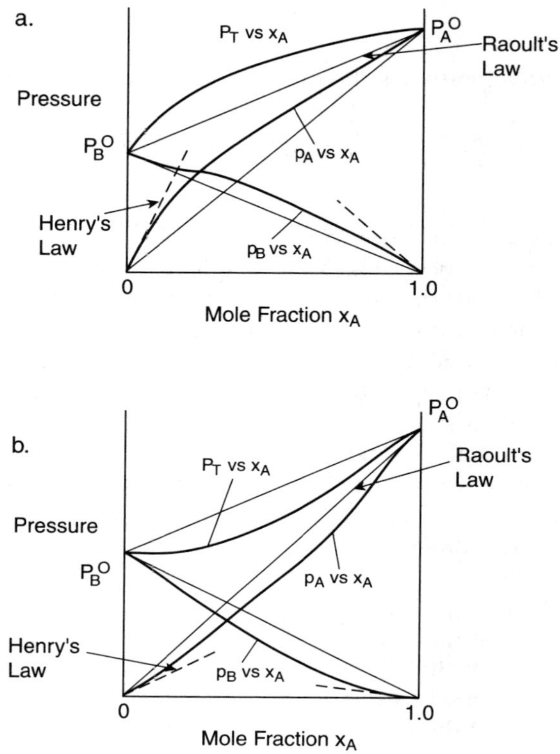

FIGURE 6.19
Vapor-pressure diagrams for nonideal systems: (a) positive deviations from ideality and (b) negative deviations.

and

$$\left(p_B\right)_{x_B \to 1} = x_B P_B^{\ o} \tag{6.16d}$$

These two cases, which have been entered in the diagrams of Figure 6.18, provide useful approximations in the limit of low-solute concentrations or high-solvent content.

6.3.2.3 *Activity Coefficients*

Although there are more formal ways of defining activity coefficients based on chemical potentials, we prefer the following simple route.

Recall that deviations from ideal gas behavior are often expressed in terms of a correction factor referred to as a compressibility factor z. We write

$$PV = z(P_r, T_r)RT \tag{6.17}$$

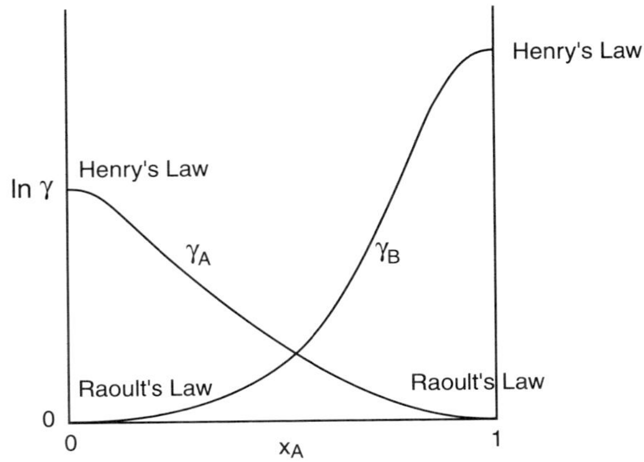

FIGURE 6.20
Activity coefficient for systems with positive deviations.

where P_r and T_r are the so-called reduced pressure and temperature, defined as the ratios of P and T to their critical counterparts P_C and T_C.

In much the same way one can define correction factors γ, termed *activity coefficients*, which describe deviations from Raoult's law for ideal solutions. The resulting expression, sometimes referred to as extended Raoult's law, has the following form:

$$p_A = \gamma_A(x_A)x_A{}^{o}P_A{}^{o} \tag{6.18a}$$

and

$$p_B = \gamma_B(x_B)x_BP_B{}^{o} \tag{6.18b}$$

where the correction factors are now functions of the binary mole fraction x. That functional relationship, expressed in the form of plots of ln γ versus x_A, is shown in Figure 6.20. Note that in the limit of high-solvent content, ln γ approaches zero (i.e., $\gamma \rightarrow 1$). At the same time, Equation 6.18a and Equation 6.18b revert to Raoult's law. At the other end of the spectrum (i.e., at low-solute content), the activity coefficients tend to constant limiting values and Henry's law results, where $H = \gamma p^{o}$.

It is evidently of considerable interest to be able to predict activity coefficients without resorting to experimentation, and immense strides have in fact been made in recent decades to accomplish this goal. Among a number of promising approaches, an analytical expression known as the UNIQUAC equation (UNIversal QUAsi-Chemical equation) has received the most widespread acceptance. In this model, the activity coefficient is decomposed into two constituents, one of which, termed *combinatorial* (C), accounts for molecular size and shape differences, while the other, denoted *residual* (R), expresses effects due to molecular interactions.

Thus,

$$\ln \gamma = \ln \gamma^C + \ln \gamma^R \tag{6.19}$$

It has become possible to express γ^C and γ^R in terms of group contributions due to molecular subunits such as CH_3, CH_2, CH_2O, and so forth. This approach has led to the successful prediction of vapor–liquid equilibria of a large number of systems. There is evidently still an ongoing need for experimentation, but this requirement is now at a much more subdued level than would otherwise be the case.

6.3.3 The Separation Factor α: Azeotropes

A convenient and consistent measure of separation is provided by the so-called *separation factor* or *relative volatility* α, which is composed of the *product* of two mole fraction ratios. These ratios are defined in such a way as to minimize changes in α with composition and are represented by the following expression:

$$\alpha = \frac{y}{x}\frac{(1-x)}{(1-y)} \tag{6.20a}$$

Here x and y refer to the more volatile component (i.e., the constituent with the higher pure-component vapor pressure or the lower boiling point). In general, the higher the value of α above unity, the greater the degree of separation or enrichment.

Note that any decrease in y/x toward 1 as y and x approach unity is neatly offset by the compensating ratio $(1 - x)/(1 - y)$. Thus, if $y = 0.99$ and $x = 0.98$, for example, α will still be considerably above unity:

$$\alpha = \frac{0.99}{0.98}\frac{1-0.98}{1-0.99} = 2.02 \tag{6.20b}$$

Consequently, the use of α predicts, correctly, that there is still substantial separation to be obtained even when the mole fractions are near unity.

The essence of the relative volatility is best understood by examining its effect on the x–y diagrams shown in Figure 6.18c and Figure 6.21c. It turns out that α bears a distinct and sensitive relation to the shape of these curves. The following features in particular stand out:

1. Separation factors greater than unity result in x–y curves that lie entirely above the 45° diagonal. The higher the value of α, the greater the distance is between the two lines. This implies that separation becomes easier as the x–y curve bulges out and away from the

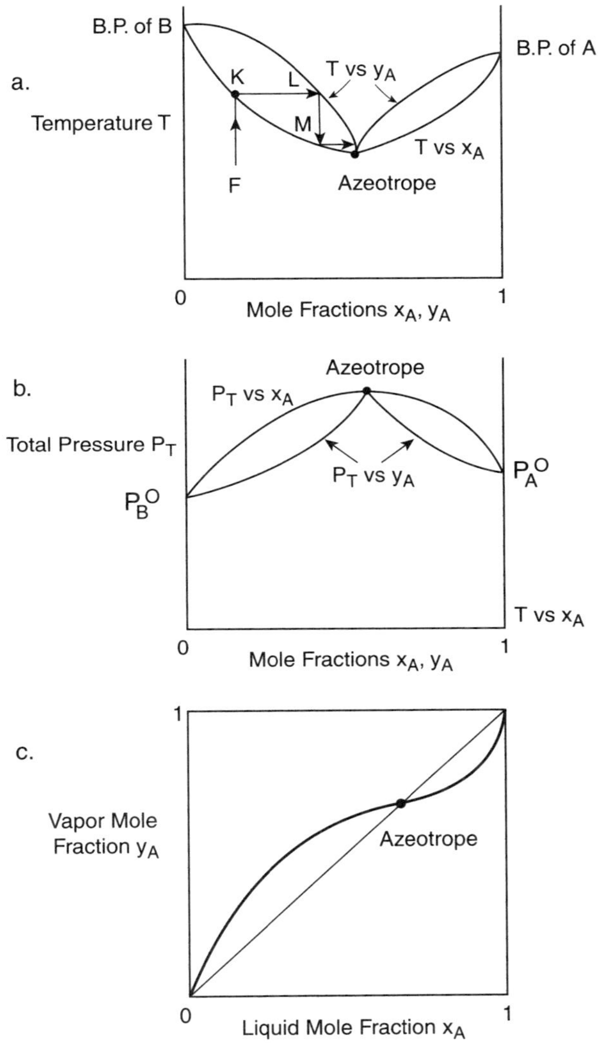

FIGURE 6.21
Azeotropic systems: (a) boiling-point diagram; (b) vapor-pressure diagram; and (c) x–y diagram.

diagonal. Systems with x–y curves close to the diagonal are by contrast difficult to separate. Figure 6.22 showing x–y plots for various values of α conveys a sense of these features.

2. Separation factors that are constant, or nearly so, result in symmetrical x–y curves. Larger variations yield asymmetrical curves, with some portions lying far above the diagonal while others are close to it.
3. When $\alpha = 1$, vapor and liquid compositions become equal (i.e., $x = y$), and the corresponding mole fraction lies on the 45° diagonal. This

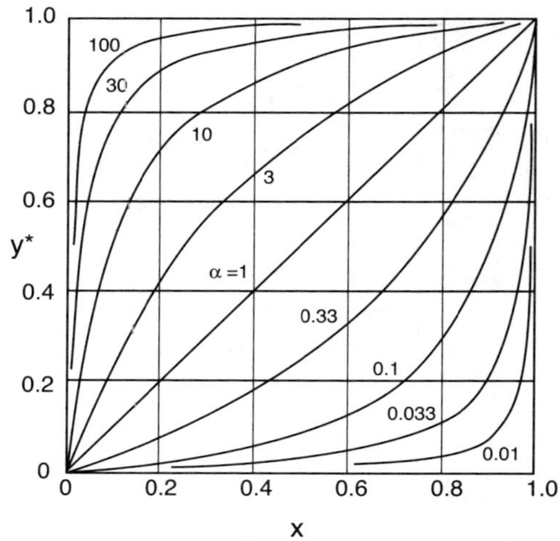

FIGURE 6.22
Vapor–liquid equilibrium compositions for various values of the separation factor . The greater the distance from the diagonal $\alpha = 1$, the easier the separation by distillation. $\alpha > 1$: More volatile component enriched in the vapor phase. $\alpha < 1$: More volatile component enriched in the liquid phase. $\alpha = 1$: No separation.

condition is referred to as an *azeotrope* and is shown in Figure 6.21c. The x–y curves for such systems will inflect across the 45° line, with portions on either side of the azeotrope lying above and below the diagonal, respectively. The portion below the line yields enrichment in the less-volatile component. Above the diagonal, the more-volatile component is enriched.

Azeotropic behavior is a common occurrence in vapor–liquid equilibria (VLE). Approximately one third of all systems listed in standard VLE handbooks exhibit azeotropes. In general, the more dissimilar the two components are, the greater the likelihood that such mixtures will be formed. Combinations of polar-nonpolar substances and those with widely differing structural features are particularly prone to azeotropic behavior. Table 6.8, which lists some of the more conventional azeotropic pairs, conveys a sense of the degree of dissimilarity which leads to the formation of azeotropes.

Several additional features of these systems are to be noted. First is the fact that mixtures that lead to azeotrope formation cannot be separated into their constituent components by simple distillation. This is best seen from the boiling-point diagram, where we indicate the pathway that results from a process of repeated vaporization and condensation. Starting with a liquid feed at F, the mixture is brought to a boil at K, and the first vapor (L) is collected and condensed. The cycle of vaporization and condensation is then repeated until the azeotropic composition at A is reached. At this point, no further enrichment by vaporization is obtained, as the compositions in the two phases are the same.

TABLE 6.8

Binary Azeotropes at 1 atm

System	Azeotrope Mole Fraction $x_1 = y_1$	Temperature, T (°C)
Water (1)–Hydrochloric acid (2)	0.889 (20.2 wt% HCl)	108.6
Water (1)–Sulfuric acid (2)	0.081 (98.4 wt% H_2SO_4)	23.5
Water (1)–Methanol (2)	—	—
Water (1)–Ethanol (2)	0.1057 (95.6 wt% EtOH)	78.15
Water (1)–Propanol (2)	0.5680	87.80
Water (1)–*i*-Propanol (2)	0.3187	80.16
Water (1)–*n*-Butanol (2)	0.75	92.95
Water (1)–Formic acid (2)	0.415	107.60
Water (1)–Acetic acid (2)	—	—
Water (1)–Dioxane (2)	0.525	87.75
Water (1)–Tetrahydrofurane (2)	0.18	63.5
Water (1)–Pyridine (2)	0.77	94.40
Methanol (1)–Acetone (2)	0.1980	54.40
Methanol (1)–Methyl acetate (2)	0.3280	53.80
Methanol (1)–Ethyl acetate (2)	0.287	62.1
Methanol (1)–Chloroform (2)	0.35	53.5
Methanol (1)–Carbon tetrachloride (2)	0.55	55.7
Methanol (1)–Hexane (2)	0.498	45
Methanol (1)–Heptane (2)	0.747	58.8
Methanol (1)–Benzene (2)	0.60	57.8
Methanol (1)–Toluene (2)	0.90	63.6
Ethanol (1)–Ethyl acetate (2)	0.48	72.1
Ethanol (1)–Chloroform (2)	Yes	
Ethanol (1)–Carbon tetrachloride (2)	Yes	
Ethanol (1)–Hexane (2)	0.345	58.0
Ethanol (1)–Heptane (2)	Yes	
Ethanol (1)–Benzene (2)	0.445	67.8
Ethanol (1)–Toluene (2)	Yes	
Acetone (1)–Chloroform (2)	0.3600	64.4
Acetone (1)–Carbon tetrachloride (2)	0.93	56.0
Acetone (1)–Carbon disulfide (2)	0.34	39.10
Acetone (1)–Pentane (2)	0.25	32
Acetic acid (1)–Octane (2)	0.675	105.1
Benzene (1)–Cyclohexane (2)	0.55	77.5

The mixture continues to boil at a constant temperature, yielding a mixture of constant composition, until the liquid charge is exhausted.

Azeotropic mixtures require special methods for their separation, which usually consist of adding a third component that has the ability to "break" the azeotrope. Perhaps the most famous case is that of ethanol–water, which has an azeotropic mole fraction in ethanol of 0.8943 at atmospheric pressure (Table 6.8). Here the added component can be benzene and results, on distillation, in the recovery of pure ethanol and a ternary azeotrope containing benzene. That mixture, on condensation, results in two immiscible aqueous and organic layers, which are separated and further processed by distillation. Changing the operating pressure P_T is yet another potentially useful way of breaking an azeotrope. This is taken up in Illustration 6.14.

We now turn to the question of the behavior of α for ideal liquid solutions. Here the Raoult's law equations (Equation 6.15a and Equation 6.15b) provide a quick answer. Dividing the two expressions, we obtain in the first instance,

$$\frac{p_A}{p_B} = \frac{x_A P_A^{\,o}}{(1-x_A)P_B^{\,o}} \tag{6.21a}$$

Replacing partial pressures by yP_T yields

$$\frac{y_A P_T}{(1-y_A)P_T} = \frac{x_A}{(1-x_A)}\frac{P_A^{\,o}}{P_B^{\,o}} \tag{6.21b}$$

and after some rearrangement,

$$\alpha = \frac{y_A}{x_A}\frac{(1-x_A)}{(1-y_A)} = \frac{P_A^{\,o}}{P_B^{\,o}} \tag{6.21c}$$

Thus, for ideal solutions, α stands in the simple ratio of the vapor pressure of the more-volatile component to that of its less-volatile counterpart. This ratio varies somewhat with temperature, but the variation is not severe and is often accounted for by composing the arithmetic or geometric average of the two end values at $x = 0$ and $x = 1$.

To provide a sense of the validity of Equation 6.21c, we composed in Table 6.9 a comparison of calculated values with the experimental range of α values taken over the span of measured mole fractions. The systems involved are those that can be expected to behave ideally (i.e., members of a homologous series, isotope pairs, and the like). For isotope mixtures, experimental α values are rare, and we have instead reported the vapor-pressure ratio, which for these substances is nearly constant and equal to the relative volatility.

Illustration 6.12: The Effect of Total Pressure on α, Part 1: Ideal Solutions

In principle, distillation can be carried out over a wide range of pressures, spanning the extremes of near-total vacuum and the critical point. While atmospheric distillation is the preferred mode of operation, reduced pressures offer the advantage of lower temperatures and energy requirements. They are often employed to avoid thermal degradation of heat-sensitive materials (ethylbenzene/styrene distillation) or are forced upon the operator by the low vapor pressure of the charge (metal distillation). The drawbacks of this mode of operation are the obvious cost and inconvenience

TABLE 6.9

Separation Factors for Some Near-Ideal Systems at 1 atm

System	Experimental Range of α	Range of P_1^o/P_2^o
Methanol (1)–Ethanol (2)	2.20–1.88	1.67–1.76
Methanol (1)–Propanol (2)	3.55–3.30	3.20–3.97
Methanol (1)–*i*-Propanol (2)	2.34–1.64	1.91–2.17
Ethanol (1)–*i*-Propanol (2)	1.15–1.21	1.17–1.27
Ethanol (1)–*n*-Butanol (2)	4.21–5.84	3.95–5.92
Ethanol (1)–sec Butanol (2)	2.48–2.54	2.20–2.37
Ethanol (1)–1-Pentanol (2)	8.96–7.45	7.15–10.9
Formic acid (1)–Acetic acid (2)	1.65–1.41	1.62–1.72
Chloroform (1)–Carbon tetrachloride (2)	1.41–2.11	1.62–1.72
Acetone (1)–Methyl ethyl ketone (2)	1.56–2.23	1.71–2.22
Benzene (1)–Toluene (2)	2.23–2.26	2.36–2.60
Nitrogen (1)–Oxygen (2)	4.03 (mean)	3.8–5
Isotopes	*T, K*	P_1^o/P_2^o
H_2–HD	20.4	1.73
$C^{12}O$–$C^{13}O$	68.3	1.0109
$O^{16}O^{16}$–$O^{16}O^{18}$	69.5	1.01
CH_4–CH_3D	90	1.0025
$N^{14}O$–$N^{15}O$	109.5	1.032
H_2O–HDO	373.1	1.026

of a vacuum process and the lowering of material throughput. Distillation above atmospheric pressures is often employed with volatile charges containing dissolved gases and is routinely used in refinery operations. It shows superior throughput but has higher operating temperatures and energy requirements than atmospheric distillation. A change in operating pressure may also be beneficial in breaking an azeotrope. This is taken up in the next illustration.

An important factor to be taken into account in any overall assessment is the effect of total pressure on the separation factor α. We can explore this dependence for the simple case of ideal solutions by first drawing on Equation 6.21c:

$$\alpha = \frac{y_1(1-x_1)}{x_1(1-y_1)} = \frac{P_1^0}{P_2^0} \tag{6.21c}$$

and then using the Clausius–Clapeyron equation (Equation 6.1g) to relate P^o to T:

$$\log \frac{P_1^0}{P_2^0} = (B_2 - B_1)/T + (A_1 - A_2)$$

It is clearly seen from the latter that the vapor-pressure ratio and hence the separation factor both increase with a decrease in boiling point or total pressure. This result has sometimes been cast into a sweeping rule-of-thumb that low-pressure distillation leads to improved separation. Many nonideal systems follow this rule, but there are also numerous exceptions, particularly in the azeotropic category. Nevertheless, low-pressure distillation is a worthwhile alternative to explore, provided some low-pressure equilibrium data are available to confirm the expected results.

Comments

Even ideal systems have been known to defy the rules. The isotope pair CH_4–CH_3D, for example, has a separation factor of 1.001 at approximately 100 mmHg, which *rises* to 1.0035 at 1 atm. This anomaly led to the suggestion that heavy water (D_2O) might be produced economically by the high-pressure distillation of liquefied natural gas, which contains the isotope CH_3D. Because the capital cost of such a plant varies inversely with $\alpha - 1$ (see Illustration 7.9), even small increases in α can be hugely beneficial. Some exploratory work took place in the 1960s, and although the results appeared promising, market forces intervened to bring the proceedings to a halt. The current large-scale use of heavy water is confined to the CANDU nuclear reactor, which faces an uncertain future.

Illustration 6.13: The Effect of Total Pressure on α, Part 2: Breaking an Azeotrope

Consider an azeotropic system of the type shown in the x–y diagram of Figure 6.22. Below the azeotropic pint, the more volatile component is enriched in the vapor phase ($\alpha > 1$); above it, the reverse is true ($\alpha < 1$). The ethanol–water system falls in this category. We can enrich dilute alcohol solutions by distillation up to the azeotropic point, but not beyond it.

When the operating pressure is changed, one of two situations can arise. The azeotropic point moves upward and, given a sufficient change in pressure, reaches a mole fraction of $x_A = 1$ in the volatile component. Or it can move downward until it reaches the position $x_A = 0$. In both cases, a complete separation of the components by distillation is then possible. For ethanol–water, the azeotropic point is closer to the position $x_A = 1$ (see Table 6.8), so one would prefer it to move up rather than down. What we wish to explore is whether this can be achieved by lowering the operating pressure (i.e., the boiling point of the mixture). The procedure consists of three steps.

We extract activity coefficients from the given azeotropic data and set a new boiling point, say 40°C, but keep liquid composition constant at $x_A = 0.8943$. This is followed by a calculation of total pressure at 40°C using the nonideal version of Equation 6.15d. The vapor composition y_A follows immediately from the ratio $y_A = p_A/P_T$ and leads directly to the new separation factor α at 40°C. If that value is greater than 1, we are on the way to success. The required pure component vapor pressures, calculated from Table 6.1, are as follows:

	T = 78.15°C	T = 40°C
p^o_{EtOH}(mmHg)	755.6	134.3
$p^o_{H_2O}$(mmHg)	329.5	55.23

Using the azeotropic data, one obtains the following for the activity coefficients:

$$\gamma_1 = \gamma_{EtOH} = \frac{y_1}{x_1}\frac{P_T}{P^o_1} = \frac{760}{755.6} = 1.006 \tag{6.22a}$$

$$\gamma_2 = \gamma_{H_2O} = \frac{y_2}{x_2}\frac{P_T}{P^o_2} = \frac{760}{329.5} = 2.307 \tag{6.22b}$$

Note that $_{EtOH}$ is close to the Raoult's law value of 1, as required (see Figure 6.21).

Total pressure P_T at 40°C is given by the sum of partial pressures — that is,

$$P_T = p_1 + p_2 = \gamma_1 x_1 P_1^o + \gamma_2 x_2 P_2^o \tag{6.22c}$$

and, therefore,

$$P_T = 1.006 \times 0.8943 \times 134.32 + 2.307 \times 0.1057 \times 55.23$$

$$= 122.843 + 13.467 = 134.31 \text{ mmHg} \tag{6.22d}$$

The new ethanol mole fraction at 40°C is then given by

$$y_1 = \frac{p_1}{P_T} = \frac{122.843}{134.31} = 0.900 \tag{6.22e}$$

from which there follows a value for the separation factor of

$$\alpha = \frac{y_1}{x_1}\frac{(1-x_1)}{(1-y_1)} = \frac{0.9}{0.8943}\frac{0.1057}{0.1} \tag{6.22f}$$

$$\alpha = 1.064 \tag{6.22g}$$

Comments

Although we achieved our goal of raising α above 1, the low value of α would probably make the process unattractive for large-scale applications. We have, however, confirmed the feasibility of breaking an azeotrope with a modest change in operating pressure.

Illustration 6.14: Activity Coefficients from Solubilities

Activity coefficients are generally determined either from Equation 6.18a or Equation 6.18b using measured vapor- and liquid-equilibrium compositions or are estimated from the UNIQUAC equation mentioned earlier. A third method arises when the two components in question have low mutual solubilities. γ values are then confined to the Henry's law regions near $x = 0$ and $x = 1$ (Figure 6.20) and cease to exist outside those regions. This provides a way of calculating activity coefficients using classical chemical thermodynamics.

Suppose an organic solvent with a low solubility is equilibrated with water. The aqueous phase will then contain small amounts of the solvent, while water will appear in similar amounts in the organic phase. Because the concentrations involved are vanishingly small, we can approximate the chemical potential of each major constituent by its *pure component* chemical potential μ^{o} $(T.P)$. We can write the following:

$$\mu^{o}_{H_2O}(T.P) \cong \mu^{o}_{H_2O}(T.P) + RT(\ln \gamma\, x_s)_{H_2O} \tag{6.23a}$$

and

$$\mu^{o}{}_{org} \cong \mu^{o}{}_{org}(T.P) \cong \mu^{o}{}_{org}(T.P) + RT(\ln \gamma\, x_s)_{org} \tag{6.23b}$$

This leads to the result

$$\gamma_{org} \cong 1/(x_s)_{org} \tag{6.23c}$$

and

$$\gamma_{H_2O} \cong 1/(x_s)_{H_2O} \tag{6.23d}$$

where x_s represents the saturation solubility, expressed as a mole fraction.

We illustrate these results by examining the system of carbon tetrachloride–water. The solubility of CCl_4 in water is listed in Table 6.3 as 0.8 g/L ≅ 800 ppm. The solubility of water in carbon tetrachloride is almost 10 times lower at 84 ppm. Converting to mole fraction, we obtain

$$(x_s)_{\text{CCl}_4} \cong 800(18/154)10^{-6} = 9.4 \times 10^{-5}$$

and

$$(x_s)_{\text{H}_2\text{O}} \cong 804(154/18)10^{-6} = 7.2 \times 10^{-4}$$

Taking the inverse of these results, we obtain the following for the activity coefficients:

In the aqueous phase:

$$\gamma_{CCl_4} \cong 11,000$$

In the organic phase:

$$\gamma_{H_2O} \cong 1400$$

These are enormously high values, which indicates that the equilibrium vapor mole fraction will by far exceed that of the liquid (see Practice Problem 6.13).

There is another simplification to be made. Because γx_s is very nearly unity, partial pressures will be closely approximated by the pure component vapor pressures, which can be computed from the Antoine constants of Table 6.1. We have the following:

For the aqueous phase:

$$p_{\text{CCl}_4} = (\gamma\, x_s\, P^o)_{\text{CCl}_4} \cong P^o{}_{\text{CCl}_4} = 86.8\,\text{mmHg}$$

For the organic phase:

$$p_{H_2O} = (\gamma\, x_s\, P^o)_{\text{H}_2\text{O}} \cong P^o{}_{\text{H}_2\text{O}} = 23.6\,\text{mmHg}$$

The results for CCl_4 are particularly noteworthy in an environmental context. The high partial pressure indicates that a water basin contaminated with carbon tetrachloride will quickly lose most of the contaminant to the atmosphere by evaporation. We return to this topic in Illustration 7.4.

Practice Problems

6.1 The Vapor Pressure of Ice and Snow

Ice has a substantial vapor pressure, which causes it to evaporate at surprisingly high rates. Most of the disappearance of snow during a lull in precipitation is due to evaporation rather than melting.

a. Suppose that a shallow puddle of water with a surface temperature of 20°C evaporates in 2 h during a dry, windy day. How long would it take the same amount of ice at 5°C to evaporate under identical wind and humidity conditions? The vapor pressures of the water and the ice at their respective temperatures are 17.5 and 3.0 mmHg.
b. Estimate the rate of evaporation of ice.

Answer:

11.7 h, 0.032 g/m^2s

6.2 Critical Points of a Binary Liquid Mixture

Apply the same physical reasoning as in Section 6.1 to establish the behavior in the critical region of a two-component system.

6.3 Variation of Latent Heat of Vaporization with Temperature

Show, by means of thermodynamic arguments, that the latent heat of vaporization must in general vary with temperature.

6.4 Enrichment of Oxygen by Absorption and Adsorption (PSA)

Calculate the enrichment in oxygen that results when air is absorbed in water at 25°C and 20 atm and is subsequently recovered by depressurization.

In a commercial process known as *pressure swing absorption* (PSA), water is replaced by a zeolitic adsorbent that has the capability of almost completely excluding nitrogen from its pore space. Discuss the procedures to be used in such a process and its advantages.

6.5 Sea Salt by Evaporation

Sea salt contains 454 mmol/kg NaCl and 9.6 mmol/kg KCl. What percentage of the water has to be evaporated for the first crystals of KCl to precipitate? The total solids content of seawater is approximately 2%. (*Hint*: Consult Table 6.3.)

Answer:

57.1%

6.6 More about Adsorption: The Toth Equation Applied to Air Purification

The Langmuir isotherm, while adequate for many purposes, has since its inception been superseded by more refined equations with a broader range of applications. One such equation is that due to Toth, which is given by

$$X = \frac{ap}{(b + p^C)^{1/C}}$$

Consider a factory air space of 10^5 m^3 that is regularly contaminated with 10 ppm by volume hydrogen sulfide, a toxic and foul-smelling gas. It is proposed to use adsorption to purify the air. Activated carbon does not lend itself to this purpose because of its low capacity, but the zeolite *mordenite* has a suitably high loading factor represented by the following Toth isotherm:

$$X = \frac{4.5675\,p}{[0.3252 + p^{0.2425}]^{1/0.2425}}$$

where p is in units of kPa, and q is in moles/kg adsorbed. Calculate the minimum weight of mordenite required to purify the air.

Answer:

561 kg

6.7 Ion-Exchange Resins

The process of ion exchange relies on the use of synthetic organic resins, which carry ionic groups capable of exchanging cations or anions with similar species in an aqueous solution. The resins come in the form of small beads (2 to 3 mm) made up of a polymeric skeleton into which the ionic groups are introduced by suitable chemical reactions. In one version of this process, styrene is polymerized in emulsion to provide the resin matrix, which is then sulfonated, thereby introducing sulfonic acid groups $-SO_3H^+$ into the polymer skeleton. We can describe this process in the following form:

$$RSO_3^-H^+ + M^+ \underset{\rightarrow}{\overset{\leftarrow}{}} RSO_3^-M^+ + H^+$$

The hydrogen ion of this group can be readily released and its place taken by another cation from a neighboring aqueous solution. Anion exchangers use amine groups R_3N^+OH to affect a similar exchange of anions. Thus,

$$-R_3N^+OH^- + A^- \underset{\rightarrow}{\overset{\leftarrow}{}} -R_3N^+A^- + OH^-$$

Ion-exchange resins find extensive use in the deionization of water and in the recovery of valuable ions from aqueous solutions. In water-softening processes, for example, calcium ions contained in the hard water are exchanged for hydrogen ions released by the sulfonic acid groups, while the anions are replaced by hydroxyl ions provided by the amine groups. The net result is the replacement of the calcium salts by water molecules and an attendant softening of the process water. Ion-exchange resins behave in much the same way as adsorbents, with ion uptake increasing in proportion to the concentration in the contacting solution and ultimately leveling off at a saturation value when all ions in the resin have been exchanged. The phase equilibrium is consequently well-described by a Langmuir-type isotherm. To account for the ionic nature of the species, it is customary to express both resin loadings and aqueous concentrations as milliequivalents rather than the actual weight of ions. Suppose that the uptake of copper ions Cu^{++} by a particular resin is described by the Langmuir form

$$X = \frac{11.2\,C}{1 + 3.0\,C}$$

where C is in milliequivalent Cu^{2+}/L and X in milliequivalent Cu^{2+}/g resin. Use this equation to calculate the resin loading X for an aqueous solution containing 0.01 by weight of copper ions.

Answer:

0.322 meq/g

6.8 Special Types of Liquid–Liquid Equilibria

Consider the following two cases:

a. A solute distributes itself between two partially soluble solvents and is itself only partially soluble in one of these solvents. Draw the triangular diagram for this case and locate the plait point.
b. A solute distributed between two partially soluble solvents reverses selectivity at some intermediate concentration level. Draw the distribution curve for this case and identify the corresponding behavior in vapor–liquid equilibria.

6.9 Repartitioning of a Solute between Compartments

A solute is partitioned among three compartments with volumes of $V_1 = 100$, $V_2 = 50$, and $V_3 = 10$ (arbitrary units) and partition coefficients of $K_{12} = 10^2$ and $K_{13} = 10^3$. Degradation is extremely slow and the system is considered to be at a quasi-equilibrium. If the concentration in compartment 1 undergoes a sudden increase from 10^3 to 10^2, what will be the ultimate new equilibrium concentrations, assuming no degradation takes place?

Answer:

$C_1 = 2.13 \times 10^3$

6.10 Vapor–Liquid Separation Factors for Light and Heavy Water

Monodeuterated water HDO has an abundance in natural water of approximately 600 ppm. One of the methods used in early attempts to separate the two isotopes was fractional distillation, for which the separation factor at 1 atm was found to be 1.026. Although this procedure was ultimately superseded by the more efficient chemical exchange processes for larger-scale production of heavy water, distillation remains the separation method of choice for upgrading small amounts of heavy water that have been contaminated by atmospheric water vapor. Distillation is in this case carried out at reduced pressure in order to take advantage of the higher separation factor. Suppose we wish to carry out the distillation of H_2O–HDO at ambient temperatures (i.e., at subatmospheric pressures). Given that the separation factor at 40°C is 1.056, calculate its value at a temperature of 25°C.

Answer:

1.069

Note: The actual distillation process at high D_2O concentrations involves the species HDO–D_2O. Separation factors of this binary pair are nearly identical to those for H_2O–HDO.

6.11 Separation Factors in the Henry's Law Region

Derive separation factors for two-component absorption and adsorption in the Henry's law region.

6.12 Separation Factors from Separation Factors

Consider a system made up of three components 1, 2, 3. If two binary separation factors are known, for example, $\alpha_{1,2}$ and $\alpha_{1,3}$, is it possible to calculate the third?

6.13 Multiple Azeotropes

Is it possible for the *x*–*y* equilibrium curve of Figure 6.21c to inflect more than once across the diagonal?

6.14 More on Nonideal Systems

Consider the case of a nonideal liquid solution in which attractive forces outweigh the repulsive. Describe how this changes the diagrams shown in Figures 6.19a and b.

6.15 Steam Distillation

It often happens, particularly in the food and pharmaceutical industries, that a heat-sensitive liquid has to be purified by distillation, or a dissolved substance has to be concentrated by boiling off solvent without risking thermal degradation. Vacuum distillation is one way of lowering the boiling point, but this process suffers from low production rates and other disadvantages (see Illustration 6.10). A second method is steam distillation, which involves passing live steam through the solution and using it as a carrier gas to sweep off the evaporating liquid or solvent. The charge consists of partially condensed steam and a separate organic phase. The latent heat of vaporization is provided by external heating.

a. Show that the normal boiling point is always under 100°C and increases with time.
b. Consider the steam distillation of a dilute solution of a heat-sensitive substance in toluene. Estimate its initial boiling point and the initial steam consumption in kilogram steam per kilogram toluene. (*Hint*: Use Table 6.1.)

Answer:

a. 84.3°C, 0.24

6.16 Transfer of Pollutants from a Water Basin to the Atmosphere

To obtain a sense of the magnification in concentration, which occurs when a sparingly soluble substance evaporates into the atmosphere, calculate the so-called K-factor, $K = y/x$, for carbon tetrachloride in water using the data given in Illustration 6.14.

Answer:

8300

7

Staged Operations: The Equilibrium Stage

Chapter 6 laid the groundwork for the topic to be taken up here by examining in considerable detail the various phase equilibria that enter into the formulation of an *equilibrium stage*. The equilibrium stage, also termed an *ideal stage* or *theoretical stage*, plays a central role in the analysis of an important class of mass transfer operations termed *staged processes*. In these operations, two phases are brought into intimate contact in a stirred tank or its equivalent, and the desired mass transfer process is allowed to take place. The two phases are then separated, and the process is either repeated (multistage contact) or brought to a halt (single-stage contact). While there is only one mode of single-stage contact, multistage operations can be arranged in a variety of geometrical configurations. In *co-current operations*, the two phases move through the stages and cascade parallel to each other and in the same direction. This mode of contact is rarely used in practice. When the phases move in opposite directions but are still parallel, we speak of a *countercurrent operation*. Finally, when the two phases move at right angles to each other, the process is termed a *crosscurrent operation*. These modes are sketched in Figure 7.1a through Figure 7.1d.

Analysis of these processes by the equilibrium-stage model proceeds in two steps. In the first step, the two phases are visualized as entering a stirred tank or its equivalent, where they are allowed to come to equilibrium. The two phases are then conceptually withdrawn and separated, and the concentrations in each phase are calculated by means of appropriate mass balances and equilibrium data. We practice this step extensively in this chapter using both analytical expressions and specially designed geometrical constructions termed *operating diagrams*.

In the second step of the process, the two phases are contacted only for a finite time interval leading to *incomplete* equilibrium. This step, which has to be carried out experimentally, reflects conditions that prevail in an actual operation. The resulting concentration changes, which differ from those attained at equilibrium, are cast in the form of a fractional approach to equilibrium or stage efficiency E. The value of this efficiency is then grafted onto the results of the first step to arrive at an estimate of the actual prevailing concentrations.

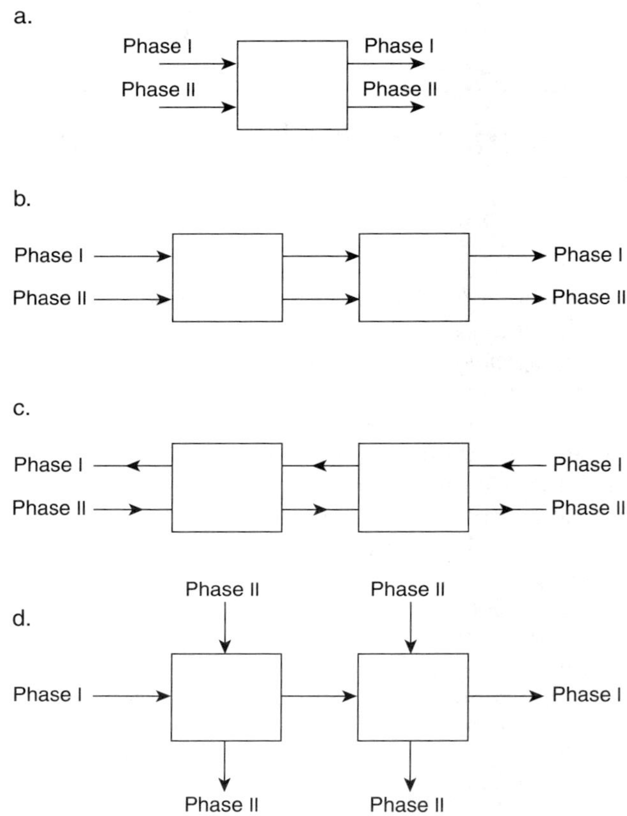

FIGURE 7.1
Staged operations: (a) single stage; (b) co-current; (c) countercurrent; and (d) crosscurrent.

The concepts we just described are illustrated in Figure 7.2a and Figure 7.2b. Figure 7.2a represents an equilibrium stage in which G kg of a carrier gas containing Y_F kg solute/kg carrier is contacted with L kg of a solvent carrying X_F kg of the solute per kilogram solvent. The solute transfers from the gas to the liquid phase until equilibrium is attained. The phases are then withdrawn and separated, with their concentrations at their respective equilibrium values X and Y^*. In Figure 7.2b, the same feed enters the stage but attains only partial equilibrium, with the exiting liquid concentration X' falling short of the equilibrium value X. We express this through the stage efficiency E by writing

$$X' = EX \tag{7.1}$$

Thus, when $E = 1$, the fractional approach to equilibrium is 1, and equilibration is consequently complete. When E is zero, no transfer takes place, and the concentration remains frozen at the level X_F of the incoming feed.

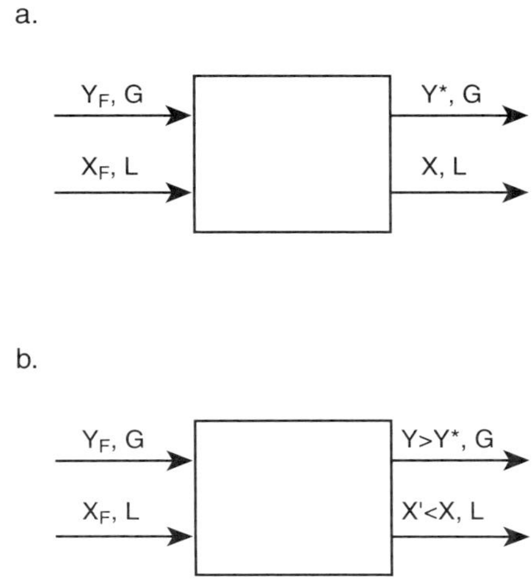

FIGURE 7.2
Staged contact of two phases: (a) equilibrium stage and (b) nonequilibrium or actual stage.

This chapter is divided into two parts. In the first, we take up the topic of equilibrium stages in their various configurations and apply them to a number of different mass transfer operations. The second part, which is less extensive, considers the effect of mass transfer resistance expressed through an appropriate stage efficiency.

7.1 Equilibrium Stages

7.1.1 Single-Stage Processes

The single equilibrium stage is the key unit on which the more complex configurations such as the crosscurrent and countercurrent cascades are based. We use it here to introduce the reader to some basic notions of equilibrium stage processes and to make a first presentation of a key tool, the operating diagram.

Figure 7.3a and Figure 7.3b show the flow diagrams for two single-stage processes. In Figure 7.3a, we display the streams entering and leaving a liquid-phase adsorption stage. S and L represent the mass of solute-free adsorbent and solvent, while X and Y are the corresponding mass ratios in units of kilogram solute per kilogram adsorbent and kilogram solute per kilogram solvent. Figure 7.3b shows a similar process of single-stage liquid extraction. The principal difference here lies in the choice of units, with concentrations expressed as mass fractions x and y, and the capitalized symbols expressing the total mass of each phase entering and leaving. The

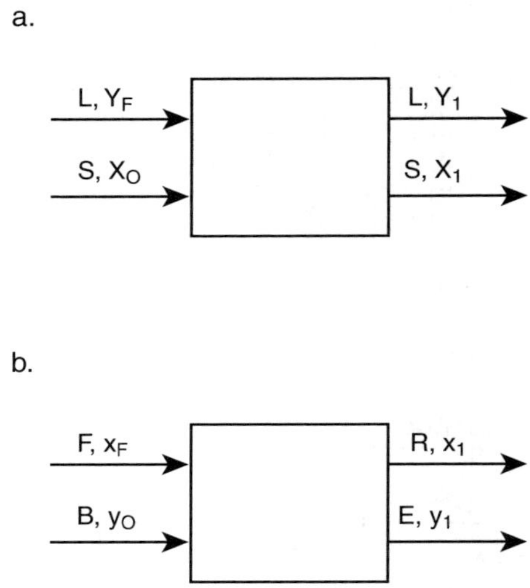

FIGURE 7.3
Single-stage equilibrium processes: (a) adsorption from a liquid solution and (b) extraction.

symbol R denotes the *raffinate* (i.e., the solution being extracted), and E stands for the *extract* containing the bulk of the solvent and the extracted solute. Our considerations here are confined to *finite* amounts of both phases. The case where differential amounts of one phase are continuously withdrawn is taken up in Section 7.1.2.

For *single-stage* operations, the following questions are to be answered:

1. With the amount and concentration of feed fixed, what is the amount of adsorbent or solvent needed to achieve a prescribed reduction in solute content?
2. Conversely, and with feed conditions again fixed, what is the concentration in the effluent if we use a given amount of adsorbent or solvent, with a known impurity level?

When dealing with *multistage* operations, the number of questions escalates. We may then ask for the number of stages required to achieve a prescribed concentration change, for example, or the most economic use of adsorbent or solvent. These will be taken up at the appropriate time.

Let us now consider the tools needed to answer the single-stage questions. In the first instance, we require a solute mass balance over the stage. For the adsorption process shown in Figure 7.3a, it takes the following form:

$$\text{Amount of solute in} - \text{Amount of solute out} = 0$$

$$(LY_F + SX_0) - (LY_1 + SX_1) = 0 \tag{7.2a}$$

Each term here is made up of the product of kilogram solvent L or kilogram adsorbent S, which do not change during their passage, and the concentrations X (kg solute/kg adsorbent) and Y (kg solute/kg solvent), which do. For the extraction process of Figure 7.3b, additional balances are required, because we do not have a convenient constant "carrier" mass available. This is taken up in Illustration 7.2.

The second tool required is a statement that the concentrations X_1, Y_1 leaving the stage are in equilibrium with each other. These values are obtained experimentally and are formally expressed by the relation

$$Y^*_i = f(X_i) \tag{7.2b}$$

where the asterisk denotes equilibrium conditions.

Laboratory data of X_i, Y_i are not always easily expressed in simple analytical form, and this has led to the development of a graphical construction termed an operating diagram, which is shown in Figure 7.4. This diagram is in essence a graphical representation of both the material balance and the equilibrium relation (Equation 7.2a and Equation 7.2b). It displays in vivid form the interrelation of the operational variables and is unsurpassed, particularly for the beginning student, in conveying an immediate sense of what needs to be done to achieve a particular result and how a change in a variable affects the outcome of the process. Many such operating diagrams will be encountered throughout the text. They are to be viewed as a vital pedagogical tool for conveying the essential features of a particular staged operation. What they are *not* is an everyday tool in the engineering world which used instead numerical or analytical solutions of the underlying behavior and equilibrium relations. It is safe to say, however, that the practicing engineer has a firm grasp of the essentials revealed by these diagrams and, more often than not, a mental image of them as well.

Illustration 7.1: Single-Stage Adsorption and Extraction: The Rectangular Operating Diagram

The first step in the construction of the operating diagram is to plot the experimental equilibrium data obtained in the laboratory in the form of an equilibrium curve. This curve is shown in Figure 7.4a and is of the Langmuir form, where solid-phase concentration now appears on the abscissa instead of the ordinate as previously shown in Figure 6.5. This is done for greater clarity of construction.

The second step is to represent the mass balance, Equation 7.2a, which is best achieved by rearranging it in the form

$$-\frac{S}{L} = \frac{\Delta Y}{\Delta X} = \frac{Y_F - Y_1}{X_F - X_1} \tag{7.3a}$$

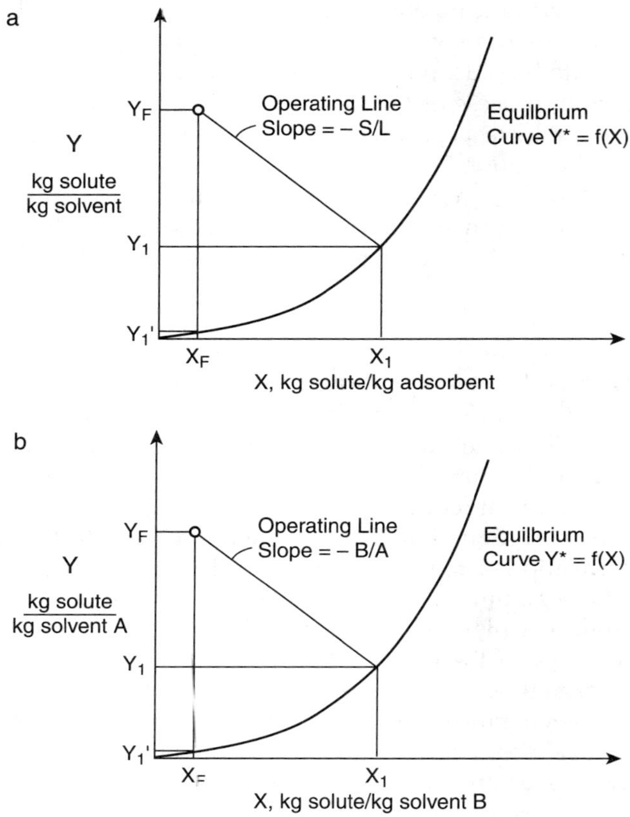

FIGURE 7.4
Single-stage process: (a) adsorption and (b) extraction (mutually insoluble solvents).

This is the equation of a straight line termed the operating line, which has a slope of S/L. In this equation, the amount of solvent L and the initial or feed concentrations X_F, Y_F are known variables; the others are either prescribed or unknown.

There are several features the diagram conveys at a glance. For example, the steeper the slope of the operating line, the greater the amount of adsorbent used and, hence, the higher the fractional recovery of solute, here given by the ratio $(Y_F - Y_1)/Y_F$. Note that when the amount of adsorbent becomes infinite, the recovery is not complete but instead stabilizes at the value $(Y_F - Y_1')/Y_F$. This is because of the residual impurity level X_0, which the incoming adsorbent carried with it. When an infinite amount of *clean* adsorbent is used, the operating line drops to the origin and recovery then becomes complete.

Suppose now that with the feed condition Y_F, X_F and L known and the amount S prescribed, we wish to establish the resulting degree of recovery. This becomes a simple matter of locating the point (Y_F, X_F) on the diagram and drawing a line of slope *S/L* through it. Its intersection with the equilibrium curve determines the concentrations of the exiting streams, Y_1 and X_1, from which we can compute the recovery $(Y_F - Y_1)Y_F$. If, on the other hand, recovery is *prescribed* and the unknown is the amount of adsorbent required, the procedure is reversed. A line is drawn through the two known points (Y_F, X_F) and (Y_1, X_1), and the unknown *S* is calculated from its slope *S/L*.

Let us turn next to the extraction process shown in Figure 7.3b. If we assume extract and raffinate solvents to be essentially insoluble in each other (e.g., *n*-octane and water), mass fractions *x* and *y* can be replaced by mass ratios *X* and *Y* based on *B* kg extract solvent and *A* kg raffinate solvent. The mass balance and operating line become identical in form to those of the adsorption process — that is, we have for the operating line

$$-B \,/\, A = \frac{\Delta Y}{\Delta x} = \frac{Y_F - Y_1}{X_F - X_1} \tag{7.3b}$$

The resulting operating diagram is shown in Figure 7.4b. It shows us, at a glance, that an increase in the extracting medium *B* will increase the slope of the operating line and hence solute recovery $(Y_F - Y_1)/Y_F$, just as it did for adsorption, but that this will be at the expense of a more dilute extract. Various other consequences can be read off with equal ease.

When mutual solubility of extract and raffinate solvents can no longer be neglected, we must turn to a representation in a triangular diagram. This is taken up in the next illustration.

Illustration 7.2: Single-Stage Liquid Extraction: The Triangular Operating Diagram

Consider the case of a feed with a composition located at *F* in Figure 7.5 being contacted with *B* kg of pure solvent, located at point *B*. *M* is the location of the mixture that results when *F* kg feed are contacted with the solvent, and $\overline{RE}$ represents the equilibrium tie-line. Composing total and component solute mass balances as previously stipulated, we obtain

Rate in Rate out

$$F + B = M = R + E \tag{7.4a}$$

$$x_F F = x_M M = x_R R + y_E E \tag{7.4b}$$

It is desired to calculate the equilibrium compositions that result, as well as the amounts of raffinate *R* and extract *E* produced. We first locate the composition x_M of the mixing point *M* by eliminating total mass *M* from the left side of the two balances. Thus,

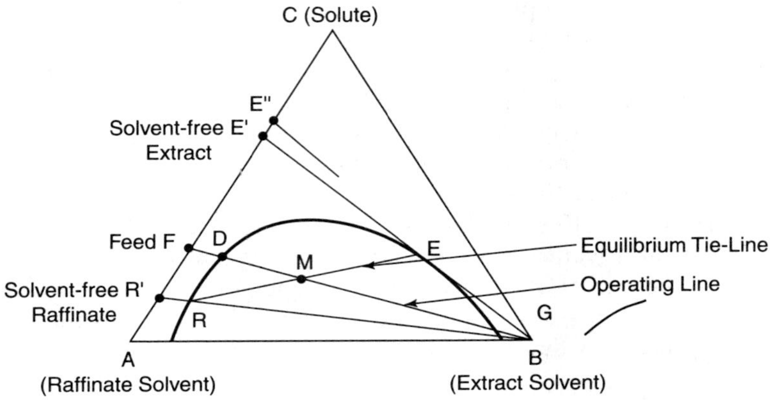

FIGURE 7.5
Operating diagram in trilinear coordinates for single-stage equilibrium extraction.

$$x_M = \frac{x_F F}{F + B} \tag{7.4c}$$

This composition is entered on the operating line connecting F and B.

Next a tie-line is drawn through M to establish the raffinate and extract compositions at R and E. This requires some visual interpolation between neighboring tie-lines obtained from experimental equilibrium data and completes the first part of the task. The second part is accomplished by again eliminating M from the two balances, this time from the right side of Equations 7.4a and 7.4b. Thus,

$$E = \frac{(F + B)(x_M - x_R)}{Y_E - x_R} \tag{7.4d}$$

and

$$R = \frac{(F + B)(y_E - x_M)}{y_E - x_R} = F + B - E \tag{7.4e}$$

Thus, the operating diagram, together with the two algebraic mass balances, has provided us with a complete answer to the problem. The reverse task, that of calculating the solvent requirement for a prescribed solute recovery, is solved by drawing both the tie-line (which is now known) and the companion operating line, and, from their intersection, establishing the value of x_M. The solvent requirement is then obtained from the left side of the two mass balances as

$$B = \frac{F(x_F - x_M)}{x_M} \tag{7.4f}$$

Comments

Several additional interesting features may be deduced from the diagram:

1. Removal of the solvent from the raffinate and extract, which can (for example) be accomplished by distillation, results in the solvent-free compositions located at R' and E'. These points lie on the straight-line extrapolations of $\overline{RB}$ and $\overline{EB}$, as required by the mixing rule. We note in addition that a maximum in the solute concentration results when the product extract is located on the tangent drawn through B to the solubility curve, leading to point E''.
2. An *increase* in the amount of solvent used results in a movement of the mixing point M toward G, while a *reduction* in solvent causes it to approach the point D. When M coincides with D, the amount of solvent is at a minimum, and the amount of extract is infinitesimally small. This follows from the lever rule, Equation 6.8f. Under these conditions, any solvent present in the system resides entirely in the raffinate phase. Conversely, the point G represents the maximum amount of solvent we can use and is attended by an infinitesimally small amount of raffinate. Evidently, the actual amount of solvent will lie somewhere between the two extremes. Note that these two limiting values are established from the intersection of the operating lines with the solubility curve and do not require any tie-line data.

7.1.2 Single-Stage Differential Operation

A special type of single-stage contact arises when one phase, which is at all times fully equilibrated with its partner, is slowly withdrawn in infinitesimally small amounts, while the second phase remains within the stage and undergoes a corresponding slow change in concentration. This process is best visualized, and most easily implemented, in the case of distillation, which is then referred to as simple distillation. The equations to be used again involve integral mass balances, but they now contain an unsteady term to reflect the slow changes in the contents of the stage.

Figure 7.6 shows the variables involved in a differential distillation process. For a binary system, there are four: the moles liquid in the still or boiler at any instant W and its mole fraction x_B, the rate of vapor withdrawal D (moles/s) and the instantaneous vapor composition y_D. The mass balances and the equilibrium relation $y_0^* = f(x_W)$ provide only three of the required equations. For the fourth we must draw on an energy balance. This stands to reason, because the rate of vapor production D will evidently depend on

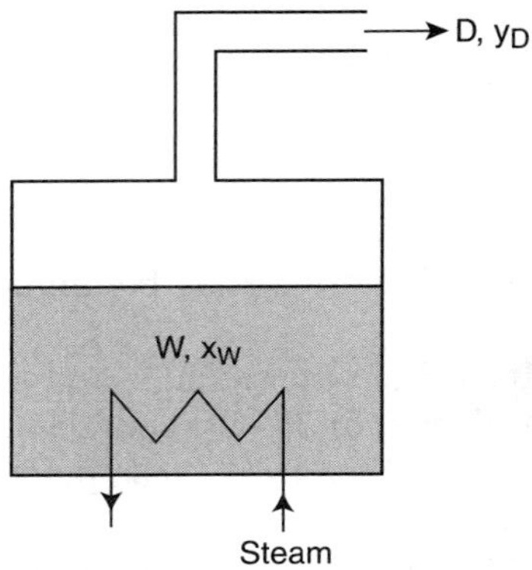

FIGURE 7.6
Differential or simple distillation.

the energy input to the system. The temperature of the still will then come into play as a fifth variable and must be accounted for by an appropriate equilibrium relation. This set of equations, which has to be solved numerically, yields the changes with time of the five variables in question.

A reduced solution to the problem can be obtained if one eliminates time as a variable and establishes the instantaneous relation between the state variables, for example, $W = f(x_W)$. This is still a highly useful result, because it can tell us how much of a given initial charge must be distilled in order to obtain a desired enrichment. It also provides us with other items of interest and falls short only by failing to establish the full time dependence of the variables.

Illustration 7.3: Differential Distillation: The Rayleigh Equation

We start the derivation of the model by composing the unsteady integral mass balances for the system:

$$\text{Rate of moles in} - \text{Rate of moles out} = \begin{matrix}\text{Rate of change}\\ \text{of contents}\end{matrix}$$

$$-D = \frac{d}{dt}(W) \tag{7.5a}$$

and

$$-y_D D = \frac{d}{dt}(x_W W) \tag{7.5b}$$

This is supplemented by the equilibrium relation

$$y_D^* = f(x_W) \tag{7.5c}$$

To eliminate the time variable, we resort to a favorite trick of ours, one that is frequently used and should be part of the analyst's tool box. The two mass balances are divided, thereby eliminating not only time as a variable but also the unknown distillation rate D. Thus,

$$y_D = \frac{d(x_W W)}{dW} \tag{7.5d}$$

which can be combined with the equilibrium relation and expanded to yield

$$f(x_W) = x_W + W\frac{dx_W}{dW} \tag{7.5e}$$

and, consequently,

$$\frac{dx_W}{f(x_W) - x_W} = \frac{dW}{W} \tag{7.5f}$$

Formal integration of this expression then leads to

$$\ln\frac{W}{W^o} = \int_{x_W^o}^{x_W} \frac{dx_W}{f(x_W) - x_W} \tag{7.5g}$$

where the superscript o denotes the initial conditions.

This expression is known as the *Rayleigh equation* and relates the amount W left in the still at any instant to its composition x_W.

Comments

We start by noting, as we have done on other occasions, that any of the variables appearing in Equation 7.5g may be regarded as an unknown. Thus, for a given boiler content W, the equation will yield the corresponding composition x_W, and conversely, we can calculate the fraction to be distilled, $1 - W/W^o$, by prescribing a desired enrichment $(x_W - x_W^o)/x_W^o$. What is often overlooked is the fact that the equilibrium relation $f(x_W)$ can also be established from experimental x_W versus W data. A particularly simple case arises when the $f(x_W)$ is linear in x_W, which holds at low values of the mole fraction x (Henry's law region). We then have

$$\ln\frac{W}{W^o} = \int_{x_W^o}^{x_W} \frac{dx_W}{(H-1)x_W} \tag{7.6}$$

and, consequently,

$$\frac{W}{W^o} = \left(\frac{x_W}{x_W{}^o}\right)^{1/H-1} \tag{7.7}$$

This expression can be exploited to derive isotope separation factors, which here equal the Henry constant. Suppose, for example, that 95% of the charge has been boiled off (i.e., $W/W^o = 0.05$), and the isotope has been depleted by 2% (i.e., $x_W/x_W{}^o = 0.98$). Then $H - 1 = \alpha - 1 = \ln 0.98/\ln 0.05$ and $\alpha = 1.0067$.

We can also use Equation 7.7 as an adjunct to the *cumulative* balances for the process, which were derived in Illustration 2.5 and which we repeat here. They are

$$W^o = W + D' \tag{2.15a}$$

and

$$x_W{}^o W^o = x_W W + x_D{}' D' \tag{2.15b}$$

where D' = moles in the distillate receiver. These can be used to derive the composite distillate concentration $x_D{}'$, which prevails at any given instant. With W prescribed, and the initial conditions known, we can calculate x_W from Equation 7.7 and by substitution into Equation 2.15b arrive at a value for $x_D{}'$. Thus, with the exception of temperature and distillation rate D, both of which require an energy balance, we have managed to establish all pertinent variables of the system. All of these equations are applied in Illustration 7.10 dealing with batch column distillation.

Illustration 7.4: Rayleigh's Equation in the Environment: Attenuation of Mercury Pollution in a Water Basin

We showed in Illustration 6.10 that substances with low solubility in water, such as hydrocarbons, chlorinated organics, and mercury, can still have high partial pressures p because of their exceptionally high activity coefficients γ. The relevant relation is given by the extended Raoult's law, Equation 6.18a:

$$p = \gamma x P^o \tag{6.18a}$$

which can be recast in the form

$$y = \gamma \frac{P^o}{P_T} x \tag{7.8a}$$

The group $\gamma P^o/P_T$ can be viewed as a partition coefficient K, which for the substances mentioned is generally quite high. The loss of even a small amount of solution by evaporation could thus be expected to result in a marked drop of pollutant concentration. This is small comfort for the environment as a whole, because the toxic substances are merely being transferred from one medium to another. Their fate needs nevertheless to be tracked.

We consider here a body of water containing dissolved mercury at the saturation level and set ourselves the task of calculating the reduction in mercury content that occurs when 0.01% of the content is evaporated. It is assumed that the vapor in question is at all times in equilibrium with a well-mixed liquid (i.e., that Rayleigh's equation applies).

The equilibrium relation (Equation 7.8a), which will be needed in the model, contains the activity coefficient γ, which is not usually known or easily measured. The difficulty was circumvented in Illustration 6.10 by relating γ to solubility x_s, using the chemical potentials of reactive phases. We obtain

$$\underset{\text{Mercury}}{\mu^\circ (T,P)} = \underset{\text{Saturated solution}}{\mu^\circ (T,P) + RT \ln \gamma\, x_s} \tag{7.8b}$$

and, hence,

$$\gamma = 1/x_s \tag{7.8c}$$

Substitution into Equation 7.8a yields

$$y_{Hg} = \frac{x_{Hg}}{x_s} \frac{P^o{}_{Hg}}{P_T} \tag{7.8d}$$

Here we must be careful to exclude the partial pressure of air from P_T because the equilibrium involved is one of the binary Hg–H_2O vapor in equilibrium with the liquid phase. We must therefore write

$$y_{Hg} = \frac{x_{Hg}}{x_s} \frac{P^o{}_{Jg}}{p_{Hg} + P^o{}_{H_2O}} \cong \frac{x_{Hg}}{x_s} \frac{P^o{}_{Hg}}{P^o{}_{H_2O}} \tag{7.8e}$$

where we omitted mercury partial pressure because it is small compared to the full vapor pressure of water $P^o{}_{H_2O}$.

Substituting Equation 7.8e into the Rayleigh equation (Equation 7.5g) and integrating yields the following result:

$$\left[\frac{P^o{}_{Hg}}{x_s P^o{}_{Hw}} - 1\right] \ln \frac{W}{W_o} = \ln \frac{x}{x_o} \tag{7.8f}$$

The data to be introduced at this stage are as follows:

Mercury solubility (Table 6.3): 3×10^{-2} mg/L = 2.7×10^{-7} mol fraction
Mercury vapor pressure (25°C): 0.173 Pa
Water vapor pressure (25°C): 3.17×10^3 Pa

Noting that $W/W_o = 1 -$ (fraction evaporated), one obtains by substitution into Equation 7.8f

$$\left[\frac{0.173}{2.7 \times 10^{-9} \times 3170} - 1\right] \ln(1 - 10^{-4}) = \ln \frac{x_{Hg}}{x_s} \tag{7.8g}$$

Taylor series expansion of the logarithmic term then yields

$$\ln (1 - 10^{-4}) \cong 10^{-4} \tag{7.8h}$$

so that

$$\ln \frac{x_{Hg}}{x_s} = -2.02 \tag{7.8i}$$

and

$$x_{Hg}/x_{Hg}^{o} \approx x_{Hg}/x_s = 0.133 \tag{7.8j}$$

Thus, nearly 87% of the original mercury was transferred to the atmosphere, a phenomenal amount considering only 1/10,000th of the solution has evaporated.

Comments

It takes astuteness to overcome some of the difficulties and avoid the pitfalls of this problem. The fact that γ is unknown and could not be located in the literature on vapor–liquid equilibria could have brought the proceedings to a halt. Instead, it had to be realized that γ could be related to solubility (which is tabulated) and the relationship established by means of elementary thermodynamics. These are considerable leaps of thought.

A second point concerns the calculation of y_{Hg}. The first instinct would be to use the conventional relation $y = p/P_T$ and set total pressure equal to 1 atm. This would have led to the wrong result, because the partial pressure of air would have been included. It required some thought to realize that y_{Hg} refers to the system H_2O–Hg, not H_2O–Hg–air.

It is of some interest to calculate the ratio or partition coefficient, $K = y/x$, which is a measure of enrichment in the vapor phase. Drawing on Equation 7.8a, we have

$$K = y / x = \frac{1}{x_s} \frac{P^o_{\text{Hg}}}{P^o_{\text{Hg}} + P^o_{\text{H}_2\text{O}}} = \frac{1}{2.7 \times 10^{-9}} \frac{0.173}{0.173 + 3.17 \times 10^3} = 2 \times 10^4$$

This value explains and confirms the astoundingly high depletion of mercury in the water basin.

7.2 Staged Cascades

7.2.1 Crosscurrent Cascades

The linkage of several single-stage contacts leads to a composite array termed a *cascade*. As mentioned earlier, the mode of contact in these configurations results in three distinct types of cascades: crosscurrent, co-current, and countercurrent. In the crosscurrent cascade, which is taken up here, the two phases enter and leave in directions that are, at least symbolically, at right angles to each other. This can take place in two different ways. In one mode of contact, the extracting agent (i.e., the solvent or adsorbent) is used in only one stage and is then removed for processing before being reused. This mode of contact, shown in Figure 7.7a for the case of adsorption, is sometimes and not very accurately referred to as co-current and has a high extraction efficiency, but also a high inventory. In the second mode of contact, the extracting agent is used *repeatedly* within the cascade before being withdrawn for regeneration. This type of configuration is shown in Figure 7.7b and has the advantage of low adsorbent or solvent inventory.

The model equations for the case shown in Figure 7.7a are now three in number: one solute mass balance each for the two stages and a companion equilibrium relation. We have the following:

BALANCE ON STAGE 1

$$\text{Mass of solute in} - \text{Mass of solute out} = 0$$

$$(LY_0 + S_1X_0) - (LY_1 + S_1X_1) = 0 \tag{7.9a}$$

BALANCE ON STAGE 2

$$\text{Mass of solute in} - \text{Mass of solute out} = 0$$

$$(LY_1 + S_2X_0) - (LY_2 + S_2X_2) = 0 \tag{7.9b}$$

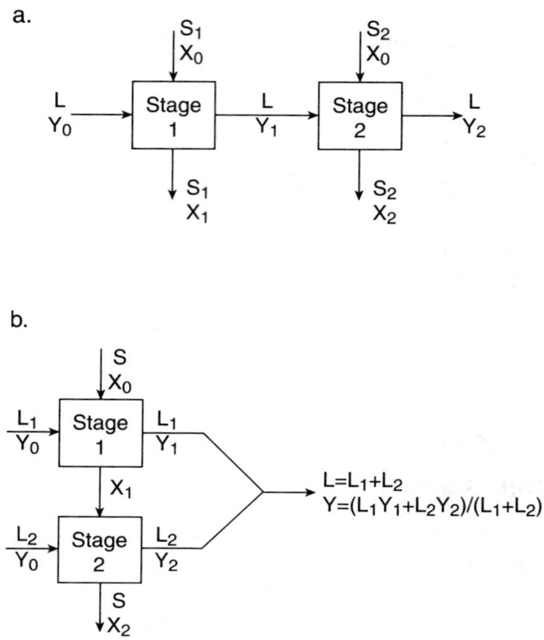

FIGURE 7.7
Two modes of crosscurrent adsorption processes: (a) use of fresh adsorbent in each stage and (b) repeated use of the same adsorbent.

EQUILIBRIUM RELATION

$$Y^* = f(X) \tag{7.9c}$$

Here again it is convenient to represent the equations in graphical form by means of an operating diagram in much the same way as was done in the case of single-stage operations. This is again best done by rearranging the two mass balances into the following form:

$$-\frac{S_1}{L} = \frac{\Delta Y}{\Delta X} = \frac{Y_0 - Y_1}{X_0 - X_1} \tag{7.9d}$$

and

$$-\frac{S_2}{L} = \frac{\Delta Y}{\Delta X} = \frac{Y_1 - Y_2}{X_0 - X_2} \tag{7.9e}$$

Suppose now that the amounts of adsorbent S_1 and S_2 entering each stage have been prescribed, and it is desired to determine the final effluent

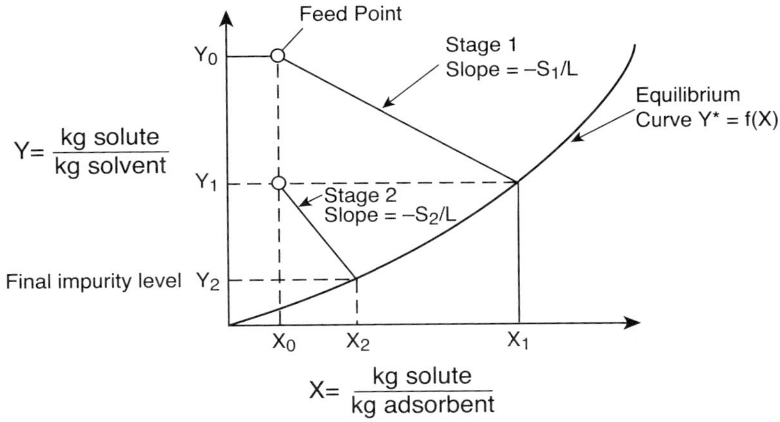

FIGURE 7.8
Operating diagram for a two-stage crosscurrent adsorption process using fresh adsorbent.

concentration $\underline{Y}_2$. The feed condition, comprising the impurity level in the incoming adsorbent X_0, the feed concentration Y_0, and the solvent content L, is known. The procedure for constructing the operating diagram, shown in Figure 7.8, is then as follows:

1. Locate the feed point (X_0, Y_0) in the diagram.
2. Draw a line of slope $-S_1/L$ through the feed point. Its intersection with the equilibrium curve fixes the concentrations (X_1, Y_1) leaving stage 1.
3. Draw a second operating line of slope $-S_2/L$ through the point (X_0, Y_1). Its intersection with the equilibrium curve defines the final impurity level Y_2.

It is clear from this construction that an infinite number of combinations of S_1 and S_2 exist that will reduce the feed-solute level Y_0 to Y_2. It is also seen that if either S_1 or S_2 is set equal to zero (i.e., a single-stage contact is used), the resulting operating line, which extends from (X_0, Y_0) to (X_2, Y_2), will always have a slope $-S/L$ greater than the combined slopes $(S_1/L + S_2/L)$ so that

$$S_{\text{single stage}} > (S_1 + S_2)_{\text{2-stage cascade}}$$

that is, the single stage always consumes more adsorbent for a given reduction in solute content than a two-stage arrangement.

It follows that between the two extremes $S_1 = 0$ and $S_2 = 0$ a combination $S_1 + S_2$ exists which represents a *minimum* total adsorbent inventory $(S_T)_{\text{Min}}$. This minimum amount can be determined by simple calculus for the case of linear equilibrium. We show this in the following illustration.

Illustration 7.5: Optimum Use of Adsorbent or Solvent in Crosscurrent Cascades

Consider a crosscurrent adsorption process in which the equilibrium is linear, a condition that holds at low solute level. We then have

$$X_i = HY_iH > 1 \tag{7.10a}$$

where H is the Henry's constant.

Assume further that the adsorbent is initially clean — $X_0 = 0$ (this is frequently the case). Substitution of Equation 7.10a into the two mass balances (Equation 7.9a and Equation 7.9b) then yields

$$S_T = S_1 + S_2 = HL\left(\frac{Y_0}{Y_1} + \frac{Y_1}{Y_2} - 2\right) \tag{7.10b}$$

The condition $dS_T/dY_1 = 0$ is now used to establish the optimum intermediate concentration $(Y_1)_{opt}$ that will yield the minimum value of S_T. We obtain

$$dS_T/dY_1 = Y_0Y_1^2 + Y_2^1 = 0 \tag{7.10c}$$

and, hence,

$$(Y_1)_{opt} = (Y_0Y_2)^{1/2} \tag{7.10d}$$

In other words, the optimum solute level from the first stage that will minimize S_T is the geometric mean of the concentrations entering and leaving the cascade. Let us see how this affects the amount of adsorbent to be used.

Backsubstitution into the material balances (Equation 7.9a and Equation 7.9b) yields

$$\frac{S_1}{L} = H\left[\frac{Y_0}{(Y_0Y_2)^{1/2}} - 1\right] \tag{7.10e}$$

and

$$\frac{S_2}{L} = H\left[\frac{(Y_0Y_2)^{1/2}}{Y_2} - 1\right] \tag{7.10f}$$

But,

$$\frac{Y_0}{(Y_0Y_2)^{1/2}} = \frac{Y_0(Y_0Y_2)^{1/2}}{Y_0Y_2} = \frac{(Y_0Y_2)^{1/2}}{Y_2} \tag{7.10g}$$

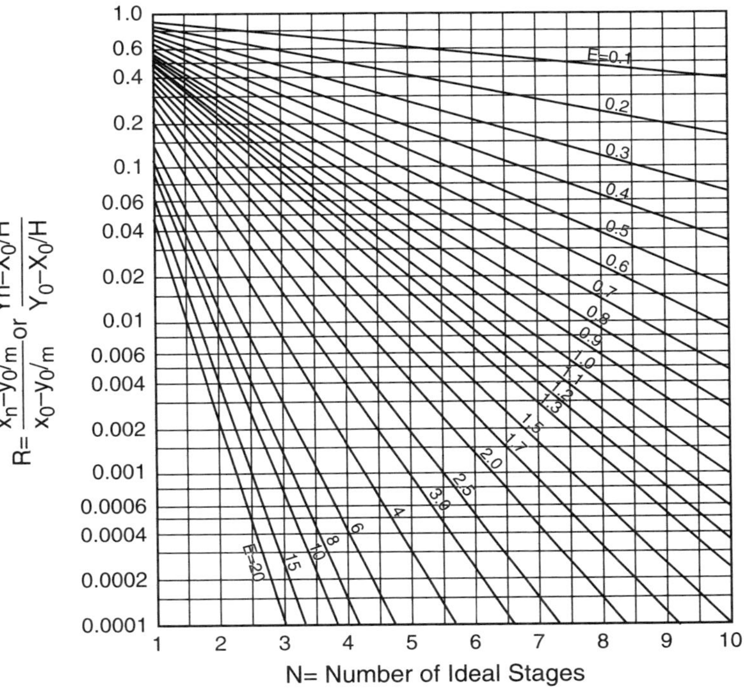

FIGURE 7.9
Crosscurrent cascades using equal amounts of adsorbent or solvent (linear equilibrium).

so that the bracketed terms in Equation 7.10e and Equation 7.10f are identical and, consequently, $S_1 = S_2$. Hence, it is seen that in this case of a two-stage cascade with linear equilibrium, optimum operation calls for the use of *equal amounts of adsorbent* given by either Equation 7.10e or Equation 7.10f.

It was shown that this principle applies to *any number of stages* — that is, the optimum use of adsorbent requires an *equal division* of that adsorbent among the stages of a crosscurrent cascade. The same principle applies to crosscurrent extraction of systems with mutually insoluble solvents. The solute recovery or removal that results in such cascades is depicted graphically in Figure 7.9. In this plot m represents the distribution coefficient for extraction or Henry's constant H for adsorption, E is the so-called extraction factor mB/A or HS/L, and Y_n or x_n is the effluent concentration from the nth stage of the solution being treated. R is the *fractional residue content* in the raffinate, and 1-R is the *fractional recovery.*

Let us apply this plot to a specific extraction process. We consider 100 kg of a feed of 1% nicotine in water that is to be extracted in a three-stage crosscurrent cascade employing 50 kg kerosene in each stage. The distribution has a slight curvature, with m varying over the range 0.80 to 0.90. We choose a mean value of 0.85 and obtain the extraction factor

$$E = mB / A = 0.85\frac{50}{99} = 0.43 \tag{7.11a}$$

Using Figure 7.9, this yields, for solute-free kerosene ($y_0 = 0$), the extraction ratio

$$\frac{x_n}{x_0} = 0.34 \tag{7.11b}$$

and, consequently,

$$x_n = 0.34 \times 0.010 = 0.0034 \tag{7.11c}$$

The same value is obtained using the graphical construction of the operating diagram.

Comments

The optimization problem considered indicates that the best way to operate a crosscurrent cascade is by equal subdivision of solvent or adsorbent. Be aware, however, that to completely optimize a plant, the cost of the stages, the cost of solvent recovery, and the value of the extracted solute must also be considered. Thus, in addition to optimum solvent use, we need to determine the optimum number of stages, and this requires bringing in all of the above-mentioned factors. Still, *equal subdivision of solvent or adsorbent is a good policy to pursue* and comes close to meeting the requirements of a global optimum for systems with linear distributions.

Illustration 7.6: A Crosscurrent Extraction Cascade in Triangular Coordinates

We consider 1000 kg of a feed containing 50% by weight of acetone in water, which is to be reduced to 10% by extraction with 1,1,2-trichloroethane in a crosscurrent cascade; 250 kg solvent are to be used in each stage. Representative tie-line data are shown in Figure 7.10.

Solution of this problem calls for the repeated application of the methods established for single-stage extraction in Illustration 7.2. We start by drawing a line connecting the solvent at B to the feed located at F and follow this by locating the mixing point M, using Equation 7.4c. We obtain

$$x_{M_1} = \frac{x_F F}{F + B} = \frac{0.5 \times 1000}{1000 + 250} = 0.40 \tag{7.12}$$

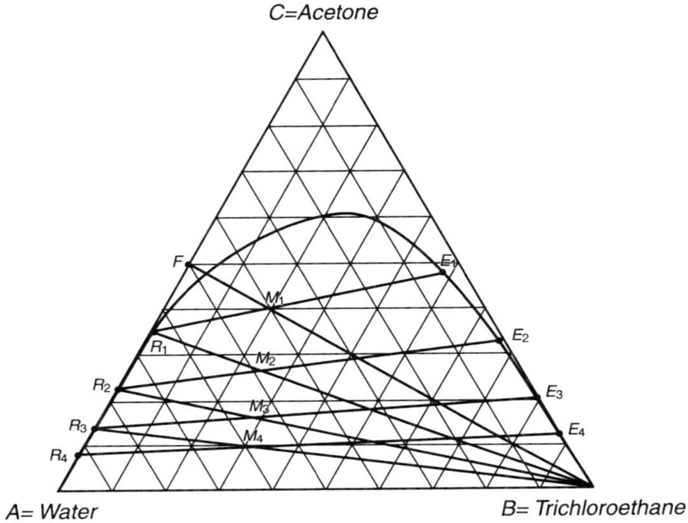

FIGURE 7.10
Operating diagram for Illustration 7.6.

The tie-line through this coordinate is shown in Figure 7.10 and yields raffinate and extract concentrations corresponding to the end points of the tie lines, i.e., $x_{R_1} = 0.35$ and $y_{E_1} = 0.475$. These values can then be used to calculate the amounts of raffinate and extract using Equation 7.4d and Equation 7.4e.

$$E_1 = \frac{(F+B)(x_{M_1} - x_{R_1})}{y_{E_1} - x_R} = \frac{(1000+250)(0.40-0.35)}{0.475-0.35} = 500\,\text{kg}$$

$$R_1 = F + B - E_1 = 1000 + 250 - 500 = 750\text{ kg}$$

The results of these calculations for four stages are given below:

Stage	x_M	x_R	y_E	E	R
1	0.400	0.35	0.475	500	750
2	0.262	0.223	0.325	382	618
3	0.159	0.134	0.204	310	558
4	0.0925	0.075	0.120	305	503

From this tabulation it emerges that three stages are too few and four stages are too many to reduce acetone content in the raffinate to $x_R = 0.10$. There are three alternatives we can pursue:

1. Use three stages and slightly more solvent per stage, determining the quantity to be used by trial and error until $x_{R_3} = 0.10$.
2. Use four stages and slightly less solvent, again with a trial-and-error procedure.
3. Use four stages with the same amount of solvent and accept a lower acetone content of $x_{R_4} = 0.075$ in the final raffinate.

The last alternative is the most convenient, as well as the most practical, because the stage inefficiency will inevitably consume the margin provided by the fourth stage.

7.2.2 Countercurrent Cascades

The countercurrent cascade is the most popular among the various existing stage configurations. It combines economy of consumption of solvent or stripping agent with a high recovery of solute, but pays for it in part with a greater number of stages than would be required in a comparable cross-current cascade. Perhaps its largest application apart from distillation operations occurs in the field of gas absorption, the flow sheet for which appears in Figure 7.11. Staged gas absorbers usually consist of vertical columns containing trays on which the two phases come into intimate contact. The gas enters from below through openings in the plate or tray and bubbles through the liquid that flows across and down to the next tray.

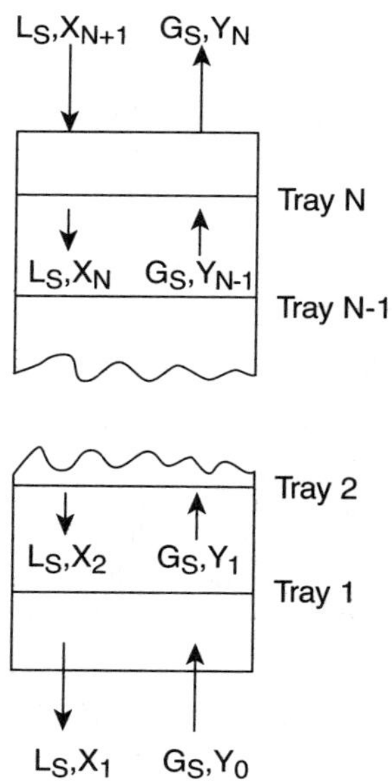

FIGURE 7.11
Countercurrent gas absorber.

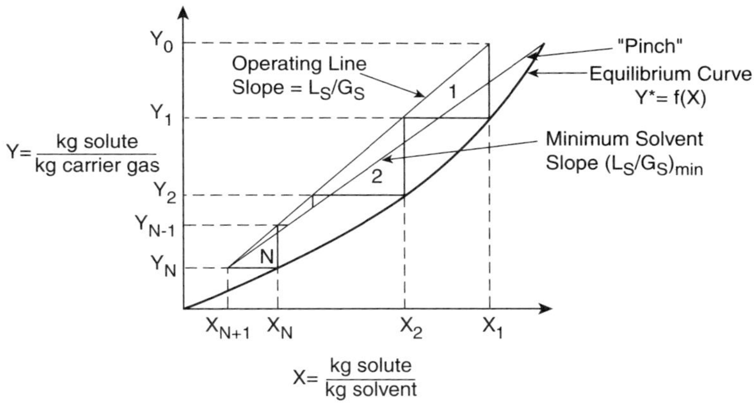

FIGURE 7.12
Operating diagram for a countercurrent absorption cascade.

In gas absorption, gas and liquid streams enter the bottom tray 1 and the top tray N at opposite ends of the cascade, carrying with them the solute concentrations Y_0 and X_{N+1}. They leave the same trays with concentration Y_1 and X_N. Note that the subscripts refer to the tray from which a particular stream exits, with the subscripts 0 and N+1 representing imaginary stages numbered 0 and N+1. Concentrations are expressed in mass ratios, and flow rates G_S and L_S in kilogram per second of solute-free carrier gas and solvent.

Our tools again comprise mass balances and equilibrium relations, which are expressed graphically in the operating diagram shown in Figure 7.12. They contain an operating line that represents the solute material balances and a staircase construction, which spans the interval between operating line and equilibrium curve and represents the various stages or trays.

A material balance over the entire gas absorber leads to the following equation:

$$\text{Rate of solute in} - \text{Rate of solute out} = 0$$

$$(L_S X_{N+1} + G_S Y_0) - (L_S X_1 + G_S Y_N) = 0 \tag{7.13a}$$

Rearrangement then gives

$$L_S \,/\, G_S = \frac{Y_N - Y_0}{X_{N+!} - X_1} \tag{7.13b}$$

Usually in the design of countercurrent absorbers, the solvent rate L_S and purity X_{N+1} as well as G_S and Y_0 are known, and the desired exit gas concentration Y_N is specified. Equation 7.13b can then be plotted in the operating diagram by drawing a line of slope L_S/G_S through the point (X_N, X_{N+1}).

Consider next a solute balance on a *single* stage, for example, tray 1. We obtain

$$\text{Rate of solute in} - \text{Rate of solute out} = 0$$

$$(L_S X_2 + G_S Y_0) - (L_S X_1 + G_S Y_1) = 0 \tag{7.13c}$$

This equation can be represented in the operating diagram as follows: We locate the point (Y_0, X_1), where the abscissa value X_1 is obtained from the intersection of the operating line with the horizontal ordinate value $Y = Y_0$. We next drop a *vertical* line from this point to the equilibrium curve $Y^* = f(X)$. The point of intersection will have the coordinates (Y_1, X_1), because these values represent the concentration levels leaving tray 1 and are known to be in equilibrium with each other. This is followed by drawing a *horizontal* line through (Y_1, X_1) all the way to the operating line. The point of intersection will have the coordinates (Y_1, X_2). This is shown by rearranging Equation 7.13c to read

$$L_S / G_S = \frac{Y_1 - Y_0}{X_2 - X_1} \tag{7.13d}$$

which is the expression of a line of slope L_S/G_S that passes through the points (Y_1, X_2) and (Y_0, X_1). But Equation 7.13b, which has the same slope, also passes through the point (Y_0, X_1). The two lines (Equation 7.13b and Equation 7.3d) must therefore coincide, and we have established the first step of the staircase construction seen in Figure 7.12. This construction of alternating vertical and horizontal lines between operating line and equilibrium curve is continued until the known concentrations at the top of the column (Y_N, X_{N+1}) are reached. A count is then made of the number of trays between gas inlet and outlet concentrations. That number represents the number of stage contacts required to reduce the feed concentration Y_0 to the prescribed value of Y_N using a fixed solvent flow rate L_S. A fractional step at the outlet end of the staircase is usually rounded off to one stage.

We now note a number of features of this construction, which are of use in both engineering calculations and analysis:

- The operating line in essence represents solute mass balances around a single stage or an aggregate of stages. Its coordinate points (Y_j, X_{j+1}) establish the relation between concentrations *entering* a stage, while the equilibrium curve relates concentrations *leaving* a stage.
- Reducing the amount of solvent L_S *lowers* the slope of the operating line and *simultaneously increases* the number of stages required for a prescribed reduction in solute content. This process can be continued until the operating line intersects the equilibrium curve. At that point a "pinch" results at the high concentration end (Figure 7.12), yielding

an *infinite* number of stages and a corresponding *minimum* flow rate of solvent or adsorbent. This is the lowest flow rate that will achieve the desired solute removal and is a useful limiting value to establish. Below that value, the prescribed effluent concentration Y_N can no longer be attained, even if an infinite number of stages were used.

- An increase in solvent flow will reduce the number of contact stages and consequently lower the capital cost of the plant. This advantage is earned at the expense of an increase in operating costs occasioned by the higher flow of solvent. There consequently exists an optimum flow rate that will *minimize* the combined operating and capital costs. That optimum can be established only with precision by a detailed economic analysis of the process. It has been found in such studies, however, that the optimum flow usually lies in the range 1.5 to 2 times the minimum value. That range is commonly used to establish the flow of solvent or adsorbent to be used, and hence the operating line, in preliminary designs of the process.
- The reverse task to the design problem considered above (i.e., the prediction of the effluent concentration of an existing or hypothetical plant (number of stages N and feed conditions known) cannot be achieved in the same direct fashion. We must resort instead to a trial-and-error procedure by drawing a series of parallel operating lines of known slope L_S/G_S until a staircase construction accommodating N stages reaches the known impurity level of the solvent or adsorbent. This presents no undue difficulties and can be accomplished quite rapidly.
- The operating diagrams we discussed are not only elegant in their simplicity but are also capable of conveying important information in rapid fashion. Thus, the effect of a change in flow rate, concentrations, or the number of stages can be quickly assessed, at least qualitatively, by visual inspection. Although in practice much of this work is now dealt with using appropriate computer packages, particularly in the case of multicomponent systems, the operating diagram remains unsurpassed, as was noted earlier, in conveying the essence of staged operations and in providing a valuable educational tool.

7.2.3 Countercurrent Cascades: The Linear Case and the Kremser Equation

In Chapter 6, attention was repeatedly drawn to the limiting case of a *linear phase equilibrium*. In the case of gas absorption and adsorption, the linearity was expressed through *Henry's law*, $p = Hx$ or $X = HY$, with an associated Henry's constant H. In the case of liquid–liquid phase equilibria, the constant distribution coefficient m played the role of a Henry's constant (cf. Equation 6.9). Compartmental models used in biology and environmental science almost invariably assume a constant "partition" or distribution coefficient, which is again akin to a Henry's constant.

In principle, linearity is confined to vanishingly small solute concentrations or pressures, often referred to as the "Henry's law region." This has been repeatedly confirmed experimentally and is in fact a thermodynamic requirement. It does, however, often extend to higher solute pressures or concentration, as we had shown in Illustration 6.4.

The case of a staged countercurrent cascade with a linear phase equilibrium was first addressed some 75 years ago in two landmark papers: A. Kremser, *Natl. Petr. News* 22(21), 42 (1930); M. Souders and G.G. Brown, *Ind. Eng. Chem.* 24, 5190 (1932). Although the equations derived in these articles apply to dilute solutions only, their importance has grown enormously in recent years. Two factors conspired to bring this about. One was the increasing necessity to deal with and remove toxic and other objectionable substances at the *trace level*. Countercurrent gas absorption, operating in the Henry's law region, is among the tools used to achieve this. The second reason is the need, in an era of diminishing resources and escalating commodity prices, to extract as much as possible of a valuable substance from a process stream. This applies particularly to the mineral processing and metallurgical industries and forces us yet again to operate, at least in part, in the Henry's law region.

The analytical expression relating the number of stages in a countercurrent cascade to the operating parameters in the Henry's law region can be obtained by performing steady-state algebraic mass balances and also involves a bit of inspired algebra. The reader is encouraged to undertake the task (Practice Problem 7.6). The result, known as the Kremser–Souders–Brown equation (or Kremser equation for short), is given by

$$1 - R = \frac{E^{n+1} - E}{E^{n+1} - 1} \tag{7.14a}$$

or in rearranged form

$$N = \frac{\log[R(1 - 1/E) + 1/E]}{\log E} \tag{7.14b}$$

A plot of Equation 7.14b appears in Figure 7.13. By suitably defining the parameters R and E, the Kremser equation can be used *for any countercurrent cascade involving linear phase equilibria*. These parameter definitions are listed in Table 7.1 and provide a convenient dictionary for use in a number of important operations.

The parameter R, which we term the *residue factor*, is a direct measure of the amount of residual solute leaving the cascade. The smaller the value of R is, the lower the effluent concentrations Y_N or X_N, and hence the higher the degree of solute recovery. The parameter E, on the other hand, is identical to the extraction ratio we defined for crosscurrent extraction (Equation 7.11a)

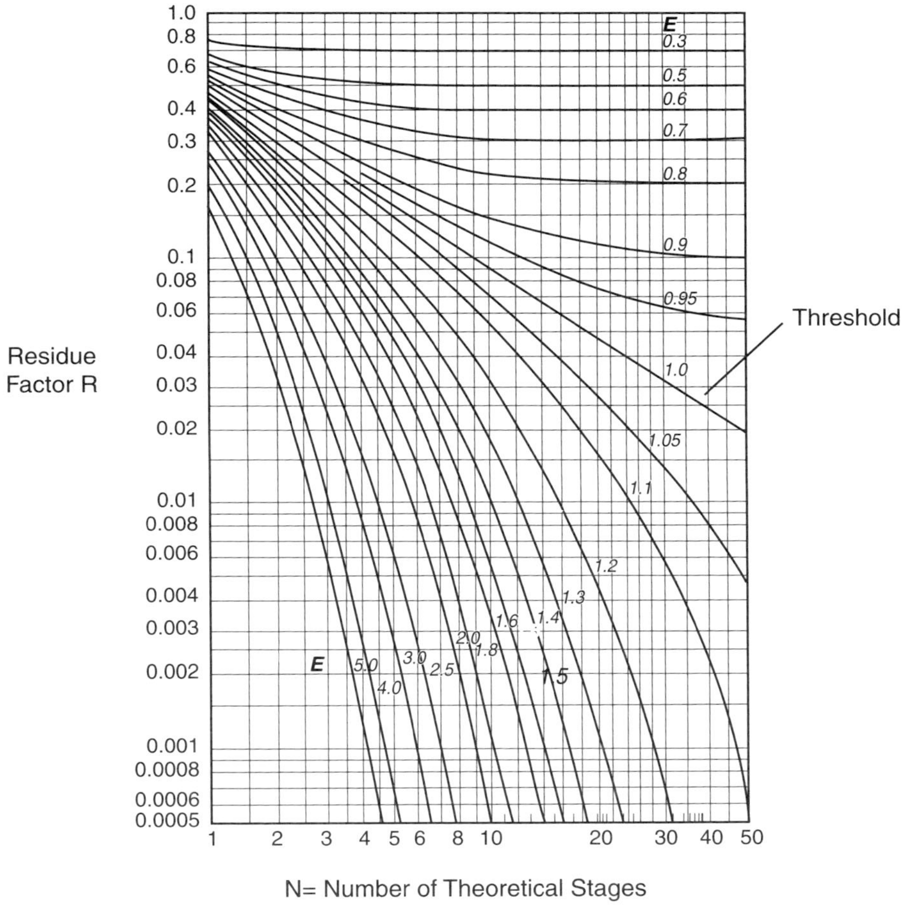

FIGURE 7.13
Countercurrent cascade with linear equilibrium: graphical representation of the Kremser equation — for definition of the residue factor *R* and extraction ratio *E*, see Table 7.1.

and varies directly with the amount of solvent or adsorbent used and their capacity. Large values of *E* lead to good recoveries and low effluent concentrations, both desirable features. Note that the subscript 0 refers to the entering feed to be treated, and that for *pure* solvents or stripping agents, $X_{N+1} = Y_{N+1} = 0$. In leaching and washing, $X = \text{kg } C/\text{kg } (B + C)$, $S = \text{kg } (B + C)$ per unit time, both referring to the solid (sludge) phase. B denotes solvent, C denotes solute.

The Kremser plot, Figure 7.13, serves several additional useful purposes. We illustrate this with the following examples.

Suppose that an existing gas absorber using clean solvent has its feed rate *G* doubled over the previous design value. A quick scan of the Kremser plot shows that to maintain the same effluent concentration Y_N or solute recovery

TABLE 7.1

Operating Parameters for Use in the Kremser Plot

Operation	R	E	Equilibrium
Gas absorption	$\frac{Y_N - HX_{N+1}}{Y_0 - HX_{N+1}}$	L/HG	Y (gas) = HX (liquid)
Gas stripping	$\frac{X_N - Y_{N+1}/H}{X_D - Y_{N+1}/H}$	HG/L	Y (gas) = HX (liquid)
Liquid extraction	$\frac{X_n - Y_{N+1}/m}{X_0 - Y_{N+1}/m}$	mB/A	Y (extract) = mX (raffinate)
Liquid adsorption	$\frac{Y_N - X_{N+1}/H}{Y_0 - X_{N+1}/H}$	HS/L	X (solid) = HY (liquid)
Solids leaching	$\frac{X_N - mY_{N+1}}{X_0 - mY_{N+1}}$	L/mS	Y (liquid) = X/m (solid)
Solids washing (CCD)	$\frac{X_N - Y_{N+1}}{X_D - Y_{N+1}}$	L/S	Y (rinse) = X (sludge)

as before, the solvent flow rate L would likewise have to be doubled. Similar considerations apply to adsorption, extraction, and leaching.

Consider next the same unit being subjected to a twofold increase in feed concentration Y_o. If no remedial action is taken, the residue ratio Y_N/Y_o will remain the same, but the effluent concentration Y_N will be doubled. To bring it down to its previous value, solvent flow rate and hence E would have to be increased. The amount of adjustment needed can once again be determined quickly through the use of Figure 7.13.

Finally, a closer examination of Figure 7.13 shows that at E values below unity, the plots veer off and asymptotically approach a constant value of R. *This implies that we cannot, under these conditions, obtain arbitrarily low effluent concentrations, no matter how high we set the flow rate of solvent or stripping agent or indeed the number of stages.* The value of $E = 1$ consequently represents an important watershed point, below which it becomes impractical or impossible to attain a desired goal. The reasons for this behavior will be addressed more fully in Illustration 8.2.

Illustration 7.7: Comparison of Various Stage Configurations

This example brings together the various stage arrangements discussed in the previous sections and compares their performances. We consider an extraction process that is to be carried out, first in a single stage and then for comparison in two-stage crosscurrent and countercurrent cascades. The two solvents involved are taken to be mutually insoluble, and the task is to determine the quantity of extraction solvent required per unit mass of raffinate solvent to reduce the solute content from a mass ratio of $X_F = 0.1$ to $X_1 = 0.01$. The distribution coefficient is constant at 3.0 (i.e., the equilibrium is linear). We obtain the following results:

1. Single stage

The solute balance for this case is given by

$$\text{Mass of solute in} - \text{Mass of solute out} = 0$$

$$AX_F - (BY_1 + AX_1) = 0 \tag{7.15a}$$

with Y_1 given by the distribution coefficient m':

$$Y_1 = m'X_1 \tag{7.15b}$$

Combining the two equations and rearranging yields

$$B / A = \frac{X_F - X_1}{mX_1} = \frac{0.1 - 0.01}{3 \times 0.01} = 3.3 \tag{7.15c}$$

and hence, $B = 3.3$ kg solvent/kg raffinate solvent.

2. Two-stage crosscurrent cascade

Here we make use of the plot of Figure 7.9 to compute the solvent quantity, which is assumed to be divided equally between the two stages for optimum operation. We have for the ordinate value, assuming pure solvent $y_0 = 0$,

$$\frac{x_n - y_0 / m}{x_o - y_0 / m} = \frac{X_1 - Y_0 / m}{X_0 - Y_0 / m} = \frac{0.01 - 0}{0.1 - 0} = 0.1 \tag{7.15d}$$

For $N = 2$, this yields the parameter value

$$E = mB/A = 3.3 \tag{7.15e}$$

and hence from Figure 7.9,

$$B/A = 3.3/3.0 = 1.1 \tag{7.15f}$$

which is the quantity of solvent used in each stage. By multiplying by two, one obtains for the solvent total

$$B_{Tot} = 2 \times 1.1 = 2.2 \text{ kg solvent/kg raffinate solvent} \tag{7.15g}$$

3. Two-stage countercurrent cascade

Making use of the Kremser plot, Figure 7.13, we have, for the residue factor R,

$$R = \frac{X_2 - Y_0 / m}{X_0 - Y_0 / m} = \frac{0.01 - 0}{0.1 - 0} = 0.10 \quad (7.15h)$$

and hence a solute recovery $r = 1 - 0.10 = 0.9$ or 90%.

The corresponding value of E, read from Figure 7.14, is 2.6. Hence,

$$E = mB/A = 2.6 \quad (7.15i)$$

and

$$B/A = 2.6/3 = 0.87$$

Therefore,

$$B = 0.87 \text{ kg solvent/kg raffinate solvent}$$

This example shows the distinct advantage that the countercurrent process holds, not only over the single-stage operation, which was to be expected, but also over its crosscurrent counterpart. Consumption by the latter is higher by a factor of $2.2/0.87 \cong 2.5$. The principal advantage of the crosscurrent cascade is that it is more easily adaptable to *batch* processing, whereas the countercurrent cascade is by necessity continuous. Cross-current operation has other advantages, which are explored in Practice Problem 7.6.

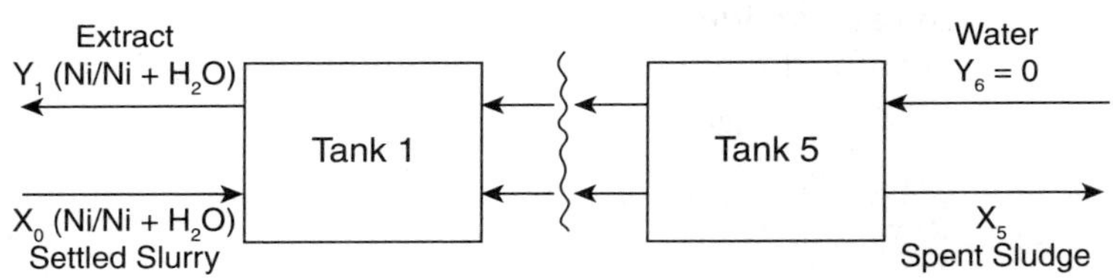

FIGURE 7.14
Countercurrent decantation cascade for nickel extraction at Goro. For results, see Table 7.2.

Illustration 7.8: Mineral Processing: Continuous Countercurrent Decantation (CCD)

The important ores of zinc, copper, and nickel generally contain the metal in the form of sulfides or oxides. Sulfides are traditionally processed by high-temperature "roasting," which converts the ore into oxides of the metallic components and sulfur, SO_x. This technology, dating back many decades, is used in the nickel extraction operations at Sudbury, Ontario, and is, in the public mind, notorious as a major cause of "acid rain" (see Practice Problem 5.3).

When the metal occurs as an oxide, the technology of choice consists of a high-temperature acid leach of the ore, followed by liquid extraction of the metal sulfate using an organic chelating compound. Sandwiched between these two operations lies a seemingly minor but, in actual fact, highly important second step — it involves the thickening and rinsing of the acid leach slurry in what is termed *continuous countercurrent decantation (CCD).* It ensures maximum nickel recovery, and it is this step we wish to address here.

Two planned nickel production facilities, both located on the Pacific island of New Caledonia, will serve to convey a sense of the scale of operations. The Goro facility, which processes the nickel oxide ore laterite, is scheduled to produce 6 tons of nickel per hour by the year 2008. The corresponding amount of sludge to be washed in the countercurrent cascade has been estimated at 5000 tons/h. Evidently this calls for the use of sizable tanks, and the available information indicates that a diameter of 70 m is to be used. The scale of operations at the second planned site, Koniambo, is similar, but the ore here is a sulfide that is processed by roasting.

The planned CCD cascade at Goro is reported to consist of six huge tanks, of which the first is likely to be a straight settler (no water added). It serves to raise the solids content of the acid leach slurry from 40% to approximately 70% (i.e., to recover 50% of the nickel in the leach). The remaining half must be recovered by washing the solids — hence, the crucial role of this step. It is assumed that the nickel content of the entering sludge is $X_0 = 0.0035$ (kg Ni/kg Ni + H_2O). All figures are tentative because of the lack of precise information.

A sketch of the washing portion of the cascade is shown in Figure 7.14. The aim now is to use the Kremser plot, Figure 7.13, to estimate the percent recovery of nickel and its final concentration in the overflow solution for various values of the ratio *L/S* (kg Ni + kg H_2O per hour in the overflow and sludge, respectively). This is done for a cascade of 4, 5, and 6 tanks in order to explore the sensitivity of the results of the parameter *N* and to find a possible justification for the actual number of tanks used. The results of the calculations are shown in Table 7.2.

The item of foremost interest and greatest economic impact is the percent nickel recovery. Using the average price of nickel in 2005 of US$14,000 per ton, the value of each percent of nickel recovered works out to $3.7 million per year. We would expect, therefore, that a preliminary design of a CCD cascade would aim at recoveries in excess of 99%.

We previously identified a ratio of *L/S* equal to 1 as a threshold between potentially poor and good performance. This value clearly does not meet the criterion for acceptable performance, irrespective of the number of stages used. An increase of *L/S* to 2 (i.e., using twice the amount of water contained in the sludge as rinse) brings an immediate improvement. A value of 3, finally, either exceeds or falls just short of 99% recovery.

TABLE 7.2

Nickel Recovery and Concentration in a Countercurrent Decanting Cascade (Kremser Equation)

Number of Rinse Tanks	L/S[a]	R[b]	Percent (%) Recovery	Final Overflow Concentration (Y_1)
$N = 4$	1	0.20	80.0	7.0×10^{-4}
	2	0.046	95.4	1.6×10^{-4}
	3	0.015	98.85	5.0×10^{-5}
$N = 5$	1	0.16	84	5.6×10^{-4}
	2	0.025	97.5	8.8×10^{-5}
	3	0.006	99.4	2.1×10^{-5}
$N = 6$	1	0.14	86.0	4.9×10^{-4}
	2	0.015	98.5	5.3×10^{-5}
	3	0.0025	99.75	8.8×10^{-6}

[a] L/S = (Ni + H_2O)/h in overflow/(Ni + H_2O)/h in underflow.
[b] $R = Y_1/0.0035$.

Superficially, at least, the use of five wash tanks can be viewed as being close to optimal. It recovers some $2 million worth more nickel than the four-tank cascade and produces a solution more than twice as concentrated as the wash from the six-tank configuration, which also has a higher capital cost. On that score, the four-tank cascade has a clear advantage over $N = 5$, and it will be a matter of a more detailed economic analysis to determine whether the savings are offset by the lower nickel recovery. In summary, while $N = 5$ appears to be an optimum, $N = 4$ cannot be ruled out as an alternative.

Comments

Note that countercurrent washing is a special case, with $m = 1$, of the more general leaching process, which has $m > 1$, indicating a higher concentration in the sludge due to adsorption. This case can also be handled by the Kremser equation provided m = constant.

7.3 The Equilibrium Stage in the Real World

7.3.1 The Mixer–Settler Configuration

In the introduction to this chapter, the equilibrium stage had been conceptualized as a stirred tank in which the two phases are brought into intimate contact and allowed to come to equilibrium. They are then separated and conveyed to another stage if necessary. In actual practice, this type of operation is carried out in dual units consisting of a mixer (the stirred tank) and a settler, where the two phases are allowed to separate by gravity. A three-stage countercurrent cascade using this configuration is shown in Figure 7.15.

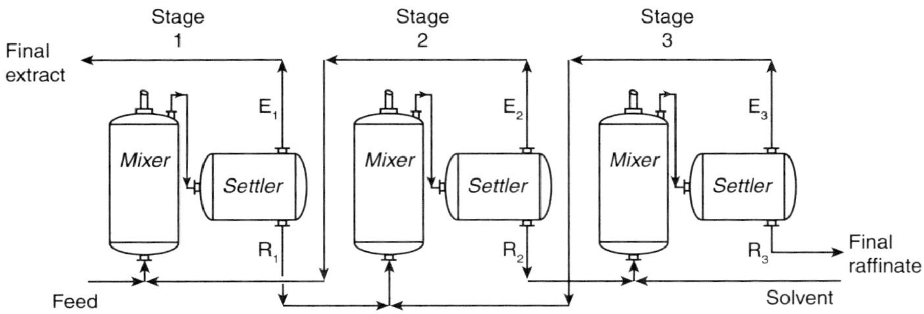

FIGURE 7.15
The mixer–settler configuration in a countercurrent cascade.

The principal use of the mixer–settler combination is in liquid extraction, and even here it is usually limited to relatively small volume operations requiring no more than two to three stages. The main impediment is the high power requirements associated with high-volume multistage configurations. On the other hand, the combination allows considerable latitude in attaining high stage efficiencies, and if the solvent has a high selectivity (large distribution coefficient m), it may become the preferred mode of operation. Similar considerations apply to liquid-phase adsorption processes.

7.3.2 Gas–Liquid Systems: The Tray Tower

Almost all staged separation processes involving gas–liquid or vapor–liquid systems employ *trays* as the principal contacting device. A vertical assembly of such trays, usually spaced at a distance of a few 10 cm apart is termed a tray tower or column, of which the fractionation towers seen in oil refineries are the most familiar example. The tray contains many perforations or openings through which the gas or vapor flows in the upward direction, while the liquid flows at right angles across the tray, contacting the gas as it does, and exchanging material with it. The gas or vapor disengages and continues upward to the next tray, while the liquid phase overflows the tray into a duct (the "downcomer") that conveys it to the next tray below.

The three principal types of trays in current use are shown in Figure 7.16a through Figure 7.16c. The simplest of these is the sieve tray (Figure 7.16b), which consists of a perforated tray and has no provision for retaining the liquid once the gas flow has ceased or dropped to a low value. The bubble-cap tray (Figure 7.16a) prevents this by providing each opening with a short length of tubing. The gas enters through it, reverses direction, and exits through the slots of an inverted cap (the "bubble cap"), bubbling through the liquid phase as it escapes. The valve tray (Figure 7.16c) acts in a similar way, but here the seal is provided by a floating disc, which lifts up and floats when the gas pressure exceeds a certain value.

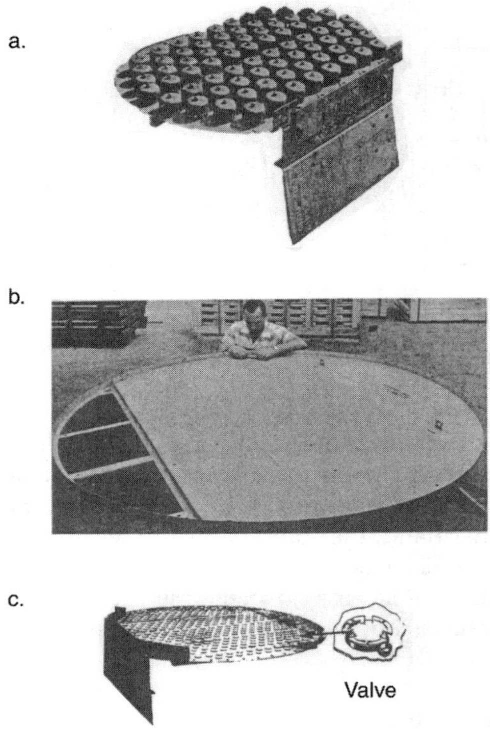

FIGURE 7.16
Types of trays: (a) bubble-cap tray — vapor rises through openings in the plate, reverses direction, and escapes through the slots of an inverted cup; (b) sieve tray; and (c) valve tray — tray openings are adjusted by means of floating disks that rise and fall with the vapor flow rate.

Additional details on trays and tray columns will be provided in Sections 7.4.5 and 7.4.6.

7.3.3 Staged Liquid Extraction Again: The Karr Column

Sieve tray columns have also found an application in staged countercurrent liquid extraction operations. The perforated plates, arranged in much the same way as in gas–liquid contact, act to break up accumulations of the dispersed phase and provide fresh surfaces for renewed mass transfer. Each of the plates, or rather the space between them, is a potential equilibrium stage, but the efficiency E is generally quite low.

E can be enhanced considerably by agitation provided by impellers mounted on a central rotating shaft and located in the space between the sieve plates. In the Karr column, shown disassembled in Figure 7.17, the spacing between the perforated plates is only of the order of a few centimeters, making it possible to attain relatively high stage efficiencies. Wire mesh packing is also sometimes used in lieu of sieve trays as a redistributor.

FIGURE 7.17
Karr extraction column plate assembly.

7.3.4 Staged Leaching: Oil Extraction from Seeds

Leaching, a hugely important and much neglected mass transfer operation, is often carried out in a staged configuration. If the solid phase is finely divided and the principal resistance resides in the external liquid, some form of agitation recommends itself. To avoid the high cost of impeller-driven mixing, use is sometimes made of *air-lifts*. In this device, air is introduced at the bottom of a hollow, slowly rotating central shaft and entrains the slurry for the top, where it is distributed and spread over the entire width of the holding tank. The bottom of the shaft carries a revolving rake, which moves the settled and leached solids toward the center for discharge, in much the same way as in conventional thickeners and decantator tanks (see next section).

In the leaching of most naturally occurring products, such as vegetable oils, the main resistance remains in the solid phase, even after preconditioning by hulling, crushing, and so forth. Agitation is in this case omitted, and contact between the two phases is by *percolation*.

An industrial unit for the countercurrent staged extraction of vegetable oils is shown in Figure 7.18. The stages consist of partitioned circular segments, each containing a charge of seeds resting on a screen bottom (Figure 7.18a). As the assembly rotates slowly, solvent is intermittently

a.

b.

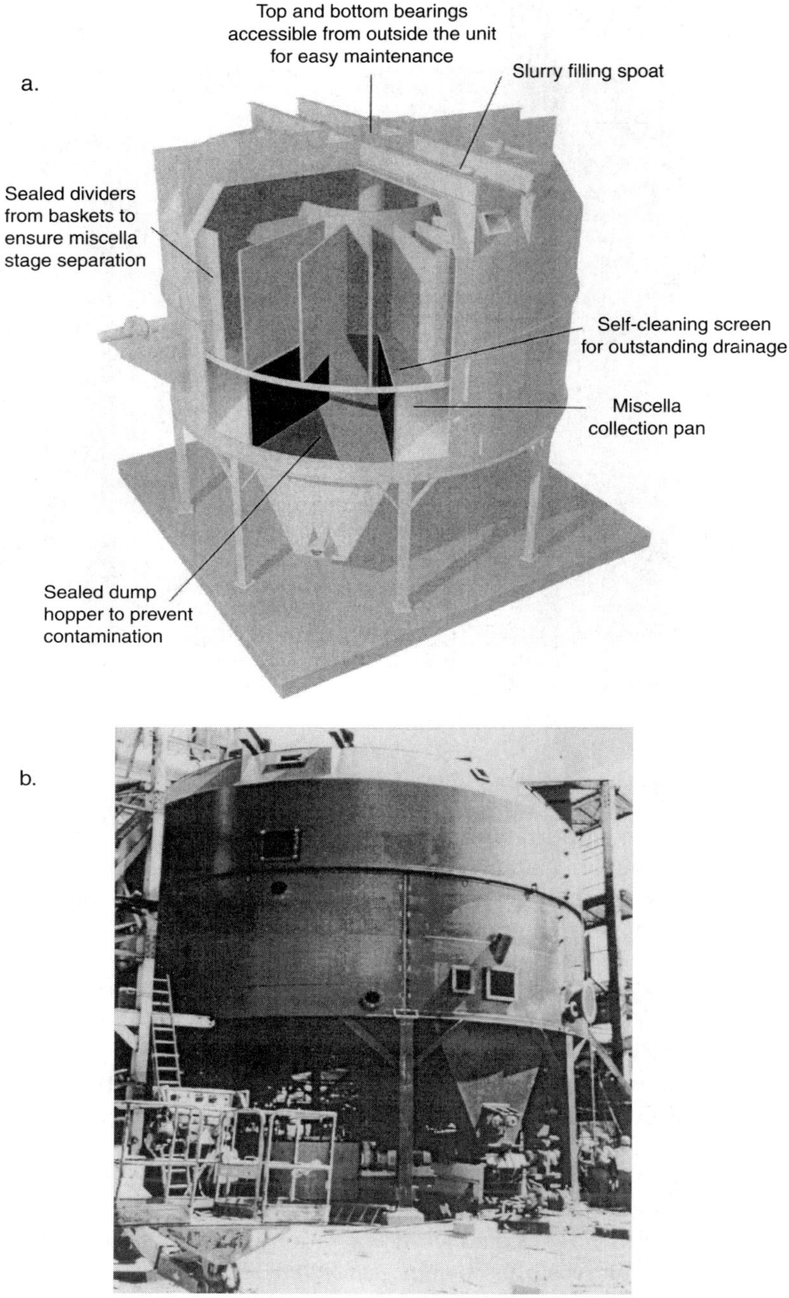

FIGURE 7.18
The reflex extractor: (a) design features and (b) installed unit. (Courtesy of De Smet S.A.)

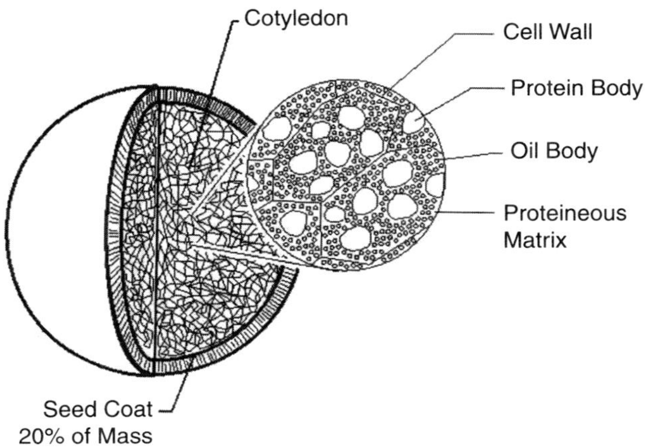

FIGURE 7.19
Mustard seed structure. (Courtesy of D. Balke and L. Diosady, University of Toronto.)

sprayed onto the seeds of a particular stage and allowed to percolate through. It is then collected at the bottom and pumped to the next stage. The final extract, termed *miscella*, contains a small amount of finely divided solid residue, which has to be removed before further processing. The actual industrial unit, shown in Figure 7.18b, is tightly encased to prevent solvent evaporation and loss.

An example of the culprits in this process, a mustard seed, is shown in Figure 7.19. Seed diameter is about 1 mm, that of the oil droplets ~1 μm. The seed coat is crushed in the pretreatment, but individual cell walls may remain largely intact.

7.3.5 Staged Washing of Solids (CCD)

The importance of washing solids after leaching and settling was demonstrated vividly in Illustration 7.8. The decantation tank in which the process is carried out is often referred to as a *thickener*, because the concentration of slurries by settling is carried out in similarly constructed tanks. Washing often entails some thickening so that the term is not misplaced.

A sketch of the device is shown in Figure 7.20a, and a picture of an actual unit appears in Figure 7.20b. Both the wash water and the slurry enter at the top through a central well and gradually flare out and settle to the bottom, where a gently rotating rate directs the sludge to a central discharge port. Washing takes place during the entire settling process, and the arrangement ensures that a clear overflow is obtained. Efficiencies are generally high although concentration levels in the tank vary with location.

Not much has changed in the design of these units over the past century, primarily because they are eminently successful in accomplishing their mission.

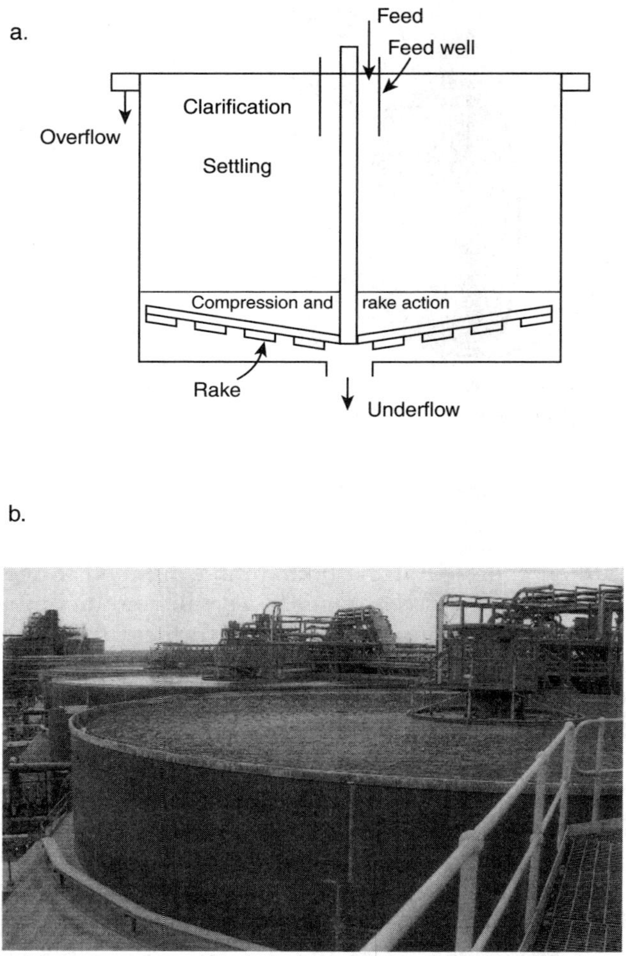

FIGURE 7.20
CCD thickeners: (a) schematic diagram and (b) units used in copper refining. (Olympic Dam, Southern Australia.)

7.4 Multistage Distillation

Although the exact origins of the art of distillation remain obscure, there is considerable evidence that its invention and early development took place in the first century in Alexandria, Egypt. A millennium later it was used for the first time in Northern Italy to produce alcoholic beverages. As may have been surmised, it was this enticing application that preoccupied the practitioners of the art in subsequent years. The early installations consisted of crude batch stills, but it was even then recognized that running them in sequence would lead to even higher enrichments. Thus, the idea of staged distillation was born. The invention of the *distillation tray column* is generally credited to the Frenchman Cellier-Blumenthal (1818), who developed the device for the purification of alcohol used in the extraction of sugar beets. The development of the chemical industry, and later of the oil and petrochemical industries, triggered an immense surge in the use of distillation columns, which remains unabated to this day. It is estimated that in 1990, some 40,000 distillation columns were in operation in the United States, accounting for 7% of the country's total energy consumption. In spite of its energy-intensive nature, distillation continues to maintain its rank among the preeminent separation technologies.

7.4.1 Continuous Fractional Distillation

The operation of a conventional tray fractionation column, and the associated variables are sketched in Figure 7.21. Although the process has most of the standard properties of a staged countercurrent operation seen before, it displays a number of new features that require special mention:

- The feed to be fractionated, which can be a liquid or vapor, or even a combination of both, enters the column at some central location rather than at one of the ends of the cascade as had previously been the case. This results in a division of the column into two parts, the *rectification* or *enriching section* above the feed tray, and the *stripping* or *exhausting section* below it. The upper section serves to enrich the vapor in the more volatile components, a portion of which is ultimately withdrawn as liquid "overhead product" or "distillate." In the lower section, residual volatile components are progressively stripped off the liquid and conveyed upward as vapor, while the downward flow of liquid becomes enriched in the heavier or less volatile components.
- The two streams entering the ends of the column are generated in separate vaporizers or "reboilers" and in condensers located at the two outlets. At the top, vapor leaving the column is condensed and returned in part to the column as "reflux," while the remainder is withdrawn as distillate. At the bottom, the exiting liquid is partly revaporized in a reboiler, and the vapor is diverted back into the column. The remainder of the stream is removed as "bottom product" or residue.
- In contrast to the processes we encountered previously, which were largely or entirely isothermal in nature, distillation has substantial heat effects associated with it. Consequently, we expect heat balances to be involved in modeling the process, as well as the usual mass balances and equilibrium relations. These balances are formulated

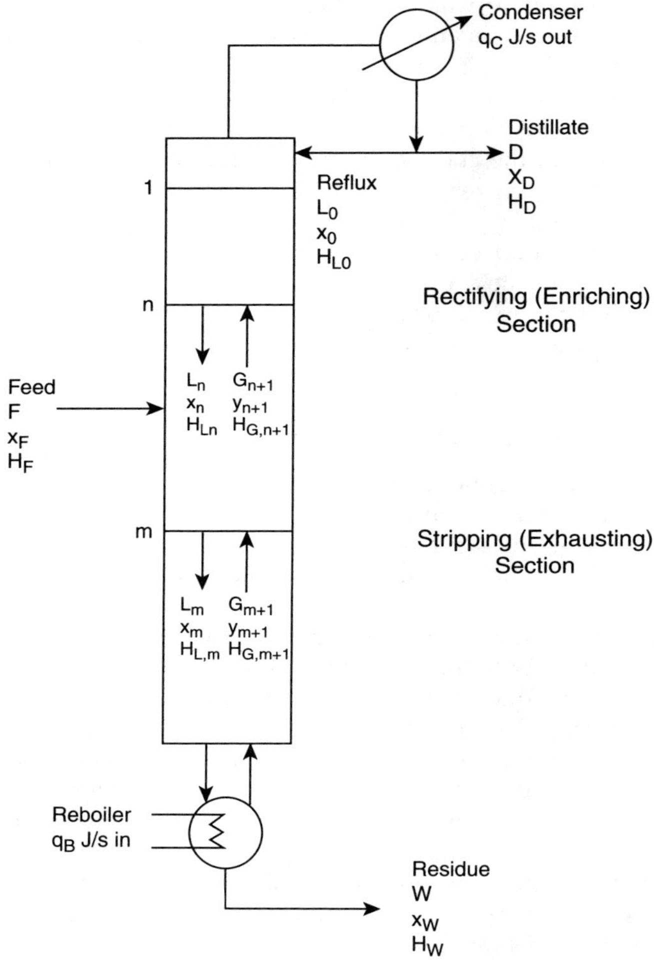

FIGURE 7.21
The fractionation column.

entirely in molar units, because the underlying equilibrium relations, such as Raoult's law and its extension, or the separation factor α, are all described in terms of mole fractions. Thus, the flow rates L and G, which appear in Figure 7.21, are both in units of moles per second, enthalpies H in units of joules per mole, and the liquid and vapor compositions are expressed as mole fractions x and y of a binary system.

- Column operation is usually taken to be isobaric, so that temperatures within the column will vary and lie in the range between the boiling points of the overhead and bottom products. The appropriate equilibrium diagrams for this case are the boiling point diagram (Figure 6.18a and b) and the x–y diagram (Figure 6.17c). The latter is used in the construction of the operating diagram to be discussed later.

7.4.2 Mass and Energy Balances: Equimolar Overflow and Vaporization

There are three balances for the binary system considered here: the total mole balance, the component mole balance, and the heat balance. It is customary to apply these in unison and, in turn, to three separate regions of the fractionation column. One set each is used to describe conditions above and below the feed tray (i.e., the rectifying and stripping sections). These balances are taken over the entire upper and lower portions and include the reboiler and condenser as well as the product stream. A third set of balances is applied to an isolated stage, the feed tray, and includes the flow rates and thermal condition of the incoming feed. Let us see how this works out in practice. For the rectifying section, we have

Rate of total moles in – Rate of total moles out = 0

$$G_{n+1} - (L_n + D) = 0 \tag{7.16a}$$

Rate of component moles in – Rate of component moles out = 0

$$y_{n+1}G_{n+1} - (x_nL_n + x_DD) = 0 \tag{7.16b}$$

Rate of heat in – Rate of heat out = 0

$$G_{n+1}H_{G,n+1} - (L_nH_{L,n} + DH_D + q_C) = 0 \tag{7.16c}$$

where q_C is the heat removed in the condenser.

Elimination of D from the mass balances leads, after some rearrangement, to the following expression:

Enriching Section

$$\boxed{\frac{L_n}{G_{n+1}} = \frac{y_{n+1} - x_D}{x_n - x_D}} \tag{7.16d}$$

This is the operating line for the enriching section. It relates the vapor composition y_{n+1} entering a tray and that of the liquid leaving, x_n, and has a slope equal to the ratio of liquid-to-vapor flow rates L_n/G_{n+1}.

The corresponding operating line for the stripping section is given by

Stripping Section

$$\boxed{\frac{\bar{L}_n}{\bar{G}_{n+1}} = \frac{\bar{y}_m - x_w}{\bar{y}_{n+1} - x_w}} \tag{7.16e}$$

where the bar is used to indicate the position *below* the feed tray to distinguish these quantities from their counterparts in the rectifying section.

To this point it has been assumed that all quantities vary from tray to tray and that, consequently, 3(N+2) balances will be required to describe the operation. Here N denotes the number of stages, with an additional 3×2 equation needed to balance the flow of mass and heat about the condenser and reboiler. These have to be further supplemented by expressions relating each enthalpy to the key state variables x, y, and T. Evidently, we are dealing with a model of considerable complexity and dimensionality, which would require a numerical solution.

We now introduce a concept that avoids this complication and drastically reduces the complexity of the model. We draw for this purpose on the heat balance, Equation 7.16c, and rearrange it to read

$$\frac{L_n}{G_{n+1}} = 1 - \frac{H_{G,n+1} - H_{L,n}}{q_C / D + H_D - H_{L,n}} \tag{7.16f}$$

where q_C is the condenser heat load (in J/s). A first simplification comes about by noting that the sensible heat of the liquid on tray n, $H_{L,n}$, is much smaller than the term $q_C/D + H_D$, which involves the latent heat of condensation. $H_{L,n}$ can, therefore, to a good approximation, be neglected in comparison to this term. If we make the further assumption that the substances have very similar molar latent heats of vaporization ΔH_v, we can approximate the numerator by the relation

$$H_{G,n+1} - H_{L,n} \doteq \Delta H_v \tag{7.16g}$$

so that the ratio of flow rates becomes a *constant* for *all* trays of the enriching section. Similar arguments can be applied to the exhausting section, with the result that

$$\frac{L_n}{G_{n+1}} = K_1 \tag{7.16h}$$

$$\frac{\overline{L}_m}{\overline{G}_{n+1}} = K_2 \tag{7.16i}$$

Consider next a total mole balance about the nth tray:

$$(L_{n-1} + G_{n+1}) - (L_n + G_n) = 0 \tag{7.16j}$$

If we now substitute Equation 7.16g into this relation, there results

$$K_1 G_n + G_{n+1} - K_1 G_{n+1} + G_n = 0 \tag{7.16k}$$

and, consequently,

$$G_{n+1} = G_n \tag{7.17a}$$

and

$$L_{n-1} = L_n \tag{7.17b}$$

We have, in other words, shown that if we assume latent heat to be the predominant thermal quantity in distillation and that this property comes close to being identical for the two components, the vapor and liquid flow can, for practical purposes, be considered constant in the rectifying section. A similar procedure applied to the stripping section leads to the conclusion that here also the flow rates will remain practically constant although different in value from those of the enriching section because of the intervening flow of feed.

What was just derived is referred to as the *Principle of Equimolar Overflow and Vaporization*.

7.4.3 The McCabe–Thiele Diagram

One immediate consequence of the principle of equimolal overflow and vaporization is a drastic simplification of the operating lines (Equation 7.16d and Equation 7.16e). Whereas previously these expressions had to be plotted laboriously step by step from tray to tray, we are now dealing with a straight line that can be drawn easily, knowing only its slope and the location of one point or, alternatively, two points. The slope L/G of the enriching operating line is referred to as the *internal reflux ratio*. It is not a convenient quantity to deal with, because it is usually not known or specified, and it is not easily manipulated in actual column operations. A more suitable parameter, and one easily controlled by appropriate valve settings, is the *external reflux ratio*, or reflux ratio R for short. This quantity represents the ratio of liquid flow L returned to the column as reflux to the flow of distillate D withdrawn as product, and it is given by

$$R = L/D \tag{7.18a}$$

Thus, when $R = 3$, 3 mol of liquid product are returned to the column for each mole withdrawn.

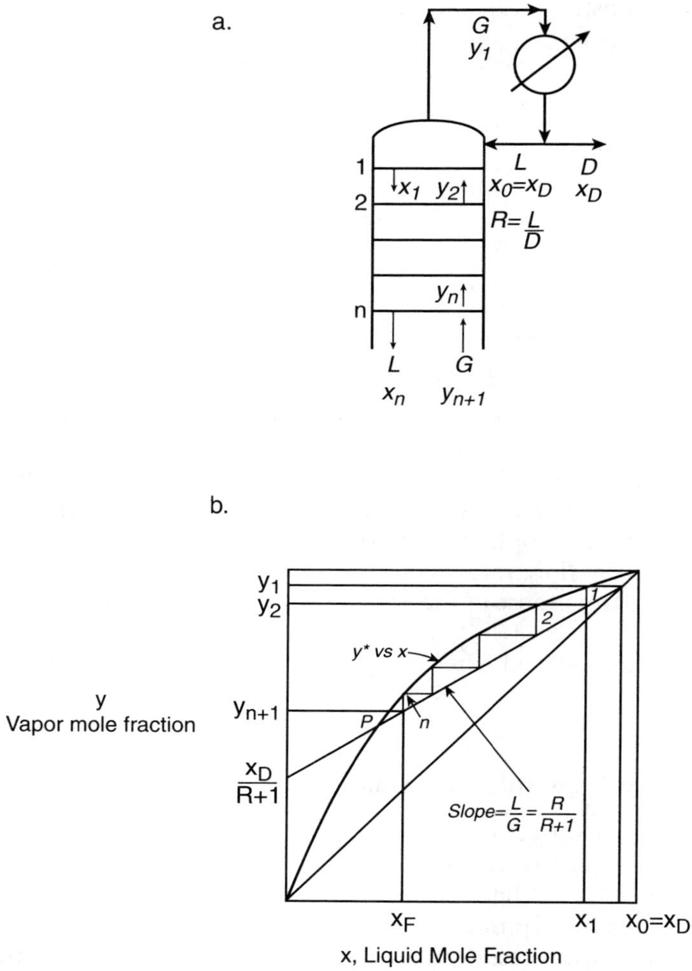

FIGURE 7.22
Details of the rectifying (enriching) section: (a) flow diagram and (b) operating diagram.

We can now recast the operating line (Equation 7.16d) in terms of this new parameter R by using a set of revised mole balances around the envelope shown in Figure 7.22a. We have, in the first instance,

$$\text{Rate of total moles in} - \text{Rate of total moles out} = 0$$

$$G - (L + D) = 0 \tag{7.18b}$$

and we obtain after combining this expression with Equation 7.18a,

$$G = D(R + 1) \tag{7.18c}$$

The corresponding component mole balance is given by

$$Gy_{n+1} - (Lx_n + Dx_D) = 0 \tag{7.18d}$$

or

$$y_{n+1} = \frac{L}{G}x_n + \frac{D}{G}x_D \tag{7.18e}$$

Consequently, using Equation 7.18a and Equation 7.18c yields the following:

OPERATING-LINE ENRICHING SECTION

$$y_{n+1} = \frac{R}{R+1}x_n + \frac{x_D}{R+1} \tag{7.18f}$$

This is the equation of a straight line with slope $R/(R+1)$, which has an intercept on the ordinate of $x_D/(R+1)$ and passes as well through the point $y = x_D$ on the 45° diagonal. This point and the y-intercept, which are usually known or prescribed, permit easy construction of the operating line, as shown in Figure 7.22b. The concentration associated with the various trays can then be derived using the "staircase" construction we employed for the countercurrent gas absorber described in Section 7.1.4 and Figure 7.12. Each point on the curve $y^* = f(x)$ represents the concentrations leaving a particular tray, (y_n, x_n), which are in equilibrium with each other, while those on the operating line relate the entering vapor composition to that of the exiting liquid. Equation 7.18f is quite general and applies to multicomponent systems of any description. The McCabe–Thiele construction, on the other hand, is confined to binary systems.

Let us next consider the corresponding balances around the exhausting section, shown in Figure 7.23a. Here there is no simple relation to the reflux ratio R, but the operating line (Equation 7.16e) is nevertheless considerably simplified because the flow rates are now constant throughout the entire section, although they differ in magnitude from those of the rectifying portion of the column. We have for the total and component mass balances, respectively,

$$\text{Rate of moles in} - \text{Rate of moles out} = 0$$

$$(\bar{G} + W) = 0 \tag{7.19a}$$

and

$$\bar{L}x_m - (\bar{G}y_{m+1} + Wx_w) = 0 \tag{7.19b}$$

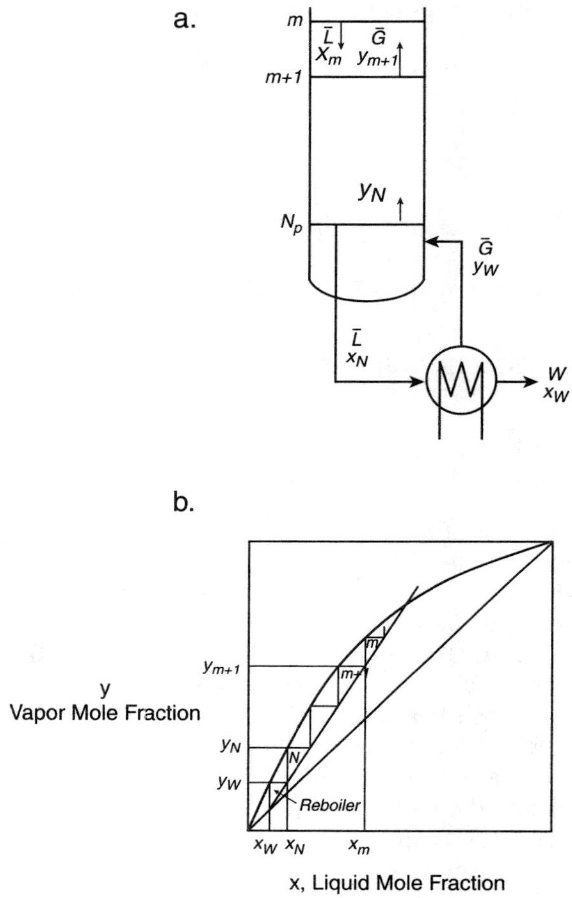

FIGURE 7.23
Details of the exhausting (stripping) section: (a) flow diagram and (b) operating diagram.

so that

$$y_{m+1} = \frac{\bar{L}}{\bar{G}} x_m - \frac{W}{\bar{G}} x_w \tag{7.19c}$$

or, alternatively,

OPERATING-LINE STRIPPING SECTION

$$y_{m+1} = \frac{\bar{L}}{\bar{L}-W} x_m - \frac{W}{\bar{L}-W} x_w \tag{7.19d}$$

This is the equation for the straight operating line shown in Figure 7.23b, which has a slope of $\bar{L}/\bar{L} - W$ and passes through the point $y = x = x_w$ on the 45° diagonal. If the vapor entering the column y_w is in equilibrium with the residue composition x_w, the reboiler can be taken to represent an additional equilibrium stage, and this is so indicated in the operating diagram of Figure 7.23b. Once the operating line is established, the usual staircase construction can be used to step off the number of stages, as shown in Figure 7.23b. It is not clear, however, how this is to be accomplished, because only one point, $y = x_w$ on the diagonal, is known or prescribed, and we have no *a priori* knowledge of either a second point or the slope of the line. In other words, the equation for the stripping operating line, as it stands, contains too many undefined variables. Some reflection will show that this must indeed be so. This follows from the fact that the liquid flow will be influenced by what comes down from the feed tray, and we must consequently draw on an additional balance, *performed around the feed tray, Figure 7.24a.*

We first compose the total mole balance and follow this up with an energy balance around the feed tray. We obtain

$$\text{Rate in} - \text{Rate out} = 0$$

$$(G + \bar{L}) = 0 \tag{7.20a}$$

and

$$FH_F + LH_L + \bar{G}H_G = GH_G + \bar{L}H_L \tag{7.20b}$$

These two equations can be combined and rearranged to yield the expression

$$\frac{\bar{L} - L}{F} = \frac{H_G - H_F}{H_G - H_L} = q \tag{7.20c}$$

The ratio of enthalpy differences, which appears in this equation and which we denoted q, represents the molar heat of vaporization of the *feed* divided by the molar heat of vaporization of the liquid in the column, assumed to be constant. Suppose, for example, that the feed consists of either saturated vapor or saturated liquid. Then the value of q will be 0 or 1, respectively, and for a partially vaporized feed it will be somewhere between these two limits. The quantity q is therefore a dimensionless measure of the thermal quality of the feed. It turns out that q also enters into the construction of the locus of the points of intersection of the two operating lines of the enriching and stripping sections. That locus is represented by the expression

$$y = \frac{q}{q-1}x - \frac{x_F}{q-1} \tag{7.20d}$$

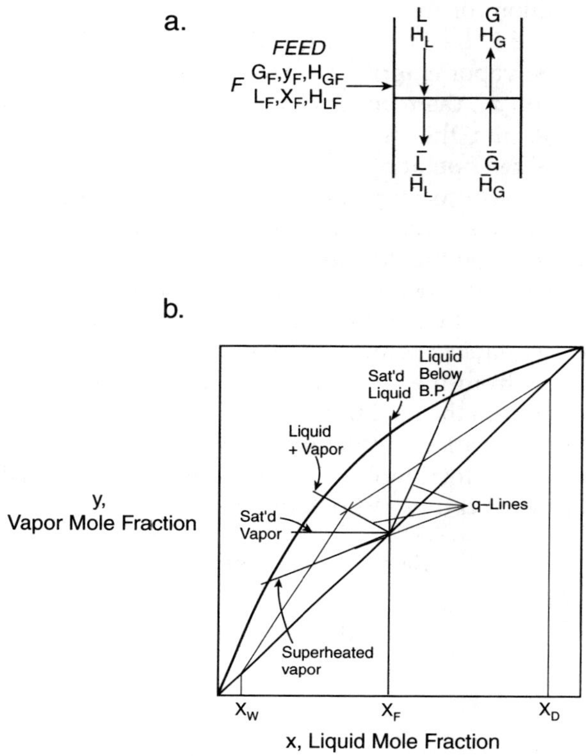

FIGURE 7.24
The feed plate: (a) flow diagram and (b) feed quality and the q-line.

which for a given q is the equation of a straight line of slope $q/(q-1)$ passing through the point $y = x = x_F$. A series of such lines for various thermal conditions of the feed is shown in Figure 7.24b. Proof of these relations is somewhat lengthy and can be found in standard texts.

7.4.4 Minimum Reflux Ratio and Number of Plates

Recall that in the discussion of the countercurrent gas scrubber, Section 7.2.2, we mention the limiting case that arises when the slope of the operating line and the associated solvent flow is progressively reduced until it intersects the equilibrium curve. This results in a condition termed a *pinch* and corresponds to a cascade with an infinite number of stages and a minimum solvent requirement. Any solvent flow rate below this value causes a rise in the effluent concentration and can therefore no longer meet the specified solute recovery.

In distillation, the liquid reflux returned to the top of the column plays, in a sense, the role of a solvent. In the course of its downward flow, it dissolves residues of the heavy component contained in the vapor phase

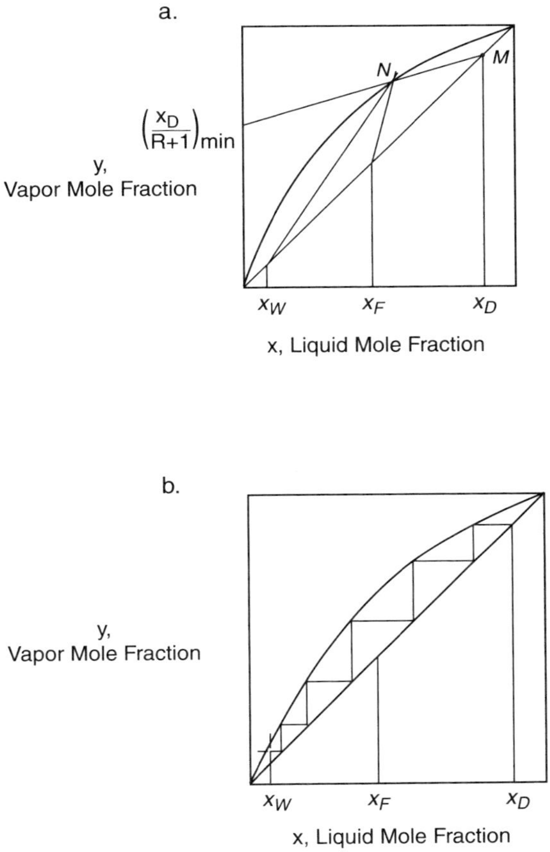

FIGURE 7.25
Two limiting conditions: (a) minimum reflux and (b) minimum number of trays (total reflux).

and thereby contributes to its enrichment in the lighter, more volatile component. A reduction of reflux, or of the reflux ratio, may therefore be expected to result in an *increase* in the required number of stages in much the same way as happened in the case of the gas scrubber. Ultimately, when the operating lines simultaneously intersect the equilibrium curve, a pinch results, and the number of stages goes to infinity. This is shown in Figure 7.25a. The reflux ratio at which this occurs can be read from the intercept $x_D/(R+1)$ and represents the minimum required to achieve the desired separation.

Let us now proceed in the opposite direction and progressively *increase* the reflux ratio. Both the logic of the preceding argument and Figure 7.25a indicate that this will lead to a *decrease* in the number of theoretical trays required. A limit is reached when no distillate is withdrawn and the entire overhead product is returned to the column as reflux. The operation is then said to be at total reflux. The reflux ratio becomes infinity, $R = L/D = L/0 = \infty$, and the operating line assumes a slope of 1 (i.e., it coincides with the 45° diagonal). This is shown in Figure 7.25b.

These two asymptotic cases are immensely useful in conveying to the analyst the lower limits of R and N, below which the desired separation will no longer proceed. In an actual operation, these values evidently must be exceeded and both R and N will assume finite values. The best or optimum value of R and N to be used will be determined by economic considerations (i.e., when the total cost consisting of fixed and operating costs is at a minimum). Let us see how this minimum comes about.

Operating costs, consisting of condenser and reboiler loads as well as pumping costs, are at their lowest value when R is at a minimum. As R is gradually increased, these costs begin to rise in almost direct proportion and ultimately tend to infinity as $R \to \infty$. There is consequently no minimum in this item. The fixed costs, on the other hand, are at first infinite at R_{Min} because of the infinite number of trays required. They then sharply drop as R is slowly increased above its minimum value because the number of trays has now become finite. As this process is continued, however, the size and cost of the condenser, reboiler, and reflux pump begin to creep up, eventually overtaking the cost reduction occasioned by the decrease in the number of plates. The fixed costs consequently, and in contrast to the operating costs, pass through a minimum, and the sum of the two results in an optimum reflux ratio, R_{Opt}. It has been found in practice that this optimum lies close to the minimum reflux ratio and is frequently, but not always, found in the range 1.2 to 1.5 times R_{Min}. This set of values is invariably used in the preliminary design of fractionation columns.

Refer to Figure 8.2 for a similar optimization problem that arises in the operation of gas scrubbers.

7.4.4.1 Comments

The McCabe–Thiele diagram is rarely used in engineering practice except for the occasional simple binary system. It remains, however, a powerful pedagogical tool for conveying the essence of staged distillation processes and is used by engineers as an important reference point. The landmark paper describing the construction appeared over 80 years ago (W.L. McCabe and E.W. Thiele, *Ind. Eng. Chem.*, *17*, 605 [1925]). Thiele, a student of McCabe's at Massachusetts Institute of Technology, later went on to develop the first model for the simultaneous diffusion and reaction in a catalyst particle (see Thiele modulus, Chapter 4).

Illustration 7.9: Design of a Distillation Column in the McCabe–Thiele Diagram

The intent in this example is to acquaint the reader with the principal steps involved in designing a distillation column and to demonstrate their

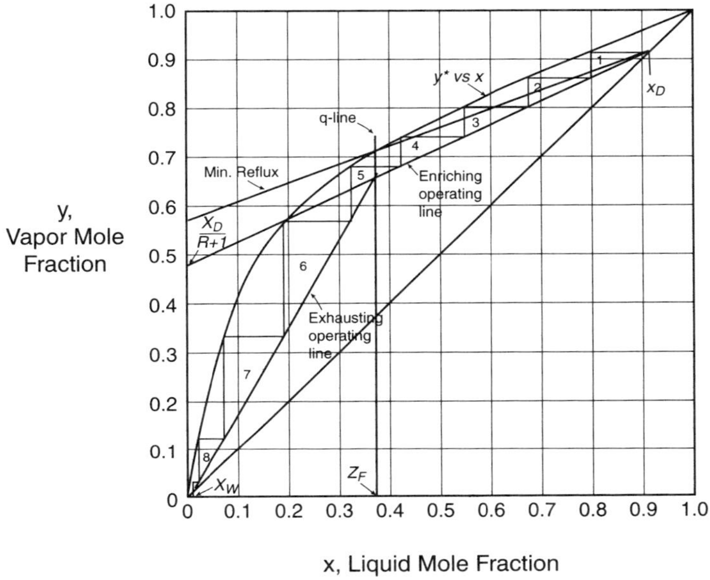

FIGURE 7.25–7.9
Operating diagram for Illustration 7.9.

implementation by means of the McCabe–Thiele diagram. We start by listing the parameters, which in the usual course of the design are either prescribed or else known *a priori*:

1. Feed rate F and composition x_F
2. Thermal condition of feed
3. Distillate composition x_D and bottoms composition x_w
4. Reflux ratio, usually specified as a multiple of the minimum reflux ratio
5. Thermal condition of the overhead and bottoms product

The parameters or quantities to be calculated are as follows:

1. Number of theoretical stages
2. Heat load for the condenser and reboiler
3. Recovery of overhead product

We consider a hypothetical system with an x–y equilibrium relation as shown in Figure 7.25–7.9. The feed, entering at the rate of 10 mol/s, is known to have a composition $x_F = 0.38$ and consists entirely of saturated liquid at its boiling point. Overhead and bottoms compositions are specified at $x_D = 0.92$ and $x_w = 0.01$, respectively. A value 1.5 times the minimum reflux ratio is to be used.

The first step in the procedure is to establish the q-line. This is a straightforward matter because the feed consists of saturated liquid and the q-line is consequently vertical.

In the second step, an operating line is drawn from distillate composition $y = x = x_D$ through the point of intersection of the q-line with the equilibrium curve, which results in a pinch and represents minimum reflux conditions. The intercept of this line on the ordinate is given by $x_D/(R_{Min} + 1)$ and establishes the minimum reflux ratio. We have

$$\frac{x_D}{R_{Min} + 1} = 0.575 \tag{7.21a}$$

and, consequently,

$$R_{Min} = \frac{0.92}{0.575} - 1 = 0.6 \tag{7.21b}$$

The actual reflux ratio to be used is 1.5 times this minimum — that is, $R = 1.5 \times 0.6 = 0.9$.

We now repeat the second step but this time draw the line through the new intercept — that is, $x_D/(R + 1) = 0.92/(0.9 + 1) = 0.48$. This is the actual operating line to be used in stepping off the number of theoretical trays.

In the fourth step, we start the staircase construction at the distillate composition $y = x = x_D$ and proceed downward, alternating between equilibrium curve and the enriching operating line. When the feed composition is reached, a crossover is made to the exhausting operating line, with the feed tray 5 straddling the q-line. The construction is then continued until we reach the bottoms composition $y = x = x_w = 0.01$. This occurs after eight stages have been stepped off. Because the liquid bottoms product and the vapor returning to the column are usually at or near equilibrium, it is customary to consider the reboiler as an additional theoretical stage. The column will consequently comprise eight theoretical stages plus the reboiler.

Let us next turn to the computation of condenser and reboiler heat loads. We assume that the relevant enthalpies have been computed and are given by

$$H_G = 50 \text{ kJ/mole}, \ H_{L,D} = 10 \text{ kJ/mole}, \ H_{L,W} = 12 \text{ kJ/mole}$$

To compute the heat loads, we require the vapor flow rates into the condenser and out of the reboiler. These are obtained by a series of total and component balances, starting with those taken over the entire column. Thus,

$$\text{Rate of moles in} - \text{Rate of moles out} = 0$$

$$F - (W + D) = 0 \tag{7.21c}$$

and

$$x_F F - (x_W W + x_D D) = 0 \tag{7.21d}$$

or

$$10 - (W + D) = 0 \tag{7.21e}$$

and

$$0.38 \times 10 - (0.01\ W + 0.92\ D) = 0 \tag{7.21f}$$

from which

$$W = 5.93 \text{ mol/s and } D = 4.07 \text{ mol/s} \tag{7.21g}$$

We then obtain from Equation 7.18c,

$$\text{Boil-up rate} \quad \overline{G} = D(R + 1) = 4.07\ (0.9 + 1) = 7.73 \text{ mol/s} \tag{7.21h}$$

Because the feed enters entirely as a liquid, the vapor flow rate will remain constant over the entire column, so that $\overline{G} = G$.

We consequently have for the heat loads,

$$q_C = G(H_G - H_{L,D}) = 7.73\ (50 - 10) \tag{7.21i}$$

$$q_C = 309 \text{ kJ/s} \tag{7.21j}$$

and

$$q_B = \overline{G}(H_G - H_{L,w}) = 7.73(50 - 12) \tag{7.21k}$$

$$q_B = 294 \text{ kJ/s} \tag{7.21l}$$

The recovery r of overhead product is given by

$$r = \frac{x_D D}{x_F F} = \frac{0.92 \times 4.07}{0.38 \times 10} = 0.985 \textit{ or } 98.5\% \tag{7.21m}$$

Comments

This example was used to illustrate the use of the McCabe–Thiele diagram and the elegant and simple way in which it conveys the design information.

In practice, the systems tend to be more complex than the simple binary example used here, and the computations are done using appropriate computer packages. Today, these packages are quite powerful and are able to handle mixtures of many components without recourse to the simplifying assumption of equimolal overflow and vaporization. The McCabe–Thiele diagram nevertheless remains a valuable tool for visualizing the principal features of the fractionation process and for providing the student an entry into the treatment of more complex systems.

Illustration 7.10: Isotope Distillation: The Fenske Equation

Vapor–liquid equilibria of isomeric or isotopic mixtures come perhaps closest to showing perfect ideal behavior and a separation factor, which for all practical purposes remains constant over the entire range of compositions. Distillation is the most commonly used technique for the separation or enrichment of several important isotopes. Both C^{13} and $O^{16}O^{18}$ are produced by the low-temperature distillation of carbon monoxide and oxygen in small commercial installations. Although these processes are generally carried out in packed columns, all preliminary design questions are settled by deducing the number of theoretical plates for a required separation, starting with the minimum number needed to achieve the desired result. These values are then translated into packing heights using the concept of the height equivalent to a theoretical plate (HETP), which was addressed in Chapter 5, Section 5.4.

Suppose we wish to gain an idea of the requirements for distilling the isotopic pair $C^{12}O/C^{13}O$, for which the value of α is 1.01 (see Table 6.9) with the equilibrium curve very close to the diagonal, which represents total reflux conditions. Here, a graphical construction on the McCabe–Thiele diagram is not recommended because the operating and equilibrium curves are too close together to allow a precise determination of the number of theoretical plates. Fortunately, for systems with a constant separation factor and operating at total reflux, a simple analytical treatment is possible, which leads to a relation between the minimum number of plates, the separation factor α, and the product compositions. This expression, known as the Fenske equation, is derived as follows.

We start by applying the definition of the separation factor, Equation 6.22a, to the reboiler and obtain

$$\frac{y_w}{1-y_w} = \alpha \frac{x_w}{1-x_w} \tag{7.22a}$$

At total reflux the operating line coincides with the 45° diagonal so that $y_w = x_N$. Equation 7.22a becomes

$$\frac{x_N}{1-x_N} = \alpha \frac{x_w}{1-x_w} \tag{7.22b}$$

A similar scheme can be applied to the *N*th plate, yielding

$$\frac{y_N}{1-y_N} = \alpha \frac{x_N}{1-x_N} = \alpha^2 \frac{x_w}{1-x_w} \tag{7.22c}$$

Continuing the procedure up the column we ultimately obtain

$$\frac{x_D}{1-x_D} = \alpha^{N+1} \frac{x_w}{1-x_w} \tag{7.22d}$$

or, equivalently,

$$N+1 = \frac{\ln \dfrac{x_D}{1-x_D} \dfrac{1-x_w}{x_w}}{\ln \alpha} \tag{7.22e}$$

This is the expression due to Fenske.

A further simplification results for separation factors close to 1, for then we have by a Taylor-expansion of the denominator,

$$\ln \alpha \cong \alpha - 1 \tag{7.23a}$$

and, consequently,

$$N+1 = \frac{1}{\alpha - 1} \ln \left[\frac{x_D}{1-x_D} \frac{1-x_w}{x_w} \right] \tag{7.23b}$$

Let us apply this expression to the distillation of the carbon monoxide isotopes, with the aim of obtaining products of $x_D = 0.9$ and $x_w = 0.1$. We have

$$N+1 = \frac{1}{1.01-1} \ln \frac{0.9}{0.1} \frac{1-0.1}{0.1} \tag{7.24a}$$

and, therefore,

$$N + 1 = 439 \tag{7.24b}$$

or

$$N = 438 \tag{7.24c}$$

Thus, a *minimum* of more than 400 theoretical plates is required to achieve the desired separation; this is an enormous number. Fortunately, high-efficiency packings are now available with equivalent HETP of a few centimeters. This limits the size of the column to heights that are not excessive and can be implemented in practice.

Comments

One feature of isotope distillation that may have been noted is the extreme sensitivity of the number of plates N to the value of α. Suppose that we manage by some means — for example, by a change in operating pressure — to increase the separation factor of the CO isotopes from 1.01 to 1.02. This represents a rather modest increase of only 1% in the value of α, which would normally have no more than a marginal effect on N. For separation factors close to 1, however, the effect is dramatically enhanced as a result of the appearance of the term $\alpha - 1$ in the denominator of Equation 7.23g. The seemingly insignificant change in α of only 1% translates here into a reduction of the number of plates from 438 to fully one half that value. This will evidently result in a considerable reduction in both the size and cost of the fractionation column. Any increase in α, no matter how small in appearance, is therefore welcome in isotope or isomer distillation, and methods for achieving this increase should be fully explored before settling on a particular process. The Fenske equation was also used to calculate the minimum number of stages in the uranium isotope separation plant constructed during World War II (see Illustration 3.8).

At the opposite end of the spectrum (i.e., at high values of α), the exact reverse holds true. Here the logarithmic dependence makes N relatively insensitive to increases in α. Suppose, for example, that x_D and x_W are set at 0.99 and 0.01, respectively. Then, for $\alpha = 2$, we still require 12 plates at total reflux, but for $\alpha = 20$, the number has dropped to only 2. Systems with a higher α can then be treated in an (single-stage) evaporator (see the following illustration).

Illustration 7.11: Forget McCabe–Thiele: The Importance of Initial Planning

The elegance and simplicity of the McCabe–Thiele method make it tempting for the beginning student to view it as the centerpiece of binary distillation column design. There are, in fact, a number of key questions that need to be addressed before proceeding with plate calculations, setting reflux rations, and the like.

TABLE 7.3

Vapor–Liquid Equilibria for H_2O–H_2SO_4 at 1 atm

Wt% H_2SO_4	Boiling Point (°C)	Mole Fractions	
		Liquid	Vapor
50	120	0.155	0
60	140	0.216	0
70	165	0.300	0
75	180	0.355	~ 0 ← Incipient SO_3 formation
98	230	0.900	~ 0.25
98.4	235	0.919	0.919[a] ← Azeotrope

[a] SO_3 reported as equivalent H_2SO_4.

In Illustration 7.9 it was tacitly assumed that the product compositions and operating pressure of the column were given, or at least easily specified. The common conception is that these parameters are determined either by market demand — the customers wants a certain purity — or by the obvious desire to maximize recovery and minimize cost. In actual fact, they often hinge on quite different and mundane factors, such as the quality of available steam or the exit temperature of the cooling medium. The chemistry of the system often intrudes as well, sometimes in unexpected ways.

Another question to be addressed is the configuration of the distillation plant. A single column is often not the best choice, in binary distillations, and here again, some unexpected factors may come into play. We illustrate this with the following example.

Sulfuric acid is used in huge tonnages by the chemical and metallurgical industries, yielding even larger quantities of spent acid, which has to be distilled to recover the H_2SO_4. An initial guide is provided in Table 7.3, which lists the pertinent vapor–liquid compositions.

The phase equilibrium is somewhat unusual and complex. Up to an acid concentration of about 75 wt%, the vapor consists of pure stream. At this point (T ~ 175°C), some decomposition of the acid into SO_3 + H_2O begins to take place. This process accelerates as the boiling point rises and culminates in an azeotrope at 98.4 wt% acid. Note that the SO_3 content of the vapor in this interval is expressed in terms of an equivalent H_2SO_4 concentration.

Spent acid has a typical concentration of 20 wt% H_2SO_4. It is clear, from Table 7.3, that a single stage, or a simple evaporator, would suffice to bring the concentration up to at least 70% without any loss of acid. It is also clear that any concentration beyond this point will have to be carried out in a second unit and at a reduced pressure to avoid escalating the decomposition of the acid. To obtain a product just short of the azeotrope, say 98 wt% H_2SO_4, without exceeding the threshold temperature of ~175°C, the distillation would have to be carried out at pressures below 10^2 atm. Here, an unexpected limitation imposes itself: the overhead product, water vapor, cannot be condensed below the exit temperature of the cooling water, typically in the range 40 to 50°C. This means that the operating pressure of the column has to at

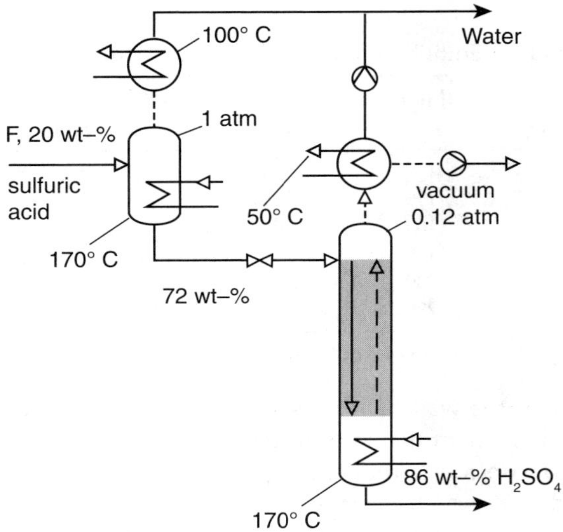

FIGURE 7.26
Recovery of H_2SO_4 from spent acid.

least equal or exceed the vapor pressure of the condensate, which is of the order of 0.1 atm. It is customary, in these cases, to set a lower limit on column pressure of 0.12 atm, corresponding to the vapor pressure of water at 50°C.

The column used to carry out this second step receives an overhead feed of 70% acid from the evaporator (see Figure 7.26). The water is stripped from the acid and ultimately passes into the overhead condenser while any small amounts of SO formed are absorbed by the acid flowing downward. In principle, bottoms concentrations of 90 wt% acid or more could be produced in this fashion without forming excessive amounts of SO. Here, however, a second limitation imposes itself. If only medium-pressure steam is available (15 atm, 200°), the reboiler temperature will be constrained to about 170°. This corresponds to an acid strength (at 0.12 atm) of only 86 wt%. This is a satisfactory, if not ideal, result. Any further increase in concentration would have to be brought about by contacting the acid with an SO_3-bearing gas in a conventional absorber.

This rude intrusion of practical consideration is not meant to detract from the McCabe–Thiele method but rather to place it in proper perspective.

7.4.5 Column and Tray Parameters

There are many features to the construction of both columns and trays, which range from the detailed (e.g., bubble-cap spacing) to the more general (type of tray, internal structure of the column). Figure 7.16 shows the three major types of trays in current use. We supplement it here with a sketch of the internal structure of a column, shown in Figure 7.27. Most of the features

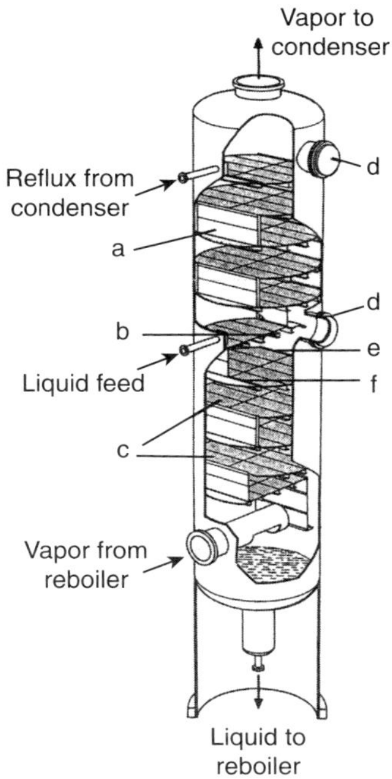

FIGURE 7.27
Structural details of a sieve tray distillation column: (a) downcomer; (b) tray support; (c) sieve trays; (d) manhole; (e) outlet weir; and (f) inlet weir.

seen there are obvious, but one or two need perhaps to be singled out for greater emphasis.

Liquid flow proceeds *across* a tray into a "downcomer" that has the cross section of circular segments and continues downward onto the next tray where it resumes its horizontal direction. Vapor moves upwards through perforations in the tray and is in crossflow with the liquid on it. A feature not always known is the provision of weirs at both the inlet and outlet to a tray. They serve to control the flow and height of liquid in the downcomer. Manholes are provided at intervals to allow internal inspection. (The uninitiated sometimes need to be reminded that it should not be sized for three-year-olds.)

The detailed design of a tray is beyond the scope of this text. It involves the use of a host of semiempirical correlations and it is best to draw on information provided by vendors. The reader should, however, have a sense of *scale* of the parameters involved, which we attempted to provide in Table 7.4. The choice of tray is assisted by Table 7.5, which summarizes the principal features of the three major types. The scale of column dimensions is conveyed by Table 7.6.

In the early years of tray development, bubble-cap trays predominated. Sieve trays were used much more sparingly because of their tendency to

TABLE 7.4

Typical Tray Parameter Values: Operation at Atmospheric Pressure

Tray spacing	50 cm
Weir length	$0.5 \times d_C$[a]
Sieve tray hole	cm
Bubble-cap diameter	10 cm
Valve diameter	5 cm
Free area (downcomer)	10%
Pressure drop	5 mmHg

[a] d_C = column diameter.

TABLE 7.5

Tray Features

	Advantage	Drawbacks
Bubble cap	Stable operation No weeping	Expensive High ΔP
Sieve	Low cost and Δp High efficiency	Weeping
Valve	Wide operational range	Expensive Some weeping

TABLE 7.6

Typical Values of Tray Column Characteristics

Diameter	1–10 m
Height	10–75 m
Number of trays	
Nitrogen–oxygen	100
Ethylene–ethane	70
Ethanol–water	60
Xylene isomers	130
Crude oil	50

weep. They have lately come to the fore by virtue of their low cost and pressure drop. Valve trays are a relatively recent development. They have penetrated the market rapidly and are now widely used in spite of their high cost. Bubble-cap trays have been able to weather the competition because of their rugged and reliable performance.

7.4.6 Limiting Flow Rates: Column Diameter

Flow rates, particularly those of the liquid, flowing downward, have a wide operational range. They must, however, be kept within certain limits to prevent a deterioration in performance or an outright breakdown in column operation. We discuss these limits in some detail below.

7.4.6.1 Gas or Vapor Flow Rates

The upper limit on gas-flow rates is dictated by two factors, excessive liquid entrainment and excessive frothing. Neither of these is necessarily fatal to the operation, but both bring about a serious deterioration in column performance. The upper limit is usually set in terms of the superficial vapor velocity v_g based on the cross-sectional area of the column A_C, which is given by the relation

$$v_g \leq K\left(\frac{\rho_L - \rho_g}{\rho_G}\right)^{1/2} \tag{7.25a}$$

where K is an empirical constant. A justification of the functional form of this expression can be found in the terminal velocity of a droplet (see Illustration 5.4). K depends in a complex way on the depth of the liquid on the tray, its surface tension, and on tray spacing, but is generally, for industrial columns, in the range 0.05 to 0.2 m/s.

If we set ρ_L = 800 kg/m³ and ρ_G = 5 kg/m³, representative of an organic mixture being distilled at 2 atm pressure, we obtain, with K set at 0.1 m/s,

$$v_g = 0.1\left(\frac{800-5}{5}\right)^{1/2} = 0.38\,\text{m/s} \tag{7.25b}$$

Suppose, now, that the boil-up rate Q_g is 0.2 m³/s. We can then, by dividing Q_g by the permissible upper limit to v_g, obtain an estimate of the *minimum* column diameter:

$$A_C = \pi d_C^2/y = Q_g/v_g \tag{7.25c}$$

and, hence,

$$d_C = \left(\frac{0.2}{0.38}\frac{4}{17}\right)^{1/2} = 0.82\,\text{m} \tag{7.25d}$$

7.4.6.2 Liquid Velocities

An increase in liquid flow, brought about, for example, by an increase in reflux ratio, causes the liquid to back up in the downcomer, ultimately spilling over onto the tray above. The column floods and operation essentially ceases. To avoid this fatal occurrence, an upper limit is usually set on liquid velocity in the downcomer, v_d, given by

$$v_d = Q_L/A_d = 10^{-2}\ \text{m/s} \tag{7.25e}$$

where A_d = downcomer cross-sectional area. The value of A_d should be verified to ensure it meets the criterion (Equation 7.25e). Occasionally the column diameter is increased to make this possible.

7.4.6.3 Lower Limits

Too low a liquid flow rate is not a fatal flaw but does cause channeling and maldistribution of the liquid, much as it does in packed columns. The remedy here is to set a lower limit on weir *crest height* of 0.5 cm, equivalent to 2 m^3/h of liquid flow per meter weir length.

The main incentive to setting a lower limit on vapor flow is the prevention of weeping. This is achieved by ensuring that the superficial velocity through the individual openings (e.g., hole of a sieve tray) is in excess of 0.2 m/s.

7.4.6.4 Comments

The brief outline given in this section is sufficient for preliminary design purposes. Much more needs to be done to complete that design. This is usually done by drawing on established design practices (see Selected References at the end of the text). In the case of new system or tray designs, vendor information becomes crucial, and pilot testing may become necessary.

7.4.7 Batch Fractional Distillation: Model Equations and Some Simple Algebraic Calculations

Frequently, in the production of rare and valuable substances, the material flow is too small to make continuous distillation for the purpose of separation or purification practicable. The intermediate or raw product is then accumulated and held in storage before being fed to the fractionation column in intermittent batches.

The distillation process, which is now unsteady, can be carried out in two modes. In the first mode, we allow distillation to proceed without outside intervention and at constant reflux until a prescribed fraction of the charge has been boiled off or the still contents have been concentrated to some desired value. In the course of the process, both the instantaneous and cumulative distillation composition x_D and x_D', as well as the contents of the still, undergo a slow and continuous change.

In the second mode of operation, which is more common, the reflux ratio is continually and automatically adjusted so as to maintain a constant x_D. The boiler contents still change, as does the reflux ratio, but both the instantaneous and cumulative overhead compositions remain invariant with time.

We now introduce two assumptions that allow us to draw on the principles and diagrams we previously established and used to describe *continuous* distillation. The first assumption presumes that the liquid content of the trays, or *hold-up* as it is termed, is negligible compared to the still contents. This allows us, in any material balance performed over the column, to ignore the

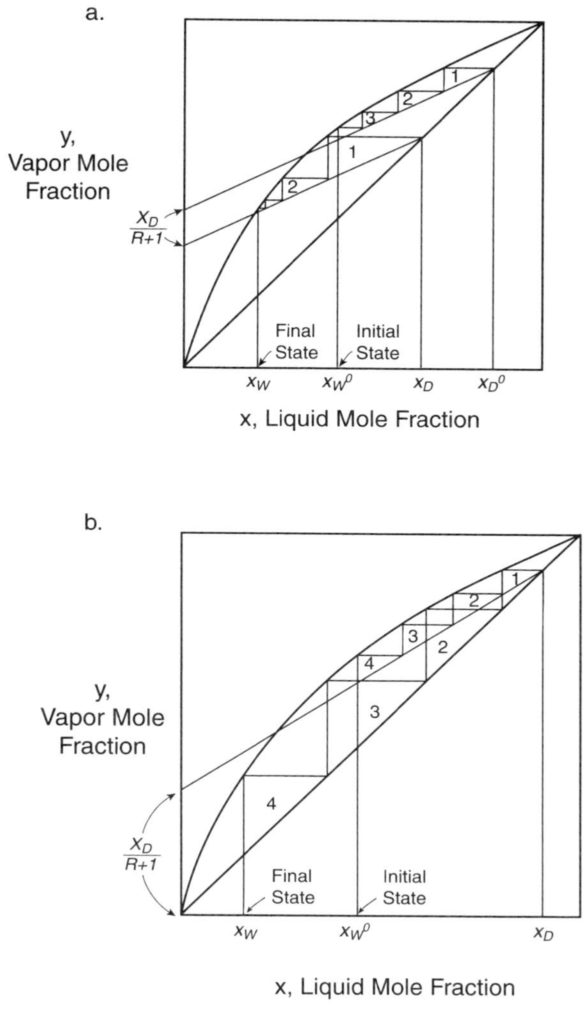

FIGURE 7.28
The two types of batch distillation processes: (a) varying distillate composition with constant reflux ratio and (b) varying reflux ratio with constant distillate composition.

contribution due to hold-up. In the second assumption, we stipulate that the process is sufficiently slow that the system has time to adjust itself to a quasi-steady-state at any instant of the operation. This highly important assumption allows us to represent both modes of operations on a McCabe–Thiele diagram, which was originally derived for steady-state operations. The difference here is that instead of a single operating line, we now have to deal with a continuous spectrum of such lines, which can, however, be accommodated on a single diagram. This has been done in Figure 7.28a and Figure 7.28b, which displays the two cases of constant and varying reflux we described.

7.4.7.1 Distillation at Constant x_D, Variable R

7.4.7.1.1 Minimum Reflux Ratio

Here we are dealing with a single operating line that extends from the desired overhead composition x_D to the batch-feed composition denoted x_F. The line is drawn through the points $x = y = x_D$ and $x_W{}^o$, y^*_W, as shown in Figure 7.28b. The minimum reflux ratio is then determined from the intercept $x_D/(R_{Min} + 1)$.

7.4.7.1.2 Number of Theoretical Plates

An actual reflux ratio is set next, which leads to an operating line with intercept $x_D/(R + 1)$. The number of theoretical plates is then stepped off between operating and equilibrium curves, as was done in the case of continuous fractionation, and yields a total of four stages between the composition x_D and $x_W{}^o$. As the distillation proceeds, the reflux ratio is gradually increased to maintain a constant overhead composition, causing the bottoms mole fraction to diminish. The process can be stopped when a prescribed composition or recovery is reached, or it can be continued until the column is at total reflux. x_W is then at its lowest point and fractional recovery is at its maximum. We reached the final state shown in Figure 7.28b.

7.4.7.1.3 Fraction Distilled and Recovery

Much of the time in actual operations, we wish to determine the fraction $f = 1 - W/W^o$ that needs to be distilled to achieve a prescribed bottoms composition x_W or recovery r defined as $r \equiv 1 - x_W W/x_W{}^o W^o = 1 - (1 - f)(x_W/x_W{}^o)$. Although it is not immediately clear how we should proceed to obtain these quantities, a good way to start is by composing total and component material balances about the entire column. We have

$$\text{Rate of moles in} - \text{rate of moles out} = \frac{d}{dt}\text{contents}$$

$$0 - D = \frac{d}{dt}W \tag{7.26a}$$

and

$$0 - x_D D = \frac{d}{dt}x_w W \tag{7.26b}$$

These are the same equations, in form at least, as those we have seen in the Rayleigh distillation of Illustration 7.3. The difference here is that x_D is *not* in equilibrium with x_W but represents an independent entity set by the operator. Proceeding as in Illustration 7.3, the two equations are divided to eliminate D and dt, and we obtain, after some manipulation, the result

$$\int_{W^o}^{W} \frac{dW}{W} = \ln\frac{W}{W^o} = \int_{x_W{}^o}^{x_W} \frac{dx_W}{x_D - x_W} \tag{7.26c}$$

Because x_D is the prescribed and constant overhead composition, Equation 7.26c can be immediately integrated to yield

$$\frac{W}{W^o} = \frac{x_D - x_W{}^o}{x_D - x_W} \tag{7.26d}$$

After some rearrangement this becomes, for the volatile component,

$$W^o x_W{}^o = (W_o - W)x_D + W x_W \tag{7.26e}$$

Initial moles	Moles distilled	Moles left over

In other words, what we derived here is nothing but a cumulative balance to time *t* of the volatile component, which could have been obtained directly by algebraic means. It can be rearranged to obtain the desired relation between fraction distilled *f* and recovery *r* in terms of the known values of x_D and $x_W{}^o$. This results in the following simple expressions:

$$r = 1 - x_W W / x_W{}^o W^o = f x_D / x_w{}^o$$

or

$$r = 1 - (1 - f)(x_W / x_W{}^o) = f x_D / x_w{}^o \tag{7.26f}$$

Note that the first part of each expression represents the *definition* of *r*, while the second part comes from the cumulative balance (Equation 7.26e). Solving the second part for *f* and back-substituting into the first, we obtain

FRACTION DISTILLED

$$f = \frac{x_W{}^o - x_W}{x_D - x_W} \tag{7.26g}$$

RECOVERY

$$r = f\frac{x_D}{x_W{}^o} = \left(\frac{x_W{}^0 - x_W}{x_D - x_W}\right)\frac{x_D}{x_W{}^0} \tag{7.26h}$$

These equations are quite generally valid, subject only to the condition that x_W cannot fall below the minimum attained at total reflux. We demonstrate their use by returning to Figure 7.25–7.9 of Illustration 7.9 and use the same prescribed overhead composition $x_D = 0.92$ and reflux ratio $R = 0.9$, but raise $x_W{}^o$ to 0.55 so that exactly three plates are accommodated on the operating diagram. To obtain the maximum recovery possible under these conditions, we carry the process to total reflux and determine the corresponding x_W. One can verify that if we step off three plates, starting at $x_D = 0.92$ and using the diagonal as operating line, we obtain $x_W = 0.175$. Substitution into Equation 7.26g and 7.26h, yields

$$f = \frac{x_W{}^o - x_W}{x_D - x_W} = \frac{0.55 - 0.175}{0.92 - 0.175} = 0.503 \tag{7.26i}$$

and for the recovery,

$$r_{\text{Max}} = f\frac{x_D}{x_W{}^o} = 0.503\frac{0.92}{0.55} \tag{7.26j}$$

$$r_{Max} = 0.84 \ or \ 84\% \tag{7.26k}$$

Thus, the *maximum* recovery we can achieve with three plates is 84%. Let us next turn to the second process shown in Figure 7.28a.

7.4.7.2 Distillation at Constant R, Variable x_D

This case, which is less frequently used, differs from the previous process in several respects. We are here no longer dealing with a *design* problem, because both x_D and x_W vary continuously and cannot be prescribed by the analyst as fixed design parameters. Rather than designing a column, we usually use the model equations to *predict* the performance of an existing or hypothetical unit with a given number of plates N_p and a prescribed value of R.

We start by establishing the operating line that will accommodate the existing number of plates and initial boiler composition $x_W{}^o$. This involves a trial-and-error procedure consisting of adjusting the operating line of constant slope $R/(R + 1)$ until it accommodates exactly N_p plates. $x_D{}^o$ of this line is the initial composition that emerges from the column (Figure 7.28a). The operation continues, with overhead and bottoms compositions decreasing steadily, until a final prescribed bottoms composition x_W is reached. The operational parameters for the entire process are then established as follows.

7.4.7.2.1 *Fraction Distilled f*

This item is obtained from Equation 7.26c, written in the form

$$\ln\frac{W}{W^o} = \ln(1-f) = \int_{w_W{}^o}^{x_W} \frac{dx_W}{x_D - x_W} \tag{7.27a}$$

Here we can no longer evaluate the integral analytically, because both x_D and x_W vary in unrelated ways. Instead, values of x_D and x_W must be read off the operating diagrams in pairs and the integral determined graphically or numerically. This is tedious but not overwhelmingly so.

7.4.7.2.2 *Average Distillate Composition* x_D'

This is a new item that must be addressed because the overhead composition is no longer constant. It is obtained from the same cumulative balance (Equation 7.26e) that was used to describe the previous case of constant x_D. This equation applies here as well with x_D' now taking the place of x_D. Equation 7.26g and Equation 7.26h remain similarly valid. Solving Equation 7.26e for x_D', we have

$$x_D' = \frac{x_W W^o - x_W W}{W^o - W} = \frac{x_W{}^o - x_W(1-f)}{f} \tag{7.27b}$$

where x_W is prescribed, and f is drawn from Equation 7.27a. Note that Equation 7.27b is identical to the cumulative composition we derived for the simple batch still in Illustration 2.9.

With both f and x_W in hand, recovery r can then be calculated from Equation 7.26f.

Let us return to the same example we considered above, with x_D and $x_W{}^o$ set at 0.92 and 0.55, and N_p at 3. Reflux ratio R is the same as before at 0.9 but is now kept constant. The aim is to calculate f and r for the same final bottoms compositions $x_W = 0.175$ used previously.

We start by drawing a series of parallel operating lines and reading off values of x_D and x_W, which are used in the graphical evaluation of the integral in Equation 7.27a. One can verify that this leads to a value of 0.913 and hence a fraction distilled f of 0.60.

The next step is to evaluate the cumulative distillate composition x_D' at the end of the process. We have from Equation 7.27b,

$$x_D' = \frac{x_W{}^o - x_W(1-f)}{f} = \frac{0.55 - 0.15(1-0.6)}{0.6} = 0.817 \tag{7.27c}$$

Using this value in Equation 7.26f yields the final result:

$$r = f\frac{x_D'}{x_W{}^o} = 0.6\frac{0.827438}{0.55} \tag{7.27d}$$

$$r = 0.89 \text{ or } 89\% \tag{7.27e}$$

TABLE 7.7

Compositions and Recoveries in the Batch Column: Distillation of an Arbitrary Ternary Mixture

Component	1	2	3
Overhead x_D	0.9	0.05	0.05
Feed $x_F = x_W^o$	0.6	0.2	0.2
Final bottoms x_W	0.1	0.45[a]	0.45[a]
Fraction distillation f[b]	0.625	0.625	0.625
Recovery r[c] (or loss)	93.5%	15.6%	15.6%

[a] Calculated from f.
[b] Calculated from Equation 7.26g.
[c] Calculated from Equation 7.26h.

Thus, although recovery has increased by 5%, the distillate composition is lower by more than 10% over the previous case. Evidently, these results can be manipulated in any number of ways by adjusting N_p and R to achieve more desirable results.

7.4.7.3 Multicomponent Batch Distillation (Forget McCabe–Thiele, Part 2)

The treatment so far has been limited, superficially at least, to binary distillation. A number of the expressions, however, have a more general character, and among them, Equation 7.26g and Equation 7.26h are particularly interesting. Their remarkable feature is that they are quite independent of the phase equilibria and number of components involved or, for that matter, of the number of plates and reflux ratio ultimately employed. They are not used to design a column but serve instead to provide important initial information required at a later stage. A similar situation was encountered in Illustration 7.11.

We consider a hypothetical ternary system with prescribed feed and overhead compositions, but of an otherwise unspecified nature. One of the final bottoms compositions can be fixed as well, while the remainder follow by calculation. No equilibrium data are available, but any existing ternary azeotrope is assumed known and fixes the overhead composition.

The aim will be to calculate the fraction distilled and the percent recovery of each component using Equation 7.26g and Equation 7.26h, as well as to fill in the missing compositions. This will provide a first and important impression of what the distillation process can accomplish.

A feed composition of $x_1 = 0.6$, $x_2 = 0.2$, $x_3 = 0.2$, is assumed, component 1 being the most volatile. Other prescribed and calculated values are summarized in Table 7.7.

Two sample calculations will serve to illustrate the process. For the fraction distilled, we have, from Equation 7.25g and the values of component 1,

$$f = \left(\frac{x_W{}^o - x_W}{x_D - x_W} \right)_1 = \frac{0.6 - 0.1}{0.9 - 0.1} = 0.625$$

The same value of f is used to calculate x_w for the second component:

$$f = 0.625 = \left(\frac{0.2 - x_W}{0.05 - x_W} \right)_2$$

and, hence, $(x_w)_2 = 0.45$.

We are still far from having a complete picture of the process, because the number of plates and minimum reflux remain unknown. This should not overshadow the considerable importance of the results.

7.5 Percolation Processes

Percolation processes refer to operations in which a fluid stream is passed through a bed of granular porous material and a transfer of mass takes place between the two phases. Such processes are seen in the purification of gases and liquids by adsorption and ion exchange, in the transfer of toxic substances from aqueous streams to surrounding soils or riverbeds, and in general whenever a fluid percolating through or over a mass of stationary porous solids exchanges material with it.

In Illustration 6.4, we introduced the reader to the notion of applying equilibrium stage concepts to operations of this type. This is at first glance a startling approach because none of the processes that fall in this category remotely resembles a stirred tank or its equivalent. They are, in fact, distributed in both time and distance and generally require partial differential equations (PDEs) for a rigorous description of the events. It has been shown, however, that if flow is sufficiently slow, we can assume local equilibrium to be established, or at least closely approached, at any point of the system. The process can then be viewed as the composite of a continuous spectrum of equilibrium stages and, therefore, that an algebraic representation becomes possible. This was shown in Illustration 6.4 for the restricted case of a linear equilibrium described by an appropriate Henry's constant.

We now extend this treatment to the more general case of a nonlinear equilibrium case of the Langmuir type. Composing a cumulative solute balance to time t, as was done in Illustration 6.4, we obtain

$$Y_F G_S A_C t = X_F \rho_b A_C t + \varepsilon Y_F \rho_f A_C t \tag{7.28a}$$

Amount introduced	Amount retained by solid	Amount left in fluid

which is identical in form to Equation 6.6a but assumes a general nonlinear relation between q_F and Y_F. Neglecting the last term, we obtain after some rearrangement

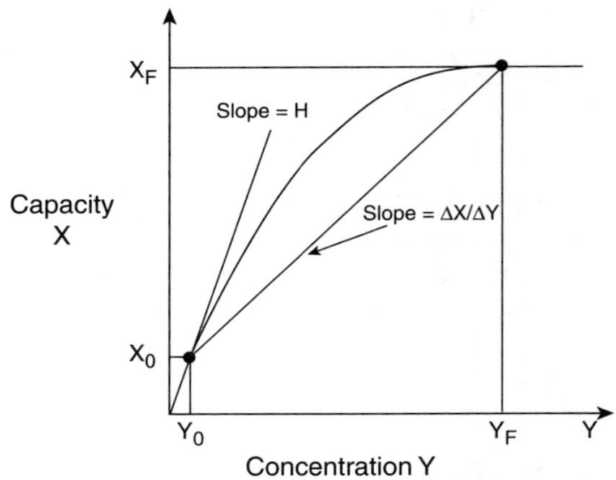

FIGURE 7.29
Operating diagram for Langmuir-type isotherm.

$$z/t = \frac{G_S}{\rho_b X_F / Y_F} \tag{7.28b}$$

where G_b is the mass velocity of the carrier fluid (in kg/m²s). When the solid phase already contains some solute at a concentration level q_0, Y_0, we can recast Equation 7.28b into the more general form:

$$z/t = \frac{G_S}{\rho_b \Delta X/\Delta Y} = \frac{v\rho_f}{\rho_b \Delta X/\Delta Y} \tag{7.28c}$$

where $\Delta X/\Delta Y = (X_F - X_0)/(Y_F - Y_0)$, and v = superficial velocity of the fluid.

An alternative form results if we solve Equation 7.28c for t and set $z = L$, where L is a particular position downstream from the inlet to the system:

$$t = \frac{\rho_b \Delta X/\Delta Y}{v\rho_f} L \tag{7.28d}$$

This confirms the intuitive notion that the greater the distance L and the greater the adsorptive loading ΔX, the longer it will take the solute front to reach the position L. Conversely, the faster the fluid flow v, the shorter the time required for the solute to "break through" at position L.

We can now construct a diagram that contains the equilibrium relation $X^* = f(Y)$ as well as an operating line, the slope of which equals the ratio $\Delta X/\Delta Y$. This is shown in Figure 7.29 and leads to the two equivalent expressions:

$$t = \frac{\rho_b \Delta X/\Delta Y}{\rho_f v} L = \frac{\rho_b L}{\rho_f v} \times (\text{Slope of operating line}) \tag{7.28e}$$

The construction has the same attractive property as previous operating diagrams in conveying at a glance the manner in which the system reacts to changes in certain operating parameters. Thus, if the feed concentration is increased, ΔY will likewise increase, lowering the slope of the operating line, and thus leading to faster breakthrough. If, on the other hand, ΔX is increased, for example, through the use of a more efficient sorbent, the movement of the solute front will slow, and breakthrough will occur much later. In typical water purification processes, ρ_b/ρ_f is of the order 1, v of the order 1 cm/s, and $L \times$ slope of the order 10^7. The on-stream time is then of the order 10^7 s $\cong$ 100 days under equilibrium conditions.

So far in our discussion, the movement of solute was assumed to be entirely in one direction (i.e., from the flowing fluid to the stationary solid). This is the case in adsorption processes or in the uptake step of ion-exchange operations. When the direction of transfer is reversed, we speak of *desorption*, *regeneration*, or, in the case of environmental systems, *clearance*. The theoretical treatment here becomes more complex and requires a more profound approach based on PDEs. We do not address this problem here and instead present the final result that emerges from that analysis for the case of complete desorption from a Langmuir-type isotherm under equilibrium conditions. The relevant equation is completely analogous in form to that for the adsorption step (Equation 7.28e), and it reads

$$t = \frac{\rho_b H}{\rho_f v} L = \frac{\rho_b L}{\rho_f v} \begin{pmatrix} \text{slope of equilibrium} \\ \text{curve at the origin} \end{pmatrix} \tag{7.28f}$$

Thus, the only change that occurred in passing from adsorption to desorption is the replacement of the slope of the operating line $\Delta X/\Delta Y$ by the slope of the equilibrium curve at the origin (i.e., the Henry constant). Let us now apply these two expressions to an environmental problem of interest.

Illustration 7.12: Contamination and Clearance of Soils and Riverbeds

When soils or sediments are exposed to contaminants contained in groundwater or in the river flow, an important question arises: What is the length of recovery period required to restore the system to its original state once contamination has ceased? Equation 7.28e and Equation 7.28f provide some important guidelines that can be used to address this question.

We start by noting that in the general case of a nonlinear equilibrium relation, the adsorption and desorption periods will always differ because of the different values of $\Delta q/\Delta Y$ and H (Figure 7.29). Because the latter quantity is invariably the greater of the two, we conclude that for nonlinear isotherms the adsorption step will proceed at a faster pace than the corresponding desorption step. This difference becomes more pronounced the steeper the equilibrium curve is at the origin.

Let us now turn to the case where the equilibrium is *linear* (i.e., the operation is entirely in the Henry's law region). This is a common, although not exclusive, occurrence in environmental systems. Inspection of Figure 7.29 shows that in this case the two relevant slopes become identical — that is, we have

$$(\Delta X/\Delta Y)_{ads} = (\Delta X/\Delta Y)_{des} = H \qquad (7.28g)$$

This leads to the surprising but also reassuring conclusion that, *for linear systems, the clearance period will always equal that for contamination*. Long recovery times arise but only as a consequence of long exposure times. The shorter the period of accidental contamination, the greater the prospects are for a fast recovery. This agrees, of course, with our physical understanding of the process, but it requires the use of Equation 7.28e and Equation 7.28f to establish that the contamination and recovery periods are identical. A similar result was seen in the uptake and depuration of toxins in animals (Illustration 5.8), and was again the direct result of the linearity of the system.

7.6 Stage Efficiencies

We turn now to the consideration of the second aspect of the equilibrium stage, that of its *efficiency*. This is an area of much greater uncertainty than we had previously seen. Evidently, the efficiency of a stage will be affected in a complex manner by an array of variables whose precise influence on the operation is difficult to quantify. In the case of trays used in gas absorption and distillation, we must consider the vigor of gas–liquid contact, the rate of liquid and gas flow, the mechanical design of the trays, as well as physical properties of the systems. For stirred tanks, the rate of stirring and the design of the stirring mechanism, the contact time allowed, and the physical properties of the systems all play a role.

There are two ways out of this dilemma. We can draw on information provided by the equipment vendor, who frequently has test facilities available for determining stage efficiencies, or we can make use of rough correlations and guidelines for a first estimate in the preliminary design of these units.

7.6.1 Distillation and Absorption

Two classical empirical correlations due to O'Connell can be used to obtain a measure of the efficiencies of bubble-cap trays for distillation and absorption (Figure 7.30a and Figure 7.30b). The principal correlating parameters in both cases are the viscosity of the liquid, μ_L, and the relevant equilibrium constants — α for distillation and Henry's constant H for absorption. High values of either of these variables adversely affect the tray efficiency.

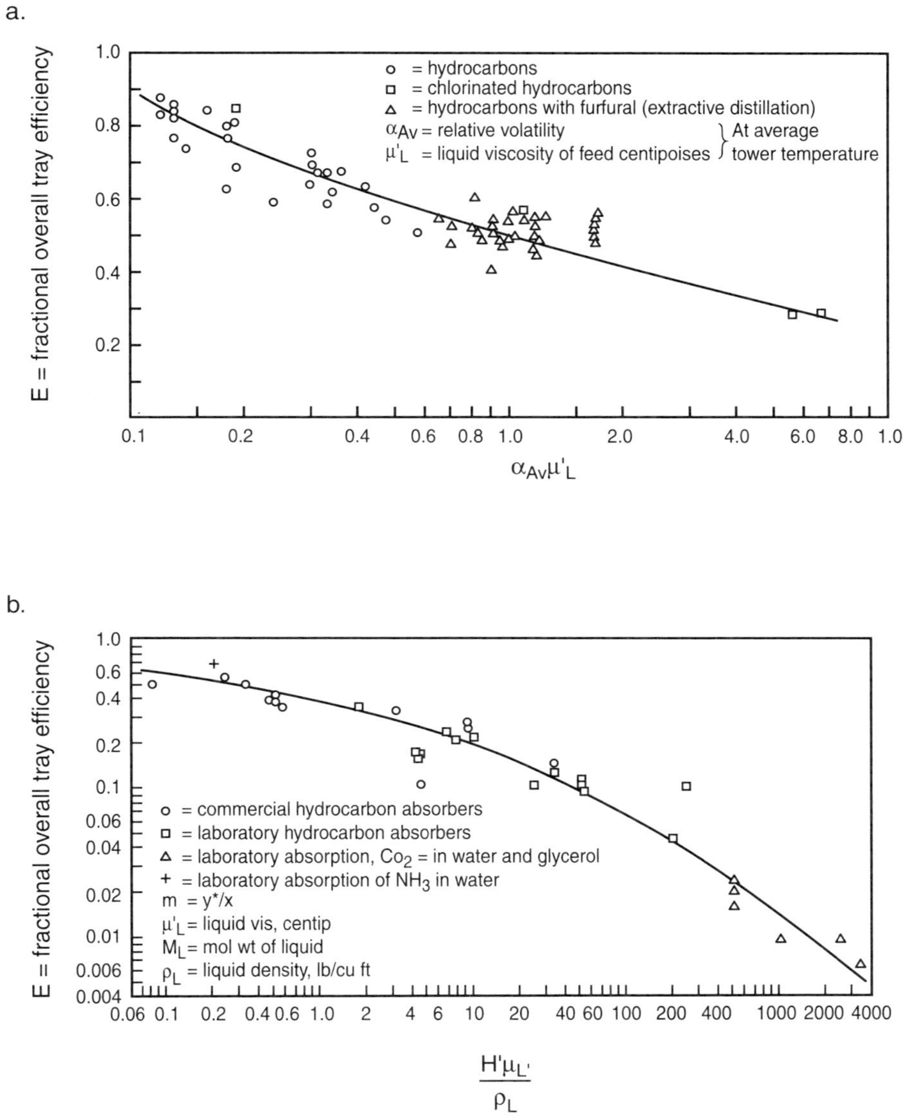

FIGURE 7.30
O'Connell's correlations for bubble-cap tray efficiencies: (a) distillation and (b) absorption.

The correlation for distillation is based on limited data for systems of hydrocarbons and chlorinated hydrocarbons and that for absorption on the performance of hydrocarbon absorbers and the scrubbing of ammonia and carbon dioxide with water. μ_L' is in units of centipoises; H' is in units of mole-fraction ratios; and ρ_{ML} is the molar density of the liquid solvent (in lb mol/ft^3). Note that the gas absorbers generally have much lower plate

TABLE 7.8

Conservative Parameter Values for Batch Extraction Efficiencies of $E > 0.8$

Residence time	$t > 25$ min
Speed of agitation	rpm > 200

efficiencies than distillation columns due primarily to their lower operating temperatures because of the higher viscosities.

7.6.2 Extraction

The extraction of solutes in agitated vessels is, at the very least, on a par in complexity with that we saw for gas–liquid contact on trays. Here again the mechanical design of the system, this time that of the impeller, enters the picture, as well as the physical properties to which we must now add the surface tension of the dispersed phase. As a result, no clear-cut correlation has emerged from the host of experimental studies reported in the literature. The studies do, however, provide some guidelines that we summarize for the convenience of the reader in Table 7.8. The parameter values listed represent "safe" lower limits designed to achieve stage efficiencies of better than 80%. Much lower values do materialize on occasion. Thus, contact times of as little as 1 min have been known to result in efficiencies of more than 75%. The table is therefore to be regarded as providing a comfortable margin of safety for conservative first estimates.

The effectiveness of staged liquid extraction columns is prone to even greater uncertainties: E may vary from a few percent to 50% or higher. Vendor information and test data become a prerequisite in this case.

7.6.3 Adsorption and Leaching

Surprisingly, it is in this area of liquid–solid contact that the stage efficiency comes closest to being quantifiable. This is due to two factors: (1) The principal resistance to mass transfer in a stirred vessel usually resides in the solid phase, and (2) one can, as a consequence, apply the "long-time" solution to Fick's equation, outlined in Chapter 4 (Section 4.1.2.2). Solution of the relevant ordinary differential equation (ODE) and conversion to efficiency results in the following expression for both adsorption and leaching (or desorption):

$$E = 1 - \exp(-10\ D_e t/R^2) \tag{7.29a}$$

Derivation of this equation is left to the exercises (Practice Problem 7.19).

The uncertainty here resides in the value of the effective diffusivity D_e, which has a typical range of 10^{-12} to 10^{-9} m²/s. Because of this uncertainty, Equation 7.29a is more fruitfully used to explore the effect of particle radius or contact time on E. This is done by converting it into the ratio form:

$$\frac{\ln(1-E_2)}{\ln(1-E_1)} = \left(R_1 / R_2\right)^2 = t_2 / t_1 \tag{7.29b}$$

Suppose, for example, that for a given particle radius R, or contact time t_1, the efficiency is 50%. Then halving the radius will raise the efficiency to 93%, but doubling the contact time only will increase it to 75%.

7.6.4 Percolation Processes

In our preceding discussion of this topic, we managed to reduce the underlying model, which consists of two PDE mass balances, to a simple algebraic cumulative balance joined to an appropriate continuous spectrum of equilibrium stages. When this restriction is removed and the mass transfer resistance is brought back into play, no alternative simplifications are possible, and we must return to the full PDE model. This model has been solved for a number of different equilibrium relations, most notably the linear case expressed by $X = HY$. The results for the latter can be expressed in terms of the following two dimensionless parameters:

DIMENSIONLESS DISTANCE Z

$$Z = k_S a(z/v) \tag{7.30a}$$

DIMENSIONLESS TIME T

$$T = k_S a \frac{\rho_f}{\rho_b} \frac{t}{H} \tag{7.30b}$$

Here ρ_f and ρ_s are the fluid and solid densities, and $k_S a$ is a volumetric mass transfer coefficient for internal diffusion based on the "long-time solution" of Fick's equation. A relation due to Glueckauf augments the factor of 10 by 50% to account for short-time contact, and this relation is commonly used in percolation processes:

$$k_S a = 15 \frac{D_e}{R^2} \tag{7.30c}$$

where D_e is the effective diffusivity in the particle, and R is its radius.

Solutions of the PDEs as a function of these parameters are given in Table 7.9 at concentration levels of 1% and 10% of the feed concentration at a distance z from the inlet. These tabulated values of Z and T can be used to calculate the time t it takes for a particular concentration level to reach the position z

TABLE 7.9

Parameters for Nonequilibrium Adsorption

	T		Efficiency $E = T/Z$	
Z	**1% of Y_F**	**10% of Y_F**	**1% of Y_F**	**10% of Y_F**
1000	900	950	0.9	0.95
800	700	740	0.88	0.93
600	520	550	0.87	0.92
400	330	360	0.83	0.90
200	150	170	0.75	0.85
100	70	83	0.70	0.83
80	52	65	0.65	0.81
60	37	48	0.2	0.80
40	22	30	0.55	0.75
20	7.8	13	0.39	0.65
10	2.5	5.0	0.25	0.50
8	1.2	3.5	0.15	0.44
6	0.38	2.2	0.063	0.37
5	0.10	1.6	0.020	0.32

(for example, the outlet of an adsorber) or, conversely, to calculate the height of an adsorber or ion-exchange column needed for it to remain functional over a prescribed period t.

The following example illustrates the use of the table. Suppose we are required to reduce the level of pollutant in water to 1% of its existing level. It is expected to keep the adsorber on-stream for 100 days before breakthrough at the 1% level occurs. What should the height of the unit be? (Data are as follows: $H = 10^4$ $v = 1$ cm/s; $\rho_f/\rho_b \approx 1$.)

$k_S a$ is estimated from Equation 7.29c to be of the order 10^2 s^1. The corresponding values of dimensionless time and distance are

$$T = 10^{-2} \times 1 \times \frac{3600 \times 24 \times 100}{10^4} = 8.64$$

$$Z \text{ (Table 7.7)} \cong 21$$

Consequently,

$$z = \frac{v}{k_S a} Z = \frac{10^{-2}}{10^{-2}} 21$$

$$z = 21 \text{ m}$$

This is somewhat excessive and would suggest the use of two 10-m columns in series. Note that when we deal with multicomponent systems, the substance with the lowest Henry constant breaks through first. This value of H

FIGURE 7.31
Dual GAC beds for water purification. (Courtesy of Adsorption Research Inc.)

must then be used in computing dimensionless time T. Figure 7.31 shows a typical dual-bed arrangement used in water purification. One unit is on-stream for up to 2 weeks or more and is loaded up to only about three quarters of its breakthrough capacity. The feed is then switched to its twin containing fresh adsorbent to allow time for unloading and replacement of the spent material. Used granulated activated carbon (GAC) is stirred to a central facility for thermal regeneration.

Table 7.7 can also be used to calculate an efficiency for the percolation process sometimes referred to as bed utilization and defined as the ratio of the minimum mass of adsorbent required under equilibrium conditions to the mass used in the actual operation (W_m/W_a). This is shown in Illustration 7.13.

Illustration 7.13: Efficiency of an Adsorption or Ion-Exchange Column

We start the procedure by composing the ratio of the dimensionless distance Z to dimensionless time T. We obtain in the first instance

$$\frac{Z}{T} = \frac{k_S a(z/v)}{k_S a(\rho_f/\rho_b)(t/H)} = \frac{z\rho_b}{v\rho_f t}H \tag{7.31a}$$

We now multiply the numerator and denominator of this fraction by the cross-sectional area of the column A_C. This has the effect of transforming the ratio into an expression representing the *actual* mass of the stationary solid per mass of fluid treated, which we term W_a:

$$\frac{Z}{T} = \frac{z\rho_b A_C}{v\rho_f t A_C} H = \left[\frac{\text{kg solid}}{\text{kg fluid treated}}\right]_a H = W_a H \tag{7.31b}$$

But as seen in the last chapter, Equation 6.7c, the corresponding *minimum* mass W_m is given by the *inverse* of the Henry constant H. Hence, $H = 1/W_m$ and one can write for the efficiency E of the process

$$E = \frac{W_m}{W_a} = \frac{T}{Z} \tag{7.31c}$$

Thus, the efficiency of a percolation process is given simply by the ratio of dimensionless time to dimensionless distance. For the process considered in the previous section, for example, we had $T = 8.64$ and $Z = 21$. Consequently,

$$E = \frac{8.64}{21} = 0.41 \text{ or } 41\%$$

This implies that 59% of the adsorber is occupied by the mass transfer zone. As we move up the columns in Table 7.7, efficiency improves dramatically. At $Z = 100$ the efficiency becomes 70%, and at $Z = 1000$ it becomes 90%, and the percentage of the bed occupied by the mass transfer zone drops to 30% and 10%, respectively. The penalty to be paid is an increase in bed height.

Practice Problems

7.1 The Operating Diagram

a. Describe what is represented in an operating diagram.
b. What is the meaning of an operating line?
c. How many operating lines are required in crosscurrent cascades, and how many are required in a countercurrent operation?
d. What are the requirements for binary distillation?
e. The first step of crude oil fractionation takes place in the so-called atmospheric tower, in which various intermediate fractions (kerosene, diesel, etc.) are withdrawn and treated in separate stripping

columns. The overhead from these side columns is returned to the main tower at appropriate feed points. How many more operating lines are required in principle if four side columns are used?

7.2 Single-Stage Adsorption: The Freundlich Isotherm

The Freundlich adsorption isotherm is a special type of equilibrium relation of the general form $Y = mX^n$. It is an empirical relation that does not converge to a saturation value at high loadings, nor does it yield the required Henry's constant at low coverages. It does, however, in many cases provide an adequate description of adsorption or ion-exchange equilibria over an intermediate concentration range. Suppose that a particular liquid–solid system is described by the following Freundlich isotherm:

$$Y = 8.91 \times 10^5 X^{1.66}$$

It is desired to reduce the impurity concentration in a given liquid from $Y_o = 9.6$ units/kg solvent to 10% of this value in a single stage or a column operated at equilibrium. Determine the minimum mass of adsorbent per 1000 kg solution required to accomplish this.

Answer:

32.0 kg

7.3 Flash Vaporization

Flash vaporization is yet another *single-stage* operation in which a liquid mixture is first heated under pressure and then allowed to flash into a chamber maintained at a lower pressure, causing partial vaporization (Figure 7.32a). Its purpose is to achieve a first crude separation of the mixture into volatile and high boiling fractions. The two equilibrated phases are then withdrawn separately for further processing.

The operating diagram for the flash vaporization of a binary mixture is shown in Figure 7.32b.

a. Define the end points of the operating line.
b. Set up an energy balance for calculating the heat load q.

7.4 Single-Stage (Rayleigh) Distillation: The Case of Constant α

A solution containing 1 mol% of a valuable (or objectionable) component is to be concentrated tenfold. The average α value for the solvent is 5. For a single-stage equilibrium distillation of such systems, the following expression applies:

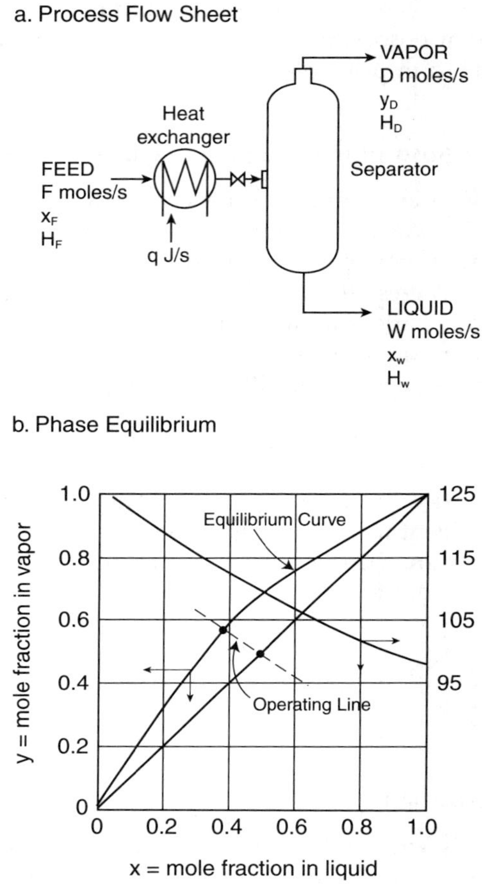

FIGURE 7.32
Flash vaporization.

$$\ln W^o / W = \frac{1}{\alpha - 1} \ln \frac{w_W{}^o(1 - x_W)}{x_W(1 - x_W{}^o)} + \ln \frac{1 - x_W}{1 - x_W{}^o}$$

a. How is this equation obtained?
b. What is the fraction to be distilled?
c. Is the process a viable one? (*Hint*: Compute the cumulative overhead loss.)

7.5 The Rayleigh Equation in Biotechnology: Ultrafiltration

Ultrafiltration is a membrane process in which a solution containing a valuable solute such as a protein is concentrated by applying pressure to it and forcing the solvent across a semipermeable membrane (i.e., a membrane more permeable to the solvent than it is to the solute). Some of the latter will usually leak through as well; that is, the process is not 100% efficient. Efficiency is here defined as $1 - C_p/C_R$, where C_p is the concentration at any instant of the solution passing through, the *permeate*, and C_R denotes the concentration in the enriched solution left behind, termed the *retentate*. In a test run of a batch ultrafiltration unit to determine leakage, it was found that the retentate concentration had doubled after 53.7% of the solution had passed through the membrane. Show in the first instance that the model is represented by the Rayleigh equation and then:

a. Determine the efficiency of the process.
b. Calculate the enrichment obtained (i.e., the ratio of final retentate concentration to the cumulative concentration of the permeate).

Answers:

a. 0.9
b. 14.5

Note: More about ultrafiltration appears in Chapter 8.

7.6 Crosscurrent Batch Extraction

Trace quantities of a valuable substance in aqueous solution are to be recovered by laboratory extraction with a solvent in which it is five times as soluble. How many extractions are needed to achieve a 99.9% recovery if (a) the amount of solvent equals that of the aqueous solution and (b) the amount of solvent is *one half* that of the solution? Which method is preferable? Would you recommend further reductions in solvent? Compare Figure 7.9 and Figure 7.13 and state the principal advantage of crosscurrent operation.

7.7 Countercurrent Operation; The Linear Case

Derive the Kremser equation (Equation 7.14), which applies to a countercurrent cascade with straight operating and equilibrium lines. Use gas absorption as an example. (*Hint*: Recall from algebra that $(x^2 + x + 1)(x - 1) = x^3 - 1$ and similarly for higher powers.)

7.8 The "Pinch"

What are the significance and consequence of a pinch in gas absorption and distillation? Describe five distinct cases that lead to a pinch.

7.9 Design of a Gas Scrubber

A gas scrubber is to be used to reduce solute content in a dilute gas feed entering at 10^4 kg/h to 1% of the incoming concentration. A solvent with a Henry constant of 0.1 kg/kg and specific gravity 0.8 is available.

a. What is the minimum solvent flow rate to be used?
b. What is the *actual* solvent flow rate you would propose, given that the plate efficiency is 10%?
c. Given a tray spacing of 40 cm, what is the minimum height of the column?
d. Given a downcomer cross-sectional area one tenth that of the tray, what is the minimum column diameter?

Check the value obtained against the maximum allowable gas velocity assuming $\rho_g = 1$ kg/m^3.

7.10 Leaching of Vegetable Oils

Set your own reasonable values for m and L/S and apply the Kremser equation to the extraction of vegetable oil from seeds described in Section 7.3.4. How many stages are required to achieve 95% recovery? Assuming a contact time of 15 min, and a nominal seed diameter of 2 mm, give an estimate of the stage efficiency. The seeds are thoroughly crushed and their interiors readily accessible. Try different values for R, L/mS, D_e.

7.11 The Reflux Ratio

a. What is the meaning of $R = 0$, $R = 1$, $R = 5$, $R = \infty$? Do the answers depend on the number of components in the feed or the relevant phase equilibrium? In what way?
b. The ratio L/V is sometimes referred to as the "internal reflux ratio." What is its relation to the "external" ratio R? What is the physical meaning of $L/V = 1$? Would you expect L/V to be higher or lower in vacuum distillation?

7.12 Effect of Feed and Reflux on Column Performance

Consider the following changes in the operating conditions of the fractionation column designed in Illustration 7.8.

a. Feed rate is doubled.
b. Feed concentration fluctuates ±10%.
c. The reflux rate is doubled.

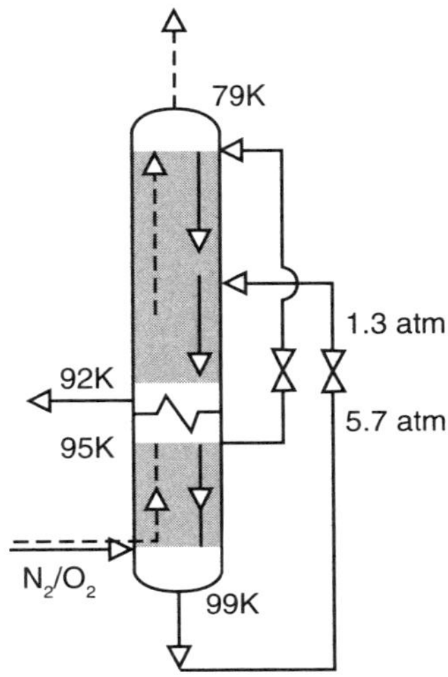

FIGURE 7.33
Air separation by cryogenic distillation: the LINDE double column.

Which variables are affected by these changes? Which remain unchanged? Support your statements with actual calculations as far as possible.

7.13 The Effect of Operating Pressure on Tray Design

Illustration 6.13 drew attention to the effect of operating pressure on the separation factor and the advantages and drawbacks of operating above and below atmospheric pressure. The task in this problem is to explore its effect on tray design.

Which two parameters would be most significantly altered by either vacuum or high-pressure operation, and in which direction?

7.14 Separation of Air by Cryogenic Distillation

The separation of air by low-temperature distillation is practiced on a large scale, primarily for the purpose of producing oxygen. The annual production of O_2 is of the order of 10^9 tons, making it second in size only to the refining of crude oil.

The method most commonly used is the Linde process, which has the peculiarity of a double or twin distillation column, shown in Figure 7.33. The reboiler and condenser, instead of being located separately at the top and bottom, are combined into a single unit for greater thermal efficiency partway up the column, separating it into two sections. The lower column, which receives partly liquefied air as feed, is operated at 5.7 atm, whereas the upper column is operated at 1.3 atm.

Nitrogen and oxygen have normal boiling points of 77 K and 90 K, respectively. The separation factor α for the system at 1 atm is approximately 4 (Table 6.9). Other temperatures of the various process streams are indicated in Figure 7.33.

a. Identify the approximate composition of each stream.
b. Why is the upper column operated at 1.3 atm? Why is the lower column operated at 5.7 atm?

7.15 Maximum Recovery in Batch Distillation

A batch still with ten theoretical plates is used to fractionate a binary mixture $x_W{}^o = 0.4$ with $\alpha = 2$ into an overhead product of constant composition $x_D = 0.95$.

a. Calculate the maximum recovery.
b. How would you proceed if the column, instead of having trays, contains a high-efficiency packing?

(*Hint*: Use the Fenske equation.)

7.16 Design of an Ion-Exchange Column

A synthetic ion-exchange resin in bead form is to be used for collecting and concentrating the copper in a dilute waste stream. Velocity of the feed of concentration 13 meq/L is 4 cm/s, bed density ρ_b = 1200 kg/m^3, and the specific gravity of the solution can be set at 1.0. Laboratory equilibrium data are available and are as follows:

meq Cu^{++}/L sol'n	0.2	2.0	4.0	8.0	12	16	20
meq Cu^{++}/g resin	2.0	4.0	4.4	4.7	4.75	4.85	4.9

If the bed is to remain on stream for 10 h, what is the minimum height of column required to treat the solution?

Answer:

3.3 m

7.17 Adsorption Purification Revisited

In Illustration 6.4 we addressed the problem of sizing a carbon bed for the removal of benzene from water. The calculations there are limited to determining the *minimum* requirements in the absence of transport resistance (i.e., under conditions of local equilibrium). We now wish to calculate the efficiency of the process by incorporating the effect of a mass transfer resistance. Carry over the pertinent variables from Illustration 6.4 and make use of Table 7.3 and the Glueckauf relation (Equation 7.30c). Assume $D = 10^{-9}$ m² /s, $R = 2$ mm, $\varepsilon = 0.3$, and $\tau = 4$.

Answer:

38%

7.18 Diameter of an Ion-Exchange or Activated Carbon Bed

The permeability K for viscous flow through a packed bed can be represented, approximately, by Ergun's correlation

$$K = 6.7 \times 10^{-3} \frac{\varepsilon^3 d_p^{\ 2}}{(1-\varepsilon)^2}$$

where ε = void fraction, d_p = particle diameter, and K is in units of square meter (m²) or darcies (1 darcy = 10^{-12} m²).

Suppose water is to be treated at the rate of 10^2 m³/s, and the height of the bed Δz was established at 4 m by the methods outlined in Sections 7.5 and 7.6. ε is estimated at 0.3, $\mu = 10^{-3}$ Pas, and d_p = 1 mm.

a. Calculate the bed diameter if flow is by gravitation only.
b. What is the diameter if the water is pumped through the bed under a pressure of 200 kPa?

(*Hint*: Use D'Arcy's law, Table 1.1.)

Note: Because of the high sensitivity of the Ergun equation to the value of ε, which typically varies over the range 0.3 to 0.6, it is often preferable to extract K values from experimental pressure drop–flow rate data or to simply set $Q/A = v$ at some "safe" value, say 0.1 m/s. The Ergun equation provides a valuable check and quantifies the dependence on ε and d_p.

7.19 Stage Efficiency for Liquid–Solid Systems

Derive Equation 7.29a.

7.20 Uranium Isotope Separation Revisited

Calculate the minimum number of stages required to enrich $U^{235}F_6$ from its natural abundance of 0.71 mol% to nuclear reactor grade (~3.5%) using a countercurrent cascade of gas centrifuges. The stripped stream is to contain no more than 0.01% of U^{235} F_6. α is taken from Illustration 1.3. (*Hint*: Draw on principles established for distillation.)

Answer:

36

8

Continuous-Contact Operations

The various staged operations taken up in the previous chapter have several features in common. The two phases involved in the transfer of mass were brought together and mixed intimately in discrete stages, which took the form of stirred tanks or some equivalent device. As a result, the concentrations were generally distributed uniformly in space and any variations they underwent were with respect to time, and not with distance. The operations were not only allowed to go to equilibrium but were actively encouraged to do so by means of agitation and the provision of sufficient contact time. Any departures from equilibrium were lumped into an entity called the stage efficiency. An efficiency of 100% signified the attainment of complete equilibrium, and values below that expressed varying degrees of nonequilibrium.

Continuous-contact operations are diametrically different in almost every aspect. The two phases are in continuous flow and in continuous contact with each other, rather than repeatedly separated and recontacted in an array of stages. Second, the attainment of equilibrium is shunned. An active driving force is maintained at all times, and its constituent concentrations vary continuously from the point of entry to the exit. The result is that the concentrations are now distributed in space and, assuming a normal steady-state operation, are invariant in time. Thus, while staged operations vary at most with time, but not at all with distance, the exact opposite holds in continuous-contact operations.

A final difference concerns the equipment used in the two cases. Continuous-contact processes are generally carried out in empty or packed columns or in tubular devices of various configurations. Staged operations may use columns, but these are usually subdivided into discrete contact stages in the form of trays. More commonly, staged processes are carried out in agitated vessels of some type, which may be used singly or in suitable arrangements.

In the following, we divide continuous-contact operations into two distinct categories. The first deals with classical packed-column operations in countercurrent flow. We revisit the packed-gas scrubber we first saw in Chapter 2 and provide a general survey of packed-gas absorption operations. Packed-column distillation and liquid extraction are addressed next, and in a somewhat unusual departure from the norm, we reexamine *coffee decaffeination* by supercritical extraction. The process involves a moving-bed configuration, which although is superficially different from the conventions of packed-column processes has identical operational parameters and can be analyzed by identical procedures.

Yet another unconventional packed-column operation is the purification of air by *biofiltration*, a relatively recent development, which is rapidly becoming a major weapon in the arsenal of environmental engineers.

The second category involves *membrane processes*, a contemporary topic of rapidly increasing importance with the promise of a bright future. Among the subcategories considered here are reverse osmosis, hemodialysis, ultrafiltration, and membrane gas separation, and the text provides some useful relations to address problems in these areas. Attention is drawn to the large number of membrane processes that take place in living organisms, of which the kidney is a particularly striking example. There are both similarities and differences compared to industrial processes, and it may not come as a complete surprise that the kidney achieves more with less than its industrial counterparts. There may be some lessons lurking here for future improvements in membrane technology.

8.1 Packed-Column Operation

Packed columns are used primarily in gas absorption and liquid extraction and in air–water contact operations such as humidification and water cooling, which we address in Chapter 9. They are found less frequently in distillation operations where their use is mostly confined to small-scale processes involving high-efficiency packing.

The analysis of packed-column operations has the same three basic goals seen in similar contexts — design, prediction of equipment performance; and, to a lesser extent, parameter estimation from experimental data. We may, for example, wish to calculate the height of a column required to achieve a certain degree of separation or level of purification. Or it may be required to assess the effect of changes in feed concentration or flow rate on column performance. In all of these problems, the same basic model equations are applied and manipulated in appropriate ways to extract the desired information.

The principal mathematical feature shared by all steady-state packed-column operations is the distribution of the concentration variables in space, principally in the direction of flow. The model will therefore have to be composed of mass balances taken over a difference element of each phase, which are then converted into ordinary differential equations (ODEs) and integrated to obtain concentration profiles and other useful information. Algebraic (integral) mass balances also make their appearance and can often be combined with the ODE balances to obtain important results.

In Illustration 2.4, the existence of this multitude of mass balances was pointed out, and we now have occasion to obtain a broader picture of their derivation and various applications. To do this, we return to the example of the gas scrubber taken up there and reexamine it in greater detail.

8.1.1 The Countercurrent Gas Scrubber Revisited

The basic model equations for countercurrent continuous mass transfer are the differential mass balances over each phase, which were derived in Illustration 2.3, and the companion equilibrium relation (Equation 2.12f). They represent a complete model for the system and are reproduced below:

GAS-PHASE MASS BALANCE

$$G_s \frac{dY}{dz} + K_{OY}a(Y - Y^*) = 0 \tag{2.12d}$$

LIQUID-PHASE MASS BALANCE

$$L_s \frac{dX}{dz} + K_{OY}a(Y - Y^*) = 0 \tag{2.12e}$$

EQUILIBRIUM RELATION

$$Y^* = f(X) \tag{2.12f}$$

Here X and Y are solute concentrations in units of kilogram solute per kilogram carrier or solvent, and the mass velocities G_s and L_s have units of kilogram (carrier or solvent) per square meter second.

The two mass balances can be manipulated in a number of ways to yield specific results of interest. One such operation consists of subtracting Equation 2.12d and Equation 2.12e, which results in the elimination of the mass transfer terms. We obtain

$$G_s \frac{dY}{dz} - L_s \frac{dX}{dz} = 0 \tag{8.1}$$

This equation can be integrated between different limits to yield the algebraic mass balances derived in Illustration 2.4 by performing integral solute balances over the column. They are as follows:

OVER PART OF THE COLUMN

$$(L_s X_2 + G_z Y) - (L_s X + G_s Y_2) = 0 \tag{2.11a}$$

and

OVER THE ENTIRE COLUMN

$$(L_sX_2 + G_sY_1) - (L_sX_1 + G_sY_2) = 0 \tag{2.11b}$$

These two expressions are then recast in a form that makes them more suitable for graphical representation. We termed these alternative formulations *operating lines* and reproduce them below:

$$\frac{Y - Y_2}{X - X_2} = \frac{L_s}{G_s} \tag{2.11c}$$

and

$$\frac{Y_1 - Y_2}{X_1 - X_2} = \frac{L_s}{G_s} \tag{2.11d}$$

We soon show how these expressions can be used to arrive at a graphical solution of the model equation.

The ODE mass balances (Equation 2.12d and Equation 2.12e) can also be tackled separately and in isolation by performing a formal integration. We obtain, for the gas-phase balance,

$$Z = \frac{G_s}{K_{OY}a} \int_{Y_2}^{Y_1} \frac{dY}{Y - Y^*} \tag{8.2a}$$

where Z = height of the scrubber, and K_{OY} is an average overall mass transfer coefficient that is obtained by experiment as an averaged constant. The integral on the right side is referred to as the number of transfer units (NTU) and the factor preceding it as the height of a transfer unit (HTU). Thus,

$$\boxed{Z = \text{HTU} \times \text{NTU}} \tag{8.2b}$$

On the surface, we do not seem to have gained much by taking this step, because the integral contains too many variables and therefore cannot be evaluated. Some thought will reveal, however, that the difference $Y - Y^*$ in the integrand can be read from a joint graph of the operating line and equilibrium curve. This is shown in Figure 8.1b, where the difference in question is given by the vertical distance between the two curves.

Let us note some features of this diagram. To begin, it has a familiar air to it. We have seen an identical representation of operating line and equilibrium curve in the operating diagram dealing with a countercurrent staged

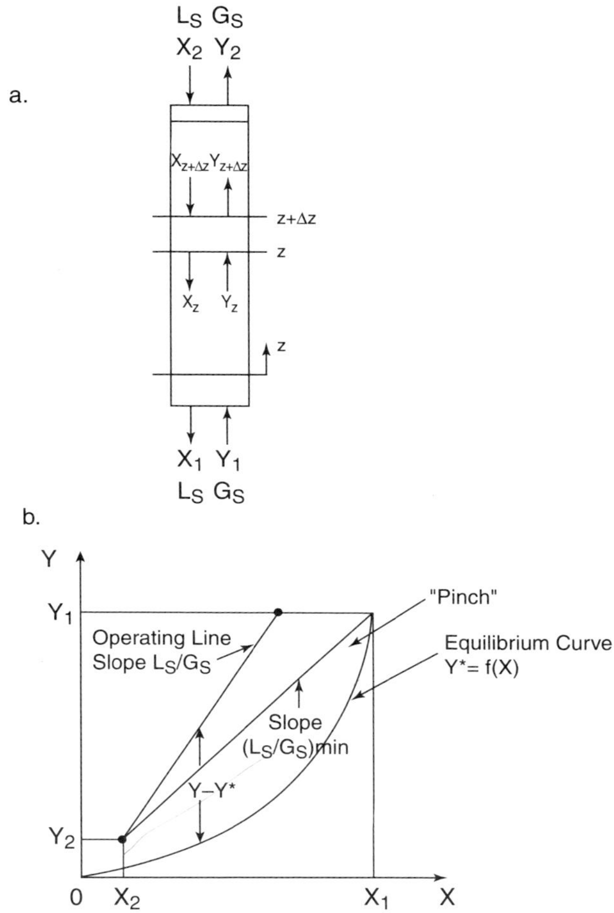

FIGURE 8.1
The countercurrent packed-gas scrubber: (a) column variables and (b) operating diagram.

cascade (Figure 7.12). However, that diagram differs from Figure 8.1 because it uses the staircase construction to establish the number of stages required to achieve a desired separation. In the present case, graphical or numerical evaluation of the NTU integral takes the place of the staircase construction, while the HTU represents, in a sense, the inherent mass transfer resistance of the process and can therefore be viewed as being the counterpart of a *stage efficiency*.

Another feature the two operations have in common is the existence of a *minimum solvent flow rate*. In both cases, this flow leads to a pinch, usually at the level of the feed concentration between the operating line and equilibrium curve. This previously led to the number of stages going to infinity. The result here is similar. The NTU integral increases rapidly as the pinch point is approached, and it ultimately diverges to infinity. This leads to a column of infinite height Z. Even though operating costs are reduced with diminishing solvent flow in both cases, the capital cost of the plant (i.e., the cascade or column) rises rapidly and goes to infinity when the minimum solvent flow rate is imposed.

The design procedure for a packed-gas scrubber parallels that of the countercurrent staged cascade. We start by fixing the point (X_2, Y_2) representing the compositions at the top of the column. Y_2 denotes the prescribed effluent concentration and X_2 the purity of the solvent. A line of slope L_s/G_s is next drawn through that point and extended to the level of the feed concentration Y_1. The slope L_s/G_s is usually set at a value in the range 1.2 to 1.5 $(L_s/G_s)_{Min}$. The vertical difference between the two graphs, the operating line, and the equilibrium curve are then used to evaluate the NTU integral. This can be done either graphically or numerically. Finally, the height of a transfer unit is established using the given carrier flow rate G_s and a volumetric mass transfer coefficient $K_{OY}a$ determined experimentally or drawn from existing correlations or tabulations (see Table 5.6). These values typically vary over the range 10 to 50 cm. With both HTU and NTU in hand, the height of the scrubber can then be determined using Equation 8.2b.

Comments

Let us first return to the basic model equations (Equation 2.12d to Equation 2.12f). This model is general and able to accommodate arbitrary equilibrium relations and system parameter values. Its solution, which is not undertaken here, must be implemented numerically and can be used for design and prediction of scrubber performance as well as parameter estimation. The reduced form of the solution we derived, which rests on the use of the operating diagram Figure 8.1b and Equation 2.2b, is not able to provide direct information on scrubber performance except by a process of trial and error. Suppose, for example, that we wish to establish the effluent concentration that results from doubling the feed rate (i.e., G_s, in an existing scrubber of height Z). The HTU value would first have to be modified to reflect the new flow rate, but more importantly, the upper limit of the NTU integral would have to be adjusted and the evaluation of the integral repeated by trial and error until the product of HTU and NTU exactly matches the given height Z of the scrubber. The numerical solution suffers from the same dilemma but can be used, once it is properly programmed and in place, to carry out a wide range of repeated calculations with a minimum of effort.

A second point concerns the solvent flow rates to be used. Low amounts of solvent carry the advantage of low solvent inventory but lead to greater scrubber heights, which in the limit of the minimum flow rate leads to an infinitely high tower. If, on the other hand, we allow an unbounded increase in solvent flow, column height and cost will be reduced to a minimum but operating costs will go to infinity. Between these two extremes there must be an optimum flow rate that will minimize the total expenditures composed of capital and operating costs.

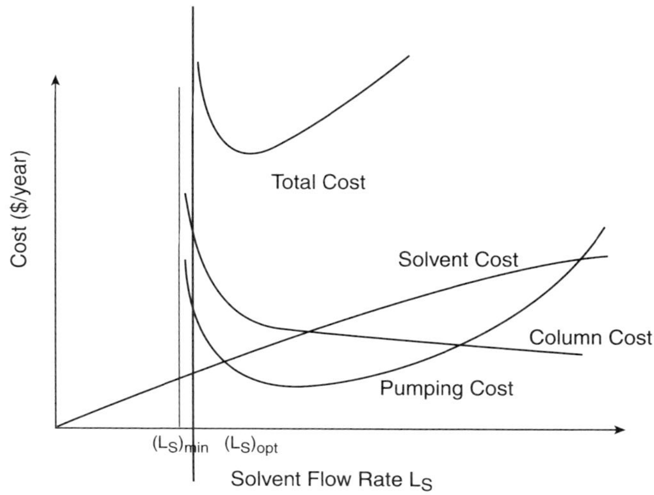

FIGURE 8.2
The countercurrent gas scrubber — cost versus solvent flow rate.

This situation is depicted in Figure 8.2, which shows a plot of total cost versus solvent flow rate. That cost is made up of three component expenditures consisting of the cost of the column, the cost of the solvent, and the cost of pumping it to the top of the tower. The column cost starts at infinity when solvent flow is at a minimum, thereafter declines sharply, and ultimately levels off to a near-constant value. The pumping cost also starts at infinity because the column is infinitely high, then declines, passes through a minimum, and gradually increases, going to infinity again as the solvent flow rate becomes unbounded. The cost of the solvent shows a much simpler relation to flow rate, rising almost in a straight line, with any departures from linearity being due to volume discounts allowed for the solvent. The total cost curve that results from the three-component expenditures shows a minimum, which, as mentioned, often lies in the range of 1.2 to 1.5 $(L_s/G_s)_{Min}$.

8.1.2 The Countercurrent Gas Scrubber Again: Analysis of the Linear Case

We consider here the case where both the operating line and equilibrium relation are linear, the latter being represented by Henry's law:

$$Y^* = HX \tag{8.3}$$

Recall the treatment of this case in Chapter 7 for a *staged* operation, which led to the Kremser equation (Equation 7.14b). The aim there had been to relate the number of *stages* N to the operational and equilibrium parameters. What we wish to do here is relate the same parameters to the NTU.

On introducing Equation 8.3 into the NTU integral of Equation 8.2a, we obtain

$$\mathrm{NTU} = \int_{Y_2}^{Y_1} \frac{dY}{Y - Y^*} = \int_{Y_2}^{Y_1} \frac{dY}{Y - HX} \tag{8.4a}$$

Drawing on the material balance (Equation 2.11a) to express X as a function of the gas-phase mass ratio Y yields

$$\mathrm{NTU} = \int_{Y_2}^{Y_1} \frac{dY}{Y(1 - 1/E) + (1/E)Y_2 - HX_2} \tag{8.4b}$$

where E is the so-called *absorption factor*, equal to the ratio of the slopes of operating line to equilibrium line, L/HG. E has its counterpart in staged operations and is called the extraction ratio there.

A standard evaluation of this integral yields

$$\mathrm{NTU} = \frac{1}{1 - 1/E} \ln \frac{Y_1(1 - 1/E) + (1/E)Y_2 - HX_2}{Y_2(1 - 1/E) + (1/E)Y_2 - HX_2} \tag{8.4c}$$

or, equivalently,

$$\mathrm{NTU} = \frac{\ln\left[R(1 - 1/E) + 1/E\right]}{1 - 1/E} \tag{8.4d}$$

Here R is given by

$$R = \frac{Y_2 - HX_2}{Y_1 - HX_2} \tag{8.4e}$$

and in the case of pure solvent ($X_2 = 0$) becomes a direct measure of the depletion Y_2/Y_1 in the solute content of the gas phase or its recovery $r = 1 - Y_1/Y_2$. It is identical to the residue factor previously defined for staged countercurrent processes (see Table 7.1). A plot of it relating it to the absorption factor E and the number of transfer units appears in Figure 8.3a. Note that the diagram is similar in appearance to the plot seen in Figure 7.13, which relates the number of stages in a countercurrent cascade to the same parameters.

We now address a number of features that both plots have in common and that define the behavior of these linear systems.

As briefly noted in Illustration 7.7, the parameter value $E = 1$ represents an important dividing line. Above it, the residue factor R converges asymptotically to a constant value with an increase in the number of transfer units. We

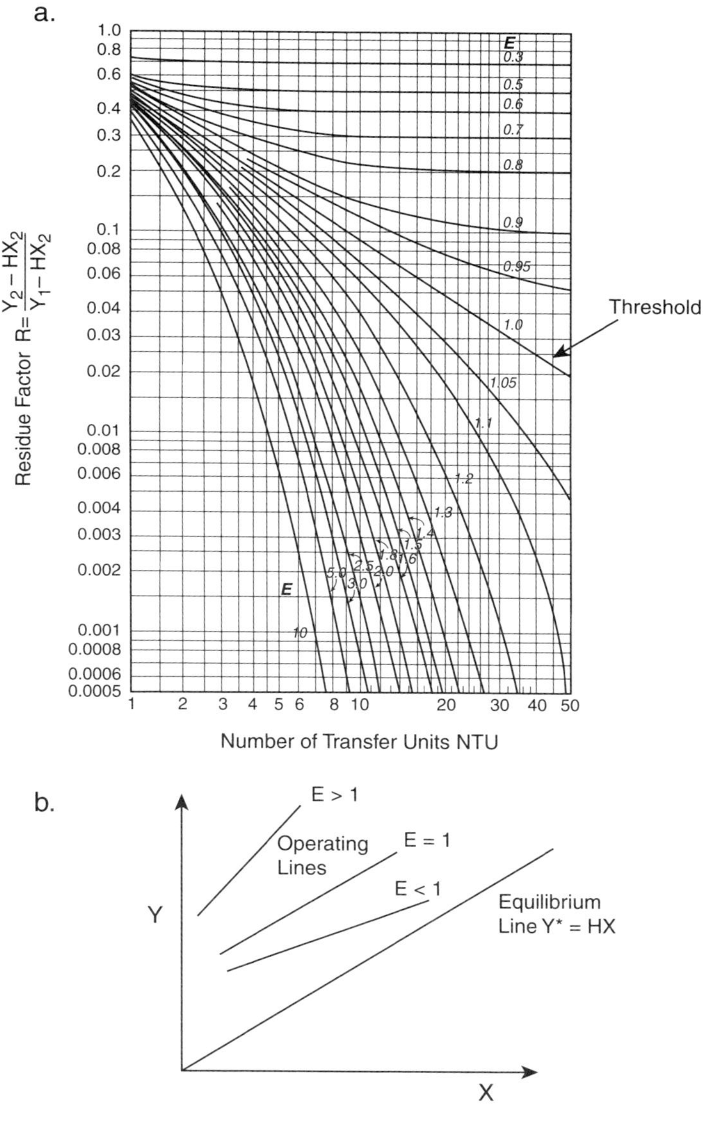

FIGURE 8.3
The gas scrubber in the Henry's law region: (a) evaluation of number of transfer units and (b) operating diagram.

cannot, in other words, absorb arbitrarily large fractions of the incoming solute by increasing solvent flow rate, as long as the value of *E* remains below unity. This becomes possible only when the absorption factor is greater than 1 — that is, whenever the slope of the operating line exceeds that of the equilibrium line. The solute fraction remaining in the outgoing gas can now be made arbitrarily small by the simple expedient of increasing the solvent flow rate.

TABLE 8.1

Packed-Column Parameters in the Linear Region

Process	R	E	Equilibrium Relation
Gas absorption	$\frac{Y_2 - HX_2}{Y_1 - HX_2}$	L/HG	Y (gas) = HX (liquid)
Gas stripping	$\frac{X_1 - Y_1/H}{X_2 - Y_1/H}$	HG/L	Y (gas) = HX (liquid)
Liquid extraction	$\frac{X_2 - Y_1/m}{X_1 - Y_1/m}$	mB/A	Y (extract) = mX (raffinate)

A complete reduction to zero solute content, however, becomes possible only in the limiting case of an infinite flow rate of virgin solvent.

A rational explanation of this behavior can be obtained by inspecting the pertinent operating diagrams, shown in Figure 8.3b. We note here that for $E < 1$, operating and equilibrium lines always *converge* and ultimately intersect in a pinch, making it impossible to set arbitrarily high solute recovery ratios. When E exceeds unity, the two lines *diverge* and no longer intersect except at the origin. It now becomes possible to fix the effluent gas concentration Y_2 at any level other than zero without risking a pinch. The dividing line for these two types of behavior is the operating line, which runs parallel to the equilibrium line. No intersection occurs in this case, and the general behavior is the same as for $E > 1$.

The NTU plot of Figure 8.3a can also be applied to other packed-column processes operating in the linear region. These include gas stripping and extraction and require the pertinent parameters to be redefined to fit each new operation. This was done in Table 8.1, which lists the different versions of R and E for each case.

Comments

This brief analysis of gas absorption with systems obeying Henry's law leads to some interesting conclusions. The lower solvent flow-rate limit here is no longer set by graphically locating the operating line, which causes a pinch. We make use, instead, of the criterion that the slope of the operating line should at least equal and preferably exceed that of the equilibrium line (i.e., Henry's constant). This results in a quick and convenient resolution of the problem of finding a "practical" lower limit of the solvent flow rate.

While operation above $E = 1$ is to be preferred, further increases in solvent flow (i.e., E) give rapidly diminishing returns. Thus, at the $R = 0.01$ level (99% recovery), a mere increase from $E = 1.1$ to 1.3 results in a decrease in NTU (and hence height of column) of about 50%. A further similar increase to $E = 1.5$ yields only a 25% drop, and so on. The range $E = 1.2$ to 2.0 therefore provides a good starting point for initial estimates of solvent requirements.

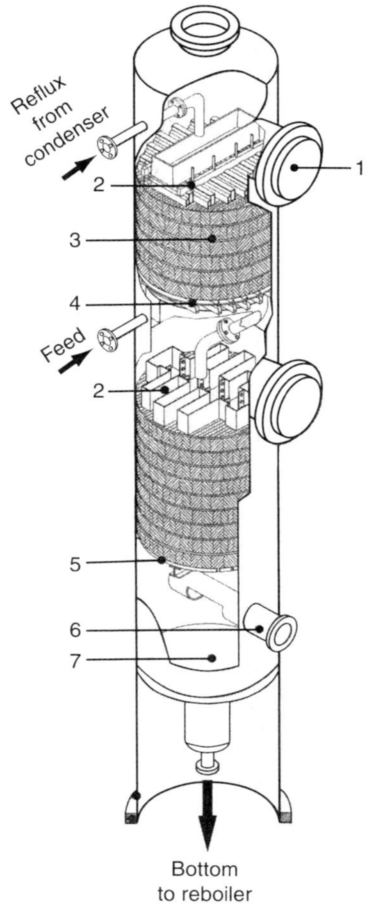

FIGURE 8.4
Main features of a packed distillation column: 1. mainways; 2. liquid distributor; 3. packing; 4. liquid collector; 5. support grid; and 6. vapor from reboiler.

8.1.3 Packed Column Characteristics

8.1.3.1 Main Features

Figure 8.4 draws attention to the principal internal components of a packed column. The particular packing shown is of the structural type displayed in Figure 5.5b. They allow the column diameter, which in distillation is usually confined to 1 m or less, to exceed that limit many times. Liquid collectors and distributors, which are shown only twice, are usually used every 3 m to counteract channeling. This undesirable effect can be particularly severe at low liquid flow rates (vacuum operation).

The internal construction of a packed-gas scrubber or absorber is similar, but the lower operating temperature leads to higher viscosities, requiring more frequent redistribution.

8.1.3.2 Relation between HTU and HETP

As noted in Chapter 5, the performance of a particular packing is often expressed in terms of a height equivalent to a theoretical plate (HETP) and is so reported by packing manufacturers. Performance data in terms of a HTU, or a volumetric mass transfer coefficient *Ka*, have a sounder theoretical basis but are relatively rare (see Table 5.6). A link between HTU and HETP can, however, be easily established for the case of a *linear equilibrium relation*. To show this, we start with the definitions of HTU and HETP and simply divide the two expressions. We obtain

$$\text{HTU} = \text{Height of Column/NTU} \tag{8.26}$$

$$\text{HETP} = \text{Height of Column/N}$$

and, consequently,

$$\text{HTU} = \text{HETP (N/NTU)} \tag{8.5a}$$

The ratio of number of plates to number of transfer units is obtained from the Kremser Equation 7.15b and Equation 8.4d. There results

$$HTU = HETP\frac{1-1/E}{\ln E}\ \text{HTU} \tag{8.5b}$$

Because the typical operating range for the extraction factor is $E = 1.2$ to 2, we find

$$\text{HTU} = (0.92 \text{ to } 0.70)\ \text{HETP} \tag{8.5c}$$

At the threshold value $E = 1$, HTU and HETP have the same value. Thus, for the linear case at least, the HTU is typically 10 to 30% lower than the HETP.

8.1.3.3 Operational Parameters

The key operational parameters for packed columns are the same as those seen in tray columns — there are three: gas and liquid loads and the pressure drop along the column. Together these three factors not only affect column performance but are the determining factors in the choice of packing size, packing type, and column diameter. They also have practical limits to their range, a feature we saw before in the case of tray columns.

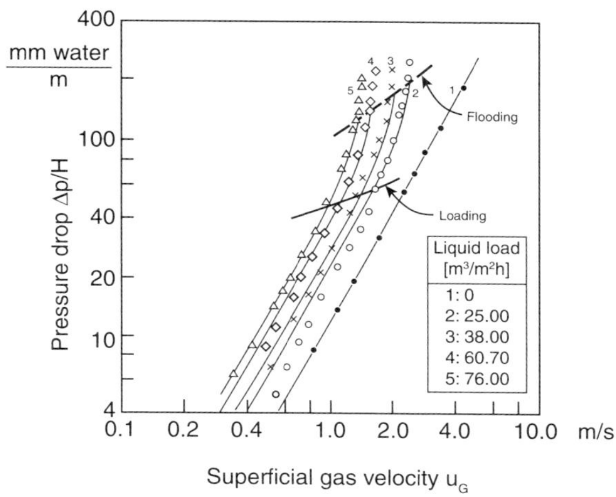

FIGURE 8.5
Pressure drop in column packed with 3.5 cm Pall rings.

Both gas and liquid loads are expressed as *superficial velocities* (i.e., volumetric flow rates divided by column cross section, $v_G = Q_G/A_C$ $v_L = Q_L/A_C$). Pressure drop is usually reported in units of millimeter H_2O per meter packing or, in the case of distillation, per theoretical plate.

The factors forcing us to set limits on the range of loads are, at the high end, the onset of flooding, and at the low end, channeling and incomplete wetting.

Flooding can be caused by both excessive gas and liquid flow rates. When gas loads exceed a certain limit, the pressure drop becomes so large that gravity flow can no longer proceed. The liquid piles up on top of the packing and is in part blown out of the tower. Excessive liquid loads lead to a similar situation.

The phenomenon is best viewed on plots of pressure Δp versus gas load v_G, with liquid load v_L as a parameter (Figure 8.5). Such plots are routinely featured in the promotional material of packing manufacturers. Abscissa values are also often reported in terms of the so-called gas-capacity factor $C = v_G[\rho_G/(\rho_L \rho_G)]^{0.5}$. This occasions a shift of the plots toward the ordinate but does not alter their appearance.

The logarithmic plots seen in Figure 8.5 initially have a slope of 2 (i.e., Δp varies with v_G^2, as it does in simple pipe flow). As v_G is increased, a location is eventually reached, termed the *loading point,* where the slope undergoes a rather abrupt increase. This is shortly thereafter followed by a *second* increase to a nearly vertical slope. This second location, termed the *flooding point,* signals the onset of liquid buildup at the top of the packing (i.e., of flooding). As a rule, we aim to operate at 50% of loading, or velocities of the order 1 m/s, because this strikes a reasonable balance between pumping cost (less at lower Δp) and column cost. These considerations may be superseded by external constraints on Δp such as the need

TABLE 8.2
Comparison of Packed and Tray Columns

Tray Column	Packed Column
Wide range of liquid loads	Wide range of gas loads
$\Delta p \sim 5$ mmHg/plate	$\Delta p \sim 0.5$ mmHg/plate
High liquid hold-up	Low liquid hold-up
Handles suspended impurities	Prone to plugging
Specialty alloys needed for corrosive fluids	Available in cheap, noncorrosive materials
Used mainly for distillation	Used mainly for gas scrubbing or small-scale distillation (<1 m diameter)

to press an existing blower into service or the outlet pressure that can be maintained in vacuum distillation.

8.1.3.4 Comparison of Packed and Tray Columns

The principal considerations that determine (or at least influence) the choice between packed and tray columns are the ability of either type to handle a given gas or liquid load, or range of loads, and the pressure drop per theoretical stage. Somewhat lower on the scale of priorities is the ability to handle corrosive fluids, liquids containing solid impurities or components with a potential to crystallize, and liquid hold-up.

Packed columns score well in terms of low Δp, robustness over a wide range of gas loads, and ready availability of corrosion-resistant components. They are less suitable for handling a wide range of liquid loads (distillation) or fluids with a tendency to deposit solids. These are the preferred domain of tray columns. Table 8.2 summarizes the principal advantages and drawbacks of the two types.

Illustration 8.1: A First Pass at Packed Column Design

The material just presented is sufficient to enable us to arrive at some first estimates of the important design and operational parameters of a packed column. Further refinements are usually undertaken in consultation with the equipment vendor and by introducing estimates of capital and operating costs. There is a trade-off between these two items, which is reflected in Figure 8.2. An increase in one of the operating parameters (say, solvent flow rate) may bring about a reduction in capital costs, but there is usually a price to be paid in terms of higher *increased* operating costs. The ideal is to operate at minimum *total cost* (i.e., with optimal solvent flow rates, column diameter, packing size, etc.). This ideal global optimum is often not attainable because of external constraints, and a proper balance can be struck only by more detailed calculations. It is reassuring that the "first-pass" estimates we wish to arrive at here are often not too far removed from the more refined results.

The process to be considered is in current use. It involves the stripping with air of low concentrations of SO_2 from a dilute acid stream and is necessary in order to avoid reaction with H_2S used in a subsequent step (see Illustration 8.2). The feed is now the liquid stream that enters the top of the column at the rate of 0.02 m^3/s while the stripping agent (air) enters at the bottom at a rate G to be determined. Operation is at 1 atm, and a 99% removal of H_2 is stipulated. Henry's constant for SO_2 is $H = 5.17$ (kg/kg air)/(kg/kg water).

The aim will be to arrive at first estimates of airflow rate G, column height, gas and liquid loading (i.e., superficial velocity), and column diameter.

We turn to Figure 8.3a for the first item. Recalling the recommended range of 1.2 to 2.0 for the extraction factor, we choose $E = 1.5$, so that

$$E = \frac{HG}{L} = 1.5 \tag{8.6a}$$

and

$$G = 1.5 \times 20/5.17 \tag{8.6b}$$

$$G = 5.8 \text{ kg/s} \tag{8.6c}$$

The corresponding value for the number of transfer units for a residue factor of 0.01 (99% removal) is NTU = 11.

The HTU value needed to calculate column height depends on the size and type of packing used. We chose 35 mm Pall rings because we happen to have a convenient pressure drop plot in Figure 8.5. Because no direct values for either HETP or HTU are immediately available, we turn to the rule of thumb given by Equation 5.14a and obtain

$$\text{HETP} = 1.5 \times 12 \times 3.5 = 63 \text{ cm} \tag{8.6d}$$

Drawing on Equation 8.5b, we obtain an equivalent value for HTU of

$$\text{HTU} = \text{HETP}\,\frac{(1-1/E)}{\ln E} = 63\,\frac{(1-1/1.5)}{\ln 1.5} \tag{8.6e}$$

$$\text{HTU} = 51 \text{ cm} \tag{8.6f}$$

and a corresponding column height of

$$Z = \text{HTU} \times \text{NTU} = 0.51 \times 11 \tag{8.6g}$$

$$Z = 5.6 \text{ m} \tag{8.6h}$$

TABLE 8.3

Packed Tower Parameters

Column Diameter, d_C (m)	Superficial Velocities v_G (m/s)	v_L (m^3/m^2h)	Δp mmH_2O/m
0.5	25	380	Flooding
1.0	6.4	96	Flooding
1.5	2.8	42	Flooding
2.0	1.6	24	40
2.5	1.0	15	18
3.0	0.71	10	8

Some space will have to be allowed for liquid collectors and distributors (at least two each), as well as bottom support, giving an overall height of about 8 m.

Because there are no external constraints on Δp, there is considerably more latitude in setting gas and liquid velocities and, hence, the column diameter. Many workers prefer to start with an assumed value of gas load ($v_V \sim 1$ m/s is the convention) and use it to calculate column diameter d_C and liquid load v_L. Δp is then read from Figure 8.5 or similar plots for other packings. We modified this approach slightly by starting with an assumed column diameter value of d_C and calculating the loads from volumetric flow rate divided by column cross-sectional area. The results are shown in Table 8.3. The volumetric flow rates used in these calculations were as follows:

Air: (5.8 kg/s)/1.2 kg/m^3 = 4.8 Nm^3/s

Liquid Feed: 0.02 m^3/s = 72 m^3/hr

It is clear that the results are unpromising up to diameters of about 1.8 m because of flooding. The performance becomes satisfactory for $d_C \geq 2$ m, and the pressure drop in each case is well below the loading point. A diameter of 2.5 m suggests itself as striking a reasonable balance between operating costs (pumping) and capital cost (column plus packing).

Illustration 8.2: Use of a Reactive Solvent: Scrubbing of H_2S with a Caustic Solution

We return here to the topic taken up in Illustration 7.11 (i.e., the processing of spent sulfuric acid). The proposal there was to recover the acid by a two-stage distillation process and return it to the plant for further use. When the acid is too dilute, the procedure may become uneconomical, and we must then consider the alternative of neutralizing the acid with lime and disposing of the calcium sulfate in a landfill site. This confronts us with a second dilemma. The spent acid often contains traces of toxic metals (Hg, Pb, Zn,

Cd) that pass into the calcium sulfate, requiring disposal in a so-called "secure" site. The fee charged at these sites can amount to many millions of dollars per year.

An alternative is to precipitate the toxic elements as sulfides using Na_2S and recover the hydrogen sulfide produced by scrubbing. Disposal of $CaSO_4$ can then take place in an *open site*. It is this latter case we wish to address here.

Packed tower scrubbing with a solution of NaOH suggests itself as a natural choice because the Na_2S produced can be recycled back for sulfide precipitation, and corrosion-resistant packings are readily available. The question now arises: How should Equation 8.2 be modified to accommodate reaction in the liquid phase? One approach suggested in Illustration 4.11 was to assume the solute to be in equilibrium with water (or dilute Na_2S solution) at the interface and apply an enhancement factor E to Ka (or HTU). The operating diagram (Figure 8.1b) can then be used as before. The second approach was to assume the reaction to be very rapid and irreversible (i.e., negligible liquid-phase resistance) and to obtain HTU (or HETP) values from the packing manufacturer.

A rapid conversion of H_2S implies that its concentration of the interface will have dropped to zero. Consequently, Y^* in Equation 8.2a vanishes, and we obtain the following by direct integration:

$$Z = \text{HTU} \times \text{NTU} = \text{HTU} \times \ln \frac{Y_1}{Y_2} \tag{8.7a}$$

For the process in question, the requirement was a reduction in H_2S content from 18,000 ppm to 0.1 ppm. We obtain

$$Z = \text{HTU} \times \ln 1.8 \times 10^5 \cong 12\ \text{HTU} \tag{8.7b}$$

This is the required height of the column.

The second question to be addressed is that of solvent flow rate, minimum and actual. The conventional method of establishing conditions for a "pinch" (i.e., a vanishing driving force $Y - Y^* = 0$) no longer applies since Y^* was declared to be zero. In fact, the entire conventional use of the operating diagram becomes invalid, a feature further explored in Practice Problem 8.4. The minimum here is dictated not by a phase equilibrium but rather by the stoichiometry of the reaction. Thus, if 1 mol/s of H_2S needs to be absorbed and neutralized, a minimum of 2 mol/s of NaOH has to be made available.

The feed rate (air and H_2S) in the process in question was set at 0.13 m^3/s = 5.3 mol/s The rate at which H_2S entered the column was therefore $18{,}000 \times 10^{-6} \times 5.3 = 9.6 \times 10^{-2}$ mol H_2S/s. This is the amount that has to be neutralized as a minimum requirement.

Suppose we opt for a 1% (0.25 molar) solution of NaOH. The minimum rate at which this solution has to enter the column is then given by

$$L_{\text{Min}} = 2 \times \frac{9.6 \times 10^{-2}\ \text{mol/s}}{0.25\,\text{mol/L}} = 0.77\ \ L\ /\ \text{s} \tag{8.7c}$$

Comments

The mundane nature of these calculations may set the teeth on edge of those looking for a more profound treatment or greater mathematical precision. This can be provided by numerical simulations at considerable cost and often only marginal gain.

Illustration 8.3: Distillation in a Packed Column: The Case of Constant α at Total Reflux

Packed-column distillation, while practiced much less frequently than gas absorption, still finds considerable use in medium- and small-scale applications and in the processing of corrosive liquids (HCl, H_2SO_4). The model equations parallel those for the absorption case and result in the same HTU–NTU relations we had seen there, with mole fractions taking the place of mass ratios as the pertinent concentration units. We obtain

$$Z = \text{HTU} \times \text{NTU} = \frac{G}{K_{OY}a} \int_{y_2}^{y_1} \frac{dy}{y^* - y} \tag{8.8a}$$

and supplement this expression with the equilibrium relation given by

$$y^* = f(x) \tag{8.8b}$$

The operating lines needed for the evaluation of the NTU integral are identical to those established in Chapter 7 and can be seen in Figure 7.17 and Figure 7.18. Either graphical or numerical procedures can be used for this purpose.

A case of special interest is that of a system with constant relative volatility α run at total reflux. The equilibrium relation is now given by

$$y^* = \frac{\alpha x}{1 + x(\alpha - 1)} \tag{8.8c}$$

Because at total reflux the operating line coincides with the 45° diagonal, we have equality of the mole fractions, $y = x$, and Equation 8.8c becomes

$$y^* = \frac{\alpha y}{1 + y(\alpha - 1)} \tag{8.8d}$$

Introduction of this relation into the NTU integral yields

$$NTU = \int_{y_1}^{y_2} \frac{dy}{\dfrac{\alpha y}{1 + y(\alpha - 1)} - y} \tag{8.8e}$$

or, equivalently,

$$\text{NTU} = \int_{y_1}^{y_2} \frac{1 + y(\alpha - 1)}{y(\alpha - 1)(1 - y)} dy \tag{8.8f}$$

This expression can be evaluated analytically using the additional integration formula

$$\int \frac{dx}{x(a + bx)} = -\frac{1}{a} \ln \frac{a + bx}{x} \tag{8.8g}$$

Its application leads to the result

$$\text{NTU} = \frac{1}{\alpha - 1} \ln \frac{y_2(1 - y_1)}{y_1(1 - y_2)} + \ln \frac{1 - y_1}{1 - y_2} \tag{8.8h}$$

This is a convenient expression to determine NTU and, from it, HTU values in laboratory experiments, which are usually run at total reflux. It also finds use in the distillation of closely boiling mixtures such as isomers or isotopes. Here α values are quite low and constant, and the operating line nearly coincides with the 45° diagonal so that Equation 8.8h becomes a valid approximation of the actual number of transfer units. An example of its application in the distillation of isotopes is found in Practice Problem 8.8.

8.1.4 Liquid–Liquid Extraction in a Packed Column

Continuous countercurrent extraction in a packed column is an attractive and frequently used alternative to the staged configurations discussed in Chapter 7. Both random packings and structured packings, samples of which were shown in Figure 5.5, are in current use. In a typical operation, the lighter *phase*, often an organic solvent, enters at the bottom of the column, while the heavier feed, often an aqueous solution, is introduced at the top. Either the light or the heavy liquid may be the dispersed phase. The packing plays its traditional role of enhancing both the degree of turbulence and the interfacial contact area. It must, however, have the additional property of being preferentially wetted by the continuous phase.

TABLE 8.4

Current Applications of Packed-Column Liquid–Liquid Extraction

Feed	Solvent
Phenol in wastewater	Isopropyl ether
H_2S in liquid propane	Amines
Methanol in hydrocarbons	Water
Acetic acid in water	Various ketones
Nickel, cobalt, zinc, copper in acid leach	Various organic chelating agents

So far, all the features mentioned parallel those seen in gas–liquid operations. Liquid–liquid systems, however, present some special challenges, which set them apart from their gas–liquid counterparts and make it much more difficult to predict the performance of an extraction column.

The first and most obvious problem is the tendency of the dispersed phase to coalesce. Packings, particularly those of a structural type, partly offset this effect. A more direct method of intervention is to pulse one of the streams, usually the dispersed phase fluid. This not only reduces droplet size but also creates a higher degree of turbulence, thus enhancing the mass transfer rates. Pulse frequencies are typically of the order 1 s^1.

A lesser known evil is the presence of axial convention currents caused by changes in density in the two phases. These reduce the overall efficiency of the column but are again partly offset by a good packing. A third difficulty, which is not as easily disposed of, is the profound effect of impurities on column efficiency. Such trace substances tend to accumulate at the interface, affecting both surface tension (i.e., interfacial area) and the mass transfer coefficient. Pulsing will almost always be a necessity in these cases, but its mitigating effect will have to be determined by pilot testing. In spite of these difficulties, liquid–liquid extraction in packed columns is now well established as a major separation technology. Table 8.4 conveys a sense of the range of industrial-scale applications. A sense of the size and structure of industrial units is conveyed in Figure 8.6.

The performance of extraction columns is described by the same HTU–NTU relations seen earlier for gas scrubbers and packed distillation columns. For the important case of mutually insoluble solvents, they take the form

$$H = \text{HTU} \times \text{NTU} = \text{HTU} \int_{X_1}^{X_2} \frac{dX}{X - X^*} \tag{8.9a}$$

and

$$H = \text{HTU} \times \text{NTU} = \text{HTU} \int_{Y_1}^{Y_2} \frac{dY}{Y - Y^*} \tag{8.9b}$$

depending on whether the raffinate or extract phase represents the controlling resistance. For mutually soluble solvents, NTU contains an additional

FIGURE 8.6
Pulsed (Bateman) columns used in copper refining.

TABLE 8.5

Range of Operational Parameters in Plant-Scale Packed-Column Liquid Extraction

Superficial velocity (either phase)	10^{-3}–10^{-2} m/s
Column diameter	0.5–2.5 m
Pulse frequency	10–100 min^{1}
HTU	0.5–2.0 m

term usually amounting to less than 10% of the total number of transfer units. Table 8.5 provides an indication of the range of HTU values and other operational parameters.

Flooding can occur in liquid extraction columns, just as it does in packed-gas scrubbers, whenever excessive flow rates cause a build-up or backing-up of one or the other phase. Pressure drop is not a convenient measure here, and other criteria are used instead. This is explored in Practice Problem 8.6.

Illustration 8.4: More on Nickel Recovery from Acid Leach: The Importance of Chemistry

The reader may by now be suffering from a surfeit of column design and performance analysis. We do not wish to abandon the topic entirely but broaden it instead into an analysis of laboratory test data, their significant features, and how they can be used in the preliminary design of a separation unit.

It will be recalled from Illustration 7.8 that the primary step of processing the nickel ore at the Goro site consisted of a high temperature and pressure acid leach, followed by the settling and countercurrent washing of the slurry. The aqueous phase from the continuous countercurrent decantation (CCD) cascade containing Ni, Co, Cu, Zn, and several other constituents then undergoes a two-step extraction process. In the first step, the solution is contacted with an organic chelating agent with a strong affinity for nickel and cobalt. Some of the other constituents are inevitably extracted to some degree as well. Because the nickel chelate complex cannot be conveniently processed further, a second extraction step is added, this time using hydrochloric acid, which converts the nickel back into the more desirable aqueous nickel chloride form. $NiCl_2$ is relatively easily pyrolyzed to nickel oxide, which is the final product of the plant. Reduction to nickel metal is usually carried out by the customer.

The proper choice of extracting solvent and the general chemistry of the system are crucial factors in arriving at a viable process. Further evidence of this appears in the reverse extraction data shown in Figure 8.7. Because the chelating agent used in the first step already has a high affinity for nickel, reverse extraction will require a solvent with even greater affinity. Under normal conditions, hydrochloric acid does not meet that requirement. It turns out, however, that an increase in temperature and acid concentrations, both of which "loosen" the chelate complex, brings about a dramatic shift in the desired direction. Here again, the proper exploitation of the chemistry of the system is critical to the success of the process.

Can the data be used to begin the process of estimating column design parameters? The answer lies in recognizing that the kinetic data of Figure 8.7 tend toward an equilibrium value with increasing time. The asymptotic ratio of the amount of nickel stripped to the amount retained by the organic phase can then be viewed as a *distribution coefficient* m for nickel. m can be used in the extraction factor $E = MB/A$ (Figure 8.2 and Table 8.1) to calculate solvent requirements B and ultimately the number of transfer units. Details of these and further calculations are left to the exercises (Practice Problem 8.5).

Comments

This illustration is meant to emphasize that major advances and improvements in separation technology do not always hinge on novel equipment or design refinements. They are more often rooted in the chemistry or physical chemistry of the system. Without the innovative use of chelating agents, metal extraction would badly flounder. Without recognition of the importance of temperature and acidity in the reverse extraction, it would be hugely inefficient. Here, as elsewhere, engineering and chemistry must be twinned. Neither should be neglected in favor of the other.

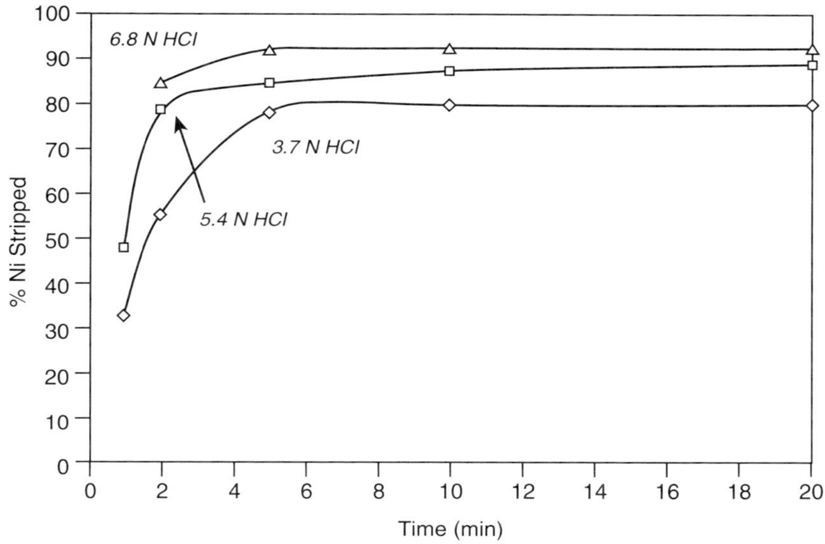

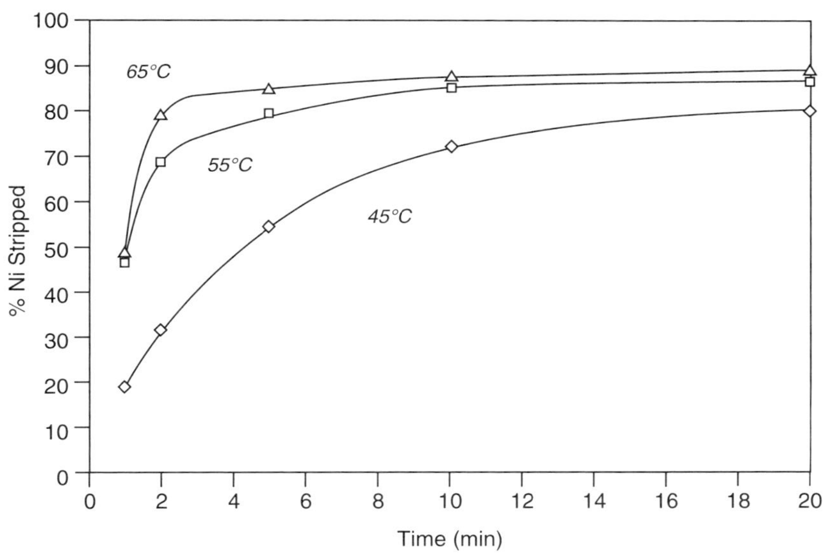

FIGURE 8.7
Nickel stripping kinetics at various acid concentrations and temperatures. Initial nickel loading 7 g/L. Ratio of acid to organic phase $A/0 = 1$. (Courtesy Jean-Jacque Parraud, Goro SAS and V. Papangelakis, University of Toronto.)

Illustration 8.5: Coffee Decaffeination by Countercurrent Supercritical Fluid Extraction

Section 6.2.5 introduced the reader to the concept of supercritical fluid extraction (SCE) and in Illustration 6.7 provided a first simple example of its application by considering the single-stage extraction of caffeine from coffee beans using supercritical carbon dioxide. The amount of CO_2 required in this elementary operation was calculated at 45.9 kg CO_2/kg coffee, an inordinately high level. Commercial decaffeination processes use a countercurrent operation carried out in tall columns (see Figure 6.16), with the coffee beans entering at the top and discharging intermittently at the bottom. This process is close enough to a continuous operation to allow us to use the principles established for such operations.

The operating diagram for the process is shown in Figure 8.4. The equilibrium curve is drawn from the relation established in Chapter 6 — that is,

$$x = 1.24\ y^{0.316}$$

The operating line will now be located *below* the equilibrium curve, because the process is one of stripping or desorption (i.e., the reverse of what we had seen in the case of gas absorption). The resistance here would evidently be in the solid phase, with the driving force given by the horizontal distance between operating line and equilibrium curve (Figure 8.8). We do not make use of the driving force in this illustration but address instead the following problem: Suppose that the caffeine content is to be reduced from 1 to 0.05%, a typical goal in commercial decaffeination and one used in the single-stage extraction process in Illustration 6.7. We first wish to establish how much the CO_2 consumption is reduced if we replace the single-stage process with a countercurrent mode of contact. Second, given a typical column height of 20 m, what would be the height equivalent to a theoretical stage (HETS)?

We start in the usual fashion by first establishing the minimum amount of "solvent" CO_2 required for the operation. The pinch here does not occur at the feed condition, as is usually the case, but much earlier at some intermediate point of the equilibrium curve. That point is located by drawing a tangent to the equilibrium curve starting from the point representing conditions at the bottom of the column (see Figure 8.8).

If we now set the actual G/S ratio at twice the minimum amount, we obtain the following result:

$$(S/G)_{actual} = 1/2(S/G)_{Min} = \Delta y/\Delta x = 0.04/0.95 \tag{8.10a}$$

The CO_2 consumption, which is the inverse of this ratio, is then given by

$$G/S = 0.95/0.04 = 23.8 \text{ kg } CO_2/\text{kg coffee} \tag{8.10b}$$

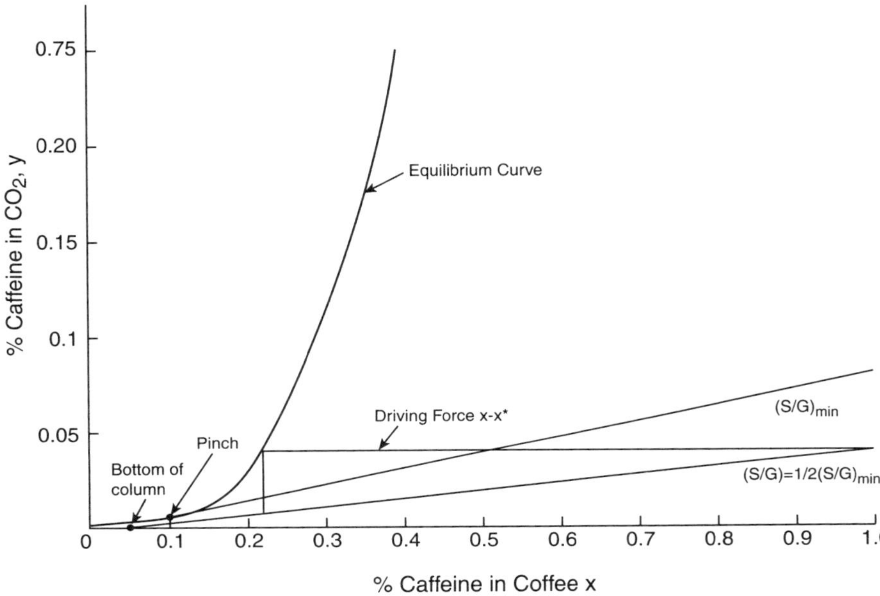

FIGURE 8.8
Operating diagram for decaffeination.

The number of stages between feed and effluent is obtained by the usual staircase construction and equals approximately four. The corresponding HETS value is then given by

$$HETS = \frac{\text{Height of column}}{\text{Number of stages}} = \frac{20}{4} = 5\,m \tag{8.10c}$$

Comments

There are a number of features in this example drawn from industry that fall outside the conventional boundaries of textbook problems. The first item of note is the location of the pinch, which occurs at a point between feed and effluent conditions. The resulting minimum CO_2 consumption is at the modest level of 11.9 kg CO_2/kg beans. Even doubling that amount for actual operating conditions still yields a consumption less than that calculated for a single-stage contact (Illustration 6.7). This is a clear-cut vindication of the countercurrent mode of operation.

The second feature of note is the small number of stages required (four) and connected to it the inordinately high value of the height equivalent to a theoretical stage (5 m). This was partly the result of the relatively generous amount of CO_2 we allowed above the minimum required. If that number is reduced to 1.5 times the minimum, the number of stages rises to approximately 10, and with it comes a corresponding drop in the HETS to 2 m. It is likely that the actual column operation is, in fact, in the range of 1.2 to 1.5 times the minimum CO_2 requirement and that the HETS is of the order of 1 m, a value much more in line with that expected in operations of this type. The reader is urged to examine these cases more closely by enlarging the relevant section of the operating diagram and stepping off the stages for each slope of the operating line.

Illustration 8.6: Design and Performance of a Biofilter

Biofilters used in the removal of objectionable components from air are, in essence, packed-bed catalytic converters, with the catalyst taking the form of an aqueous "biofilm" immobilized on the surface of the inert packing. In Illustration 4.13, an expression was derived for the effectiveness factor of the biofilm catalyst, and the task here is to convert that result into a working expression for the design and performance analysis of a biofilter.

The starting point is a conventional component mass balance over a difference element of the reactor. Thus,

$$\text{Rate of solution} - \begin{array}{c}\text{Rate of solute out}\\ \text{by flow and reaction}\end{array} = 0$$

$$QC_g\Big|_z - \left(QC_g\Big|_{z+\Delta z} + Er_i\Delta V\right) = 0 \tag{8.11a}$$

where r_i = intrinsic reaction rate = k_0, and E is the effectiveness factor. ΔV is the reaction volume of the biofilm, $\delta\Delta A = \delta a.A_C\Delta z$, where a was borrowed from conventional packed tower gas–liquid contact to denote the specific surface area. In biofilters, it has a value of the order 100 m^2/m^3 filter.

Introducing the effectiveness factor E from Equation 4.26i, we obtain, after dividing by $A_C\Delta z$ and going to the limit,

$$v\frac{dC_g}{dz} + ak_0\left(\frac{2C_gD_e}{k_0m}\right)^{1/2} = 0 \tag{8.11b}$$

where $v = Q/A_C$ = superficial velocity of the gas.

Integrating by separation of variables, we obtain the first instance

$$C_g^{\;1/2} - C_{go}^{\;1/2} = -\frac{ak_0z}{v}\left(\frac{D_e}{2k_0m}\right)^{1/2} \tag{8.11c}$$

and after rationalization,

$$C_g / C_{go} = \left[1 - \frac{az}{v} \left(\frac{k_0 D_e}{2mC_{go}} \right)^{1/2} \right]^2 \quad (8.11d)$$

or

$$z = \frac{v}{a} \left(\frac{2mC_{go}}{k_0 D_e} \right)^{1/2} \left[1 - (C_g / C_{go})^{1/2} \right] \quad (8.11e)$$

where C_{go} = inlet concentration.

Comments

Equation 8.11e is the equivalent of the z = HTU × NTU relations established for other continuous contact operations the square bracket representing the NTU. It serves the same dual purpose of column design and performance (i.e., calculation of C_g/C_{go}). It also allows us to predict the effect on performance of such parameters as air velocity, inlet concentration, or that of the partition coefficient m.

8.2 Membrane Processes

The notion of separating substances by means of membrane barriers that allow the selective passage of one or more species to the exclusion of all others has both a powerful appeal and the appearance of an impossible dream.

Consider the advantages of such a process compared to other operations:

- Elimination of thermal energy as an operating expense
- Ability to process heat-sensitive materials
- No intrusion of solvents, which merely produce an enriched phase and require further processing
- No necessity to regenerate or replace adsorbents or ion-exchange resins

In fact, the only operating costs visible to the casual observer are conveying the fluids to and (usually) through the barriers and the ultimate disposal of the product stream.

a.

b.

FIGURE 8.9
Ceramic membranes: (a) hollow-fiber elements and modules and (b) ultrafiltration unit for fruit juice clarification. (Courtesy Mempro Ceramics Corporation.)

The age-old quest for such membranes accelerated in the first half of the 20th century but was initially plagued by a lack of suitable materials, low throughput rates and selectivity, and continuous problems with fouling and plugging. The advent of synthetic polymers brought about a wide new range of materials capable of being altered and modified to suit a particular application. Low throughput rates initially persisted but were overcome with the dramatic development in 1960 of high-flux cellulose acetate membranes by Loeb and Sourirajan. This event signaled the start of a burst of activity in the field of membrane development that continues undiminished to this day. Among the more recent innovations are ceramic, stainless steel, and titanium membranes (see Figure 8.9).

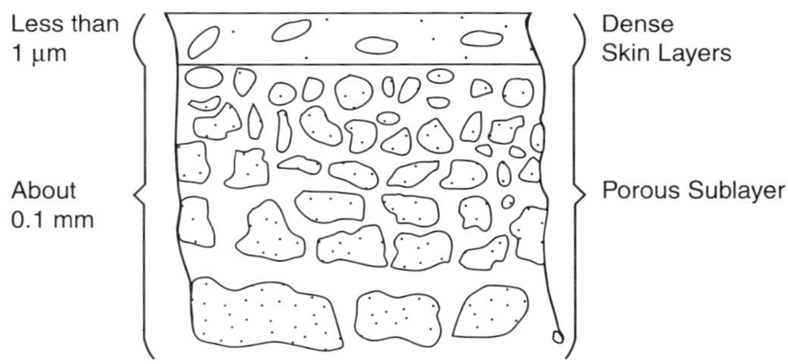

FIGURE 8.10
The asymmetric membrane.

8.2.1 Membrane Structure, Configuration, and Applications

The basic structure of present-day membranes is illustrated in Figure 8.10. It consists of a selective polymeric film and a much thicker but more porous sublayer, which provides the necessary structural support but otherwise does not actively participate in the separation. Such membranes composed of a dual layer of different materials are referred to as *anisotropic* or *asymmetric*.

The performance of a membrane process is influenced by a number of factors, chief among which are the physical structure of the membrane and the nature and composition of the feed. These factors can be used to order and classify the processes into subcategories, each with its own distinctive features. These categories have gradually assumed proportions that have turned them into self-contained operations in their own right. This is shown in Table 8.6, which summarizes the various subprocesses that have emerged, their relation to physical structure expressed through the membrane pore diameter, and the different applications that have come about as a result. Note the impressive range of the physical parameters and the actual applications. Pore diameters vary over the span of six orders of magnitude, capable of barring particles as small as sodium ions (3.7 Å) in reverse osmosis or bacteria with a diameter of 1 μm in microfiltration, while allowing free passage to water molecules. Intermediate-size particles such as proteins (100 Å) or viruses (1000 Å) can either be retained or allowed to permeate, depending on the requirements of the process.

Although fairly distinct in their capabilities, these processes have several features in common. The membrane materials used are generally polymeric in origin and are extensions, both in mode of preparation and in composition, of the cellulose acetate membranes pioneered by Loeb and Sourirajan. Another common feature shared by most of the processes listed is the use of pressure as the driving force; the exceptions are the different forms of dialysis. Thus, in hemodialysis, concentration replaces pressure as the driving force, while electrodialysis is driven by an electrical potential.

TABLE 8.6

Membrane Processes and Their Applications

Process	Pore Diameter	Applications
NF, Nanofiltration	10 Å	High flux desalination
RO, Reverse osmosis	5–30 Å	Desalination
HD, Hemodialysis	Similar to RO membranes	Removal of urea, creatinine, and other substances from blood
ED, Electrodialysis	Similar to RO membranes	Desalination in the range 500–2000 ppm salt content
GS, Gas separation	Similar to RO membranes	Recovery of H_2 in ammonia plants, separation of N_2 from air, sweetening of natural gas
UF, Ultrafiltration	50–1000 Å	Separation of small molecules (inorganic compounds, sugars, amino acids, small proteins) from macromolecules; treatment of process water and whey; clarification of fruit juices; flood filtration in the kidney
MF, Microfiltration	0.1–10 μm	Removal of microorganisms from water and beverages, or product antibiotics
CF, Conventional filtration	10 μm–100 μm	Cell harvesting

Note: 1 nm = 10 Å = 10^3 μm.

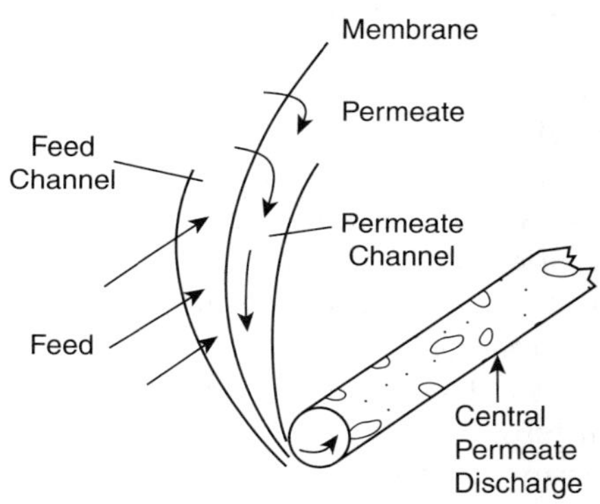

FIGURE 8.11
The spiral-wound membrane.

Geometrical configurations of the membrane modules have passed through various stages of development from which two major contenders have emerged. In the spiral-wound configuration, membrane sheets flanked by two spacer-supported channels are wound around a central perforated collector tube, which receives the permeate (Figure 8.11). Feed enters axially through one of the channels and selectively permeates radially through the membrane into the second channel, which discharges the permeate into the collector tube. In the hollow-fiber configuration, a *shell and tube* arrangement is used in which the feed enters either through the membrane fibers (core-fed) or is confined to the shell side (shell-fed). These flow geometries are illustrated in Figure 8.12a and Figure 8.12b.

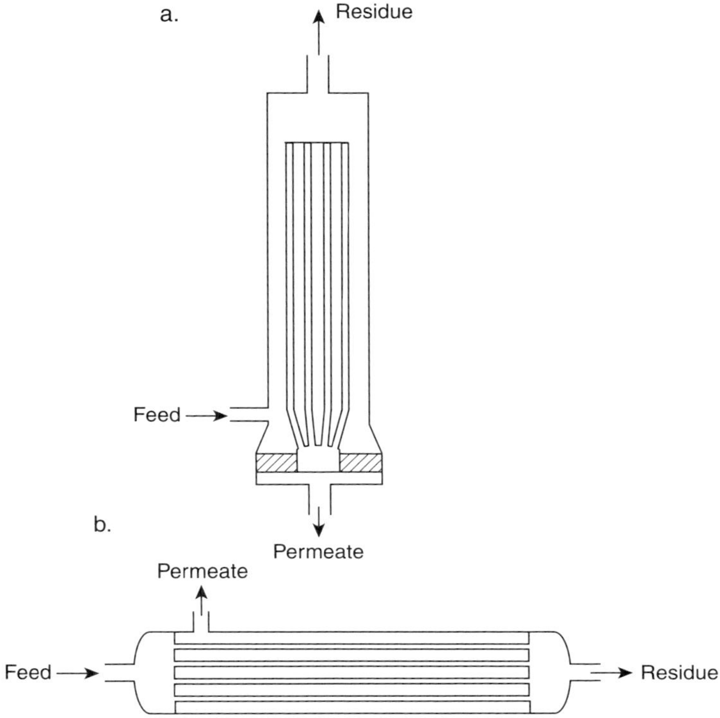

FIGURE 8.12
Hollow-fiber reverse-osmosis modules: (a) shell-side feed and (b) bore-side feed.

Let us now turn to a more detailed examination of these processes. Both *nanofiltration* (NF) and *reverse osmosis* (RO) draw on principles of osmosis for their implementation. The principal features of this phenomenon and its manifestation in reverse-osmosis operations are illustrated in Figure 8.13a. Consider a selective membrane (i.e., one that is freely permeable to water but much less so to salt) separating a salt solution from water, as shown in Part 1. In such an arrangement, water will flow from the pure-water side into the side less concentrated in water (i.e., the saltwater side). This process is referred to as normal osmosis. If, now, a hydrostatic pressure is applied to the salt side, the flow of water will be retarded, and if that pressure is sufficiently high, the flow will ultimately cease completely. At this point, we will have reached what is termed *osmotic equilibrium* (Part 2), and the hydrostatic pressure associated with this state is referred to as the *osmotic pressure*. A further increase in applied pressure will act to reverse the flow from the salt solution to the pure-water side (Part 3), and that flow will rise as the pressure is increased. This phenomenon, termed *reverse osmosis*, is exploited in all RO operations. The membranes used in these processes are so dense that no discrete pores can be discerned. Permeation proceeds instead by a solution–diffusion mechanism that makes use of a statistical distribution of

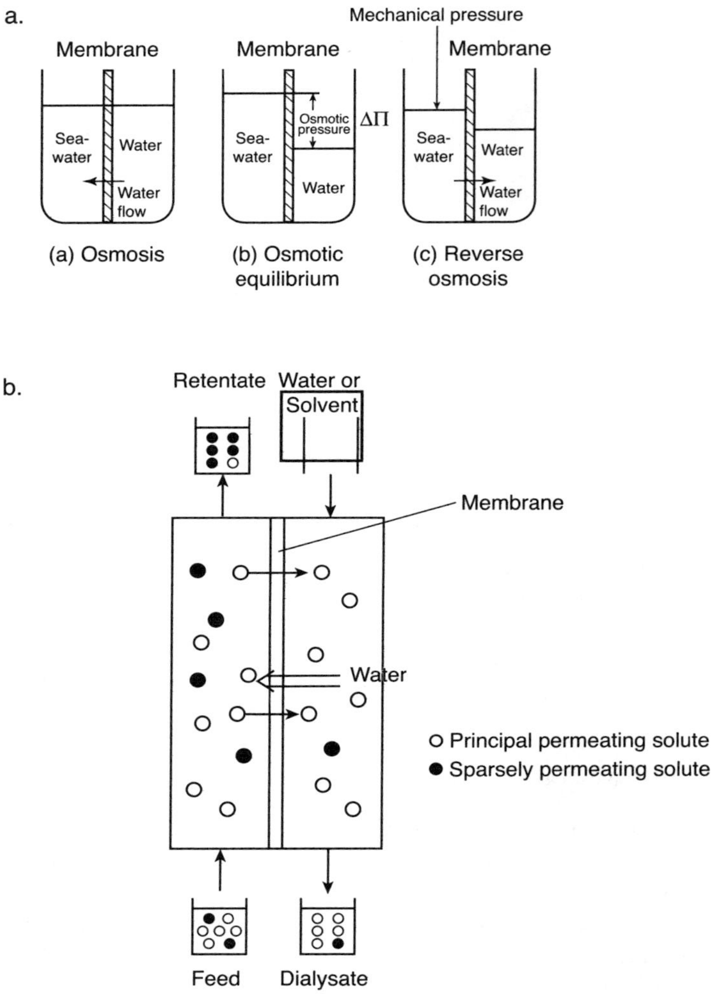

FIGURE 8.13
(a) Principles of reverse osmosis. (b) Principles of dialysis.

free volume or holes (see Chapter 3). The daily production of water using RO and NF amounts to approximately 1 billion gallons, of which 60% involves desalination of brackish and seawater. The remainder of the membrane capacity is in the production of ultrapure water for the electronics industry. Both hollow-fiber and spiral-wound configurations are employed, with pressures in the range of 50 to 70 atm for seawater and 15 to 20 atm for brackish water. Osmotic pressures for these feeds are in the range of 1 to 20 atm and are listed, along with other items of interest, in Table 8.7.

Gas separation and *hemodialysis* often employ membranes similar to those used in RO, but they differ from them in several aspects. Neither involves

TABLE 8.7

Osmotic Pressures of Various Solutions at 25°C

Solute	Concentration (mg/L)	Osmotic Pressure (atm)
NaCl	35,000	27
Seawater	32,000	23
Brackish water	2,000–5,000	1–2.7
Sucrose	1,000	1.1
Dextrose	1,000	2
Blood plasma and intracellular fluid (37°C)	—	7.2

osmotic pressure as a countervailing driving force. In gas separation, the partial pressure difference of the diffusing species acts as the driving potential, while in hemodialysis the concentration difference across the membrane acts to provide the necessary impetus. Hollow-fiber configurations are predominant in both applications.

Membrane gas separation is a rapidly evolving technology, with a strong presence in hydrogen recovery from plant off-gases and in the separation of nitrogen from air. Approximately 50% of all nitrogen used in ammonia synthesis and other applications is produced by membrane separation, the rest coming from the cryogenic liquefaction and distillation of air. Hydrogen recovery typically employs pressures as great as 150 atm on the feed or shell side of the module, with correspondingly small hollow-fiber diameters of 50 to 100 μm (Table 8.6). Nitrogen separation employs much lower feed pressures of only up to 10 atm, which can be easily accommodated in the bore side of the hollow-fiber module and can tolerate higher fiber diameters of up to 1 mm. Because of the extreme sensitivity of the separation process to small pinholes in the membrane, it has become customary to cover the basic fiber with a permeable defect-sealing layer of silicone rubber. We address questions of enrichment and selectivity attained in these processes in the next section.

Hemodialysis is a special case of a wider class of operations covered by the umbrella term *dialysis*. It involves, as is shown in Figure 8.13b, the simultaneous permeation of solutes from the higher to the lower concentration side of the membrane while water passes in the opposite direction driven by the osmotic pressure difference of the two solutions. In hemodialysis, water flux is suppressed by using iso-osmotic solutions on both sides. The nonpermeating solution is referred to as the *retentate* and contains blood cells, proteins, and other valuable components of the blood. The stream carrying away the toxic solutes, which have permeated through the membrane, is termed the *dialysate* or permeate. The use of hollow-fiber membranes in the dialysis of blood is now widespread and serves about 1 million patients worldwide involving over 100 million procedures a year. The device, often referred to as an artificial kidney, is illustrated in Figure 8.14. In it, a saline solution with an osmotic pressure nearly equal to

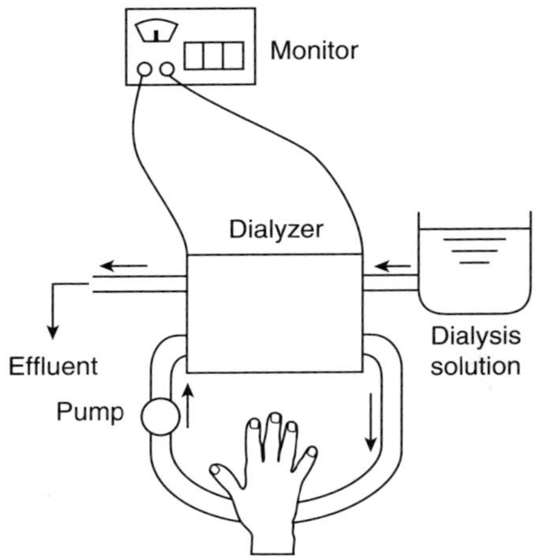

FIGURE 8.14
The hemodialyzer.

that of blood is supplied to the shell side of the module, while the blood is pumped countercurrently through the core side of the hollow-fiber module. The device contains several thousand fibers in a 5-cm-diameter tube with a total membrane area of 1 to 2 m^2. The cumulative area used worldwide for medical purposes exceeds that employed in all industrial applications.

Ultrafiltration (UF) is commonly used to strip smaller-sized molecules such as salts and sugars from solutions containing valuable macromolecules. The most prominent applications that have emerged in recent times are the concentration of automotive paint in process rinse water and the concentration of valuable proteins in whey, the supernatant liquid produced in cheese manufacture. In both cases, the application of UF technology not only helps recover valuable materials but also eliminates a bothersome disposal problem. The extension of the process to the treatment of other waste streams has been hampered by the tendency of the membrane to accumulate retained macromolecules at the surface, often in the form of adhering gel layers. This *concentration polarization*, as it is termed, which arises in all liquid-phase membrane processes but is most marked here, can be partially overcome by membrane modification, repeated cleaning, or the use of high fluid velocities. UF has a potentially huge market in the treatment of wastewaters, and current indications are that the technology has now prospered sufficiently to be adopted on a large scale for the treatment of effluents.

8.2.2 Process Considerations and Calculations

The engineer dealing with membrane processes is confronted with the usual set of tasks associated with operations of this type. In the area of design, we wish to establish the type of membrane and the total area required for a

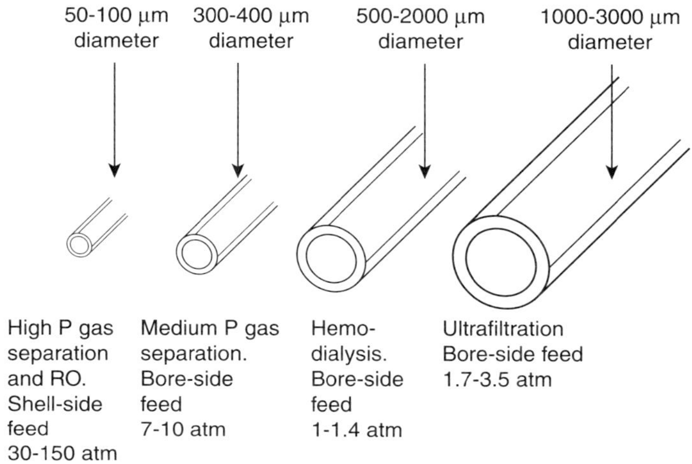

FIGURE 8.15
The principal types of hollow-fiber membranes.

specified separation or production rate. Conversely, for a given unit, real or imaginary, we would want to project performance or product concentration and recovery. Finally, it is often desirable to establish the effect of operating parameters such as feed composition, module geometry, or applied pressure on process performance. We will provide some guidelines for dealing with these tasks, aided by the survey of fiber size and operating pressure given in Figure 8.15.

Membrane processes in general involve concentration changes in at least two directions, axial and lateral or radial, and therefore inevitably give rise, at least in principle, to partial differential equations (PDEs). When both solute and water transport are involved, there will be two equations, and they will also be coupled. Rigorous numerical solutions of these models were established in the 1970s and 1980s but are generally shunned by process engineers due to their complexity. Certain simplifying assumptions can be introduced that reduce the models to the ODE or even algebraic level while providing reasonably valid estimates of the answers being sought.

In all these endeavors, rigorous or approximate, transport parameters play a key role and must be addressed first. They are dependent on a number of variables, including flow conditions, membrane and module geometry, and the specific membrane process being addressed. Thus, the flow can be laminar or turbulent, involve gases or liquids, or take place in hollow-fiber or spiral-wound geometries. We listed some of these features and the resulting transport parameters in Table 8.8.

We note, first, that transport resistance is on the fluid side in most cases; the main exception is gas permeation, which, because of the much higher diffusivities involved, is controlled by the membrane resistance. When fluid flow is laminar on the core side, mass transport will be in the entry region (see Chapter 5). This is reflected by the Sherwood number tabulated in Table 8.8. Transport coefficients on the shell side are subject to some uncertainty, which is indicated by listing a suggested range of values. Finally, we note that some of the models can be reduced to the ODE level. This occurs primarily as a result of the absence of water flux and the prevailing resistance residing in an entry-region boundary layer (hemodialysis) or in the membrane (gas separation). In the illustration that follows, use is made of these and other simplifications to provide a set of tools for the preliminary analysis of membrane processes.

Illustration 8.7: Brian's Equation for Concentration Polarization

Concentration polarization, which was briefly mentioned in our previous discussions, arises in all processes involving the passage of water or some other solvent through the membrane wall. The resulting flux will initially lead to a depletion of water and a consequent rise in solute concentration in the vicinity of the wall. This sets in motion a countervailing diffusion of solute back into the core fluid, which is exactly balanced by the rate at which solute is transported to the wall by the permeating water. We indicated these opposing flows and the resulting concentration profile in Figure 8.16, where C_b, C_w, and C_p are solute concentrations in the bulk fluid, at the wall, and in the permeate, respectively. Note that simultaneously, and in conjunction with this process, an entry-region development of the concentration profile takes place, which is described by the Lévêque-type transport relations given in Table 8.2.

This local balance of solute transport was first described by Brian in analytical form as follows:

$$\begin{matrix}\text{Rate of transport of} \\ \text{solute by water flux}\end{matrix} = \begin{matrix}\text{Rate of back} \\ \text{diffusion of solute}\end{matrix}$$

$$J_w C - J_w C_p = D\frac{dC}{dx} \tag{8.12a}$$

where J_w is the volumetric flux (in units of m^3/m^2s), and x is the distance in the direction of water flux. Note that both C and x increase in the same direction so that Fick's law takes a positive sign. This expression can be integrated over the boundary layer thickness Δx, which yields

$$\int_{C_b}^{C_w} \frac{dC}{C - C_p} = \int_0^{\Delta x} \frac{J_w\, dx}{D} \tag{8.12b}$$

TABLE 8.8

Membrane Transport Characteristics

Type of Process and Geometry	Flow Regimes	Controlling Resistance	Solute Transport Coefficients	Type of Model
Reverse osmosis spiral wound	Laminar (entry region)	Core-side fluid	$\mathrm{Sh} = 2.12\left(\mathrm{Re\,Sc}\frac{d}{L}\right)^{1/3}$	PDE
Reverse osmosis hollow fiber	Turbulent (shell-side fed)	Shell-side fluid	$k \sim 10^5$ to 10^6 m/s	PDE
Gas separation hollow fiber	Laminar	Membrane	Permeabilities	ODE
Hemodialysis hollow fiber	Laminar (entry region)	Core-side fluid	$\mathrm{Sh} = 1.86\left(\mathrm{Re\,Sc}\frac{d}{L}\right)^{1/3}$	ODE
Reverse osmosis hollow fiber	Laminar (entry region)	Core-side fluid	$\mathrm{Sh} = 1.86\left(\mathrm{Re\,Sc}\frac{d}{L}\right)^{1/3}$	PDE
Ultrafiltration hollow fiber	Laminar/turbulent	Core-side fluid and gel layer	$\mathrm{Sh} = 1.86\left(\mathrm{Re\,Sc}\frac{d}{L}\right)^{1/3}$ $\mathrm{Sh} = 0.23\ \mathrm{Re}^{0.8}\ \mathrm{Sc}^{1/3}$ (turbulent)	PDE or ODE

Note: Water permeability for contemporary RO membranes is in the range $P_w = 10^7$ to 10^8 mol/m^{2}sPa.

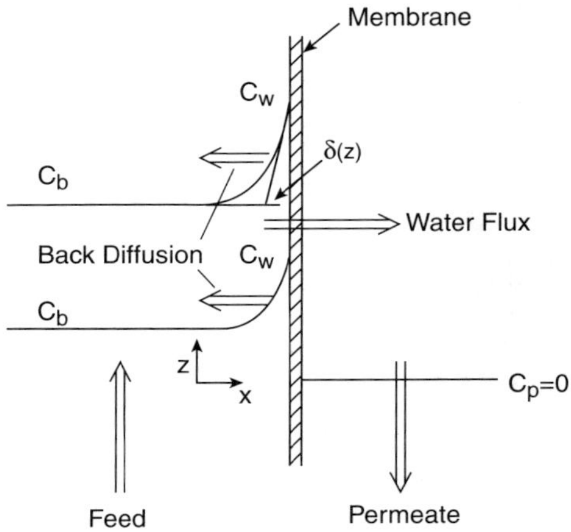

FIGURE 8.16
Concentration polarization in reverse osmosis and ultrafiltration.

or

$$\ln\frac{(C_w - C_p)}{(C_b - C_p)} = J_w \Delta x / D \tag{8.12c}$$

where the right side can be viewed as a Peclet number, given that J_w has the units of velocity.

Recognizing that the ratio $D/\Delta x$ defines the mass transfer coefficient k_c, Equation 8.11c can be recast in the form

$$\frac{C_w - C_p}{C_b - C_p} = \exp(J_w / k_c) \tag{8.12d}$$

which is the expression derived by Brian. When the permeate concentration C_p equals zero, the left side reduces to the quotient C_w/C_b. This ratio is a measure of solute accumulated at the membrane wall and is termed the *concentration polarization modulus* E_{Cp}. We use this equation in the illustration that follows and in a modified and revealing form in Practice Problem 8.6.

Illustration 8.8: A Simple Model of Reverse Osmosis and Ultrafiltration

We set as our initial task the replacement of the rigorous PDE model of reverse osmosis by a simple and approximate treatment, which makes use of certain empirical findings. Given the uncertainty of the parameters used

TABLE 8.9

Experimental Membrane Process Parameters

Process	$E_{Cp} = C_w/C_b$	Pe	E_e
Reverse osmosis	1–1.5	0.3–0.5	$< 10^2$
Ultrafiltration	70–150	5–10	$\sim 10^2$
Gas separation	~ 1	$3 \times 10^3 - 7 \times 10^2$	~ 1

in even the most rigorous models, this is no more than what good engineering sense dictates.

Our starting point is the basic flux equation for water transport through membranes driven by hydrostatic and osmotic potentials, a process alluded to in Chapter 1 (see Table 1.2). We have

$$N_w/A = P_w(\Delta p - \Delta\pi) \tag{8.13a}$$

where N_w is the molar transmembrane water flux (mol/s), and P_w is the water permeability (mol/m^{2}s Pa). A range of values for P_w is given in Table 8.9.

The osmotic pressure π is a colligative property (i.e., a property that depends on the particle molarity) and is for infinitely dilute solutions given by the van't Hoff equation:

$$\pi = Cn_pRT \tag{8.13b}$$

where C = molar concentration, and n_p = number of particles formed upon dissociation.

This equation holds well for the dilute solutions involved in nanofiltration and for brackish water. Even for seawater, with $Cn_p \sim 10^3$ moles/m^3, it predicts an osmotic pressure of $\pi = 10^3 \times 8.31 \times 298 = 2.48$ MPa = 24.6 atm, compared to an experimental value of 23 atm (Table 8.3). We will adopt this equation to replace osmotic pressure by concentration and make the assumption that the osmotic pressure in the permeate is zero (i.e., that the membrane is 100% effective in rejecting salt and the bulk concentration C_b is constant, as was deduced in Illustration 5.2 for laminar entry region flow). We can then combine the two equations and factor out the constant C_b to obtain

$$N_w / A = P_w\left(\Delta p - C_b n_p RT \frac{C_w}{C_b}\right) \tag{8.13c}$$

This expression opens up a way to introduce Brian's equation (Equation 8.12d) to eliminate the unknown polarization modulus C_w/C_b. We have

$$N_w / A = P_w\left[\Delta p - C_b n_p RT \exp\left(\frac{J_w}{k_c}\right)\right] \tag{8.13d}$$

and noting that the volumetric flux J_w (m^3/m^2s) equals 1.8×10^5 times the molar flux N_w/A (mol/m^2s), we obtain the following expression in the single variable J_w:

$$J_w = 1.8 \times 10^{-6}\, P_w \left[\Delta p - C_b n_p RT \exp\left(\frac{J_w}{k_c} \right) \right] \tag{8.13e}$$

This is a general-purpose equation applicable to both RO and UF which can be used for design (membrane area lurks in J) and predictive purpose. It is also a nonlinear equation in which J_w must be solved for numerically. We can use this equation to estimate water flux J_w, drawing on k_c values given in Table 8.8 or those for the Peclet number J_w/k_c listed in Table 8.9. Let us choose an average value for Pe = 0.4, a water permeability of 3×10^{-8} mol/m^2s Pa (see Table 8.4), and an applied hydrostatic pressure of ΔP = 1000 psig = 6.8×10^6 Pa. The particle molarity $C_b n_p$ for seawater is approximately 10^3 mol/m^3. We obtain from Equation 8.13f,

$$J_w = 1.8 \times 10^{-6} \times 3 \times 10^{-8} \,(6.8 \times 10^6 - 10^3 \times 8.314 \times 298 \exp(0.4)) \tag{8.13f}$$

or

$$J_w = 5.6 \times 10^{-7}\ m^3/m^2s = 16.8 \times 10^{-7}\ m/s \tag{8.13g}$$

This compares with a water flux of about 9×10^{-7} m/s obtained in commercial desalination processes under identical conditions.

Comments

The degree of agreement obtained with this simple model is quite remarkable but should be seen as being partly due to the judicious choice of experimental Peclet number and water permeability P_w. The range of these parameters will result in the answers extending between certain limits, which will differ by a factor of at least two, introducing uncertainties of the same magnitude in the quantity being sought. A more fruitful application of Equation 8.13d and Equation 8.13e lies in their use as a tool for exploring the effect of changes in the operating variables on the performance of an existing unit with a known water permeability. Feed rate, water salinity, and operating pressure are the more obvious candidates for such a study. We address the effect of these variables on reverse osmosis performance in Practice Problems 8.10 and 8.11. Estimation of the membrane area (i.e., design of an ultrafiltration process) is addressed in Practice Problem 8.12.

Illustration 8.9: Modeling the Artificial Kidney: Analogy to the External Heat Exchanger

We previously indicated that transmembrane transport in hemodialysis is dominated by the fluid resistance on the blood (core) side and that this resistance lies entirely in the laminar entry region. It was further seen that the dialysate concentration is adjusted to have the same osmotic pressure as that of the blood and that, consequently, no water flux is allowed to intrude on the proceedings. The only transport is that of the solutes to be removed from the blood, which is characterized by a core-side mass transfer coefficient listed in Table 8.3. We are, in other words, dealing with a straightforward combination of convective flow in the axial direction and radial transport through a film resistance. This combination can be expressed in terms of simple ODEs. Assuming co-current flow, the following results:

For the core-side mass balance:

Rate of solute in – Rate of solute out = 0

$$\begin{bmatrix} Q_B C_B \big|_{x+\Delta x} \\ K_0 \pi d (C_B - C_D)_{\text{avg}} \Delta x \end{bmatrix} = 0 \tag{8.14a}$$

For the shell-side mass balance:

Rate of solute in – Rate of solute out = 0

$$Q_D C_C \big|_{x+\Delta x} = 0 \tag{8.14b}$$

On dividing by Δx and letting $\Delta x \to 0$, we obtain the two ODEs:

$$Q_B \frac{dC_B}{dz} + K_0 \pi d (C_B - C_D) = 0 \tag{8.14c}$$

and

$$Q_D \frac{dC_D}{dz} + K_0 \pi d (C_B - C_D) = 0 \tag{8.14d}$$

where the subscripts *B* and *D* refer to blood and dialysate, respectively, and K_0 (m/s) is an overall mass transfer coefficient that includes the effect, usually minor, of the membrane resistance.

The two ODEs are solved simultaneously by an extension of the *D*-operator method outlined in Appendix A1 or by Laplace transformation. The results can be arranged into the following dimensionless form:

FOR THE CO-CURRENT CASE

$$E = \frac{1 - \exp[-N_T(1+Z)]}{(1+Z)]} \tag{8.15a}$$

FOR THE COUNTERCURRENT CASE

$$E = \frac{1-\exp[-N_T(1-Z)]}{Z-\exp[-N_T(1-Z)]} \tag{8.15b}$$

where the parameters Z, N_T and E are defined as follows:

$$Z = Q_B/Q_D = \text{Flow rate ratio} \tag{8.15c}$$

$$N_T = K_0A/Q_B = \text{Number of transfer units} \tag{8.15d}$$

$$E = (C_{Bi} - C_{Bo})/(C_{Bi} - C_{Do}) = \text{Extraction ratio} \tag{8.15e}$$

and the subscripts i and o refer to inlet and outlet conditions, respectively.

In most practical applications, dialysate flow is large compared to blood flow ($Z \cong\approx 0$) and its solute concentrations are much lower than those prevailing in the blood. Under these conditions, we obtain for both the co-current and countercurrent cases,

$$E = 1 - \exp(-N_T) \tag{8.15f}$$

or, equivalently.

$$C_{Bo}/C_{Bi} = \exp(-K_0A/Q_B) \tag{8.15g}$$

This final expression, an important one, allows us to calculate the exit concentration from the dialyser C_{Bo} as a function of membrane area A and blood flow rate Q_B.

If the task is to assess the response of the *body* to dialysis, it becomes necessary to integrate the events in the blood compartment with those taking place in the dialyzer. The configuration then becomes identical to one that is well known in the field of heat transfer. It involves heating the contents M_t of a well-stirred tank by pumping them through an external heat exchanger at a flow rate F and returning the contents to the tank. The tank represents the blood compartment and the dialyzer takes the place of the heat exchanger. The analogy is illustrated in Figure 8.17.

The model is based on the assumption that the heat exchanger is at a quasi-steady state because flow through it is much faster than the rate at which incoming temperature changes with time. The solution, which is classical and well known, is given by

$$t = \frac{M_t}{F}\left[1-\exp\left(-\frac{UA}{FC_p}\right)\right]^{-1} \ln\frac{T_s - T_t^o}{T_s - T_t} \tag{8.16a}$$

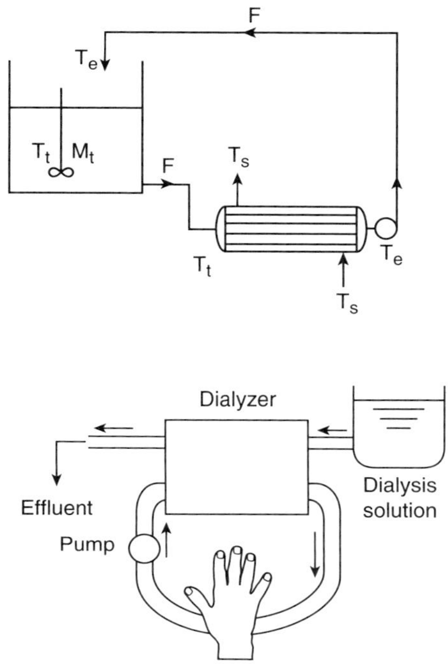

FIGURE 8.17
Analogy between an external heat exchanger and the dialyzer.

where the steam temperature T_s corresponds to the external dialyzer concentration, which is usually taken to be vanishingly small. To convert this expression to represent the duration of the dialysis procedure, we drop the heat capacity C_p and replace the heat transfer coefficient U by the mass transfer coefficient K_o. The time t it takes to reduce the toxin concentration from C^o to C is then given by

$$t = \frac{V}{Q}\left[1 - \exp\left(-\frac{K_o A}{Q}\right)\right]^{-1} \ln \frac{C^o}{C} \tag{8.16b}$$

where V and Q are now the blood volume and volumetric flow rate, respectively. Derivation of this expression is left to the exercises.

A typical value for K_oA is 1 cm^3/s; that for the blood flow rate is 5 cm^3/s. Blood volume for an adult is about 5 L. We then have, for a tenfold reduction in toxic concentration,

$$t = \frac{5000}{5}\left[1 - \exp\left(-\frac{1}{5}\right)\right]^{-1} \ln 10 \tag{8.16c}$$

that is,

$$t = 12{,}690 \quad \text{s} = 3.52 \text{ hrs} \tag{8.16d}$$

This is the duration of a typical dialysis procedure.

Equation 8.16b is a highly useful expression to assess the effect of the percent toxin to be removed, of the value K_oA for different dialyzers, and the impact of flow rate Q on the duration of the precedent. Note that K_o varies with Q as well, but the dependence is of a weak one-third power (see Table 8.8).

Illustration 8.10: Membrane Transport in Biology: The Human Kidney

Of the many transmembrane processes that take place in the human body, the formation of urine in the kidney stands out as an example of intricately interwoven mechanisms and a highly sophisticated use of transmembrane driving forces. Consider the requirements: The blood has to be stripped of toxic substances and excess water without depriving the body of valuable substances or disturbing the balance of vital substances. This has to be done in the face of a widely varying intake or loss of water and salt, variations in the rate of production of toxins, and, in general, any variations due to changes in human activity. This is achieved in a series of transmembrane transport processes along a tubular assembly known as a *nephron* (Figure 8.18).

A first crude separation takes place in the *glomerulus,* which removes about 10% of the blood flow as filtrate containing ions, toxins such as urea and creatinine, and the smaller proteins. Only about 0.1% of the water is ultimately discharged in the urine, the rest being "reabsorbed" (i.e., returned to the body) by osmotic transport in the "proximal" and, to varying degrees, in the "distal" tubule of the U-tube, known as the Loop of Henle (Figure 8.18). Additional transport takes place in the collecting duct. Just how much water is passed back is regulated by the antidiuretic hormone (ADH) that acts to increase membrane permeability to water whenever a more concentrated urine is needed.

The second species of importance transported across the tubular wall is Na^+, followed by Cl^-. The mechanism is largely non-Fickian and relies on "active transport" to move the solutes "uphill" against a gradient of increasing concentration (see Illustration 1.3 and Figure 1.4).

Active transport takes place in the proximal and distal tubules, and most notably in the ascending limb of the Loop of Henle. This transport is almost exactly balanced by back-diffusion of Na^+ from the tissue into the descending limb. The net result is the creation of a standing external concentration gradient, which allows water to pass from the tubular urine into the surrounding tissue by *osmosis* (i.e., to be "reabsorbed").

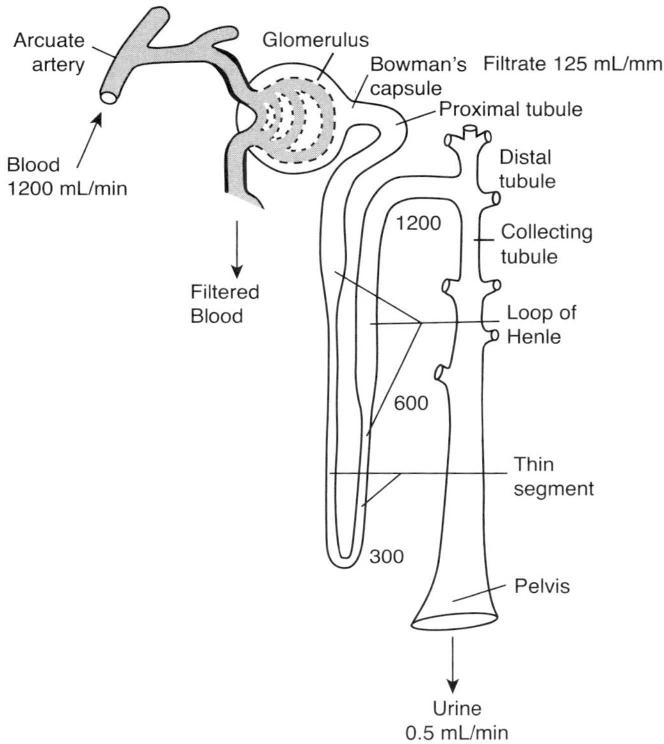

FIGURE 8.18
Part of nephron of the human kidney. There are approximately 2,500,000 such units per kidney. Numbers indicate Na^+ concentration in millimoles per liter.

TABLE 8.10

Parameters for Glomerular and Industrial Ultrafiltration (Order of Magnitude)

	Glomerular Filtration	Ultrafiltration of Whey
Membrane area, m^2	0.1	100
Filtration rate, L/s	10^3	1
Filtration flux, L/m^2s	10^2	10^2
Permeability, m^3/m^2s Pa	10^9	10^{11}
$(\Delta p - \Delta\pi)$, kPa	1	100

To convey a sense of what nephrons are able to accomplish, we make a comparison in Table 8.10 of the characteristics and performance of the glomeruli with those of an industrial ultrafiltration process (e.g., the treatment of whey, a by-product of cheese manufacture). In both cases, water and solutes pass through the membrane wall — salts, glucose, toxins, and some small proteins in the glomerulus, lactose in the case of whey.

It is clear from these results that while filtration rate per square meter is of the same order for both cases, the net pressure driving force available to the glomeruli is only 1/100 that of the industrial process. The organ makes up for this by having a 100-fold higher permeability.

It is interesting to use some of the kidney parameters cited to calculate the time required to clear the body of toxins, resulting, for example, from a drug overdose. It is assumed that the toxin is rapidly distributed in the extracellular fluid "compartment" (which includes plasma) of 10-L volume, and that none is metabolized or reabsorbed during its passage through the kidney. A toxin mass balance over the compartment then yields:

Rate out = Rate of change

$$-QC = V\frac{dC}{dt} \tag{8.17a}$$

where Q = glomerular filtration rate = 0.125 L/s. Integration by separation of variables then yields

$$\ln\frac{C_1}{C_2} = \frac{Q}{V}t \tag{8.17b}$$

The time needed to reduce the concentration by a factor of ten is then given by

$$t = \frac{1.0}{0.125}\ln 10 = 188\,\text{min} \tag{8.17c}$$

or approximately 3 h.

To shorten this time, "forced diuresis" (i.e., stimulation of urine formation) is often used for treating a drug overdose.

Comments

The number of species transported through various segments of the nephron wall is considerably higher than indicated. Sulfate, phosphate, and biocarbonate ions, as well as various amino acids, vitamin C, and lactate, among others, transfer in significant amounts, often to be reabsorbed into the body. The principal events, however, are those given in the outline and are three in number:

- Reabsorption of more than 99% of the original filtrate
- An attendant 100-fold increase in the concentration of toxins in the filtrate, which are insoluble in the membrane wall
- Active transport of Na^+ to maintain a standing external concentration gradient needed for osmotic water transport

Illustration 8.11: Membrane Gas Separation: Selectivity α and the Pressure Ratio ϕ

It was previously indicated (see Table 8.2) that membrane-based gas separation processes, while still of considerable complexity, can be modeled at the level of ODEs. This is because the principal transport resistance resides within the membrane and because the mass balances need be concerned only with concentration changes in the direction of flow. These equations, however, must be supplemented by force balances to take account of the nonlinear pressure drop associated with permeation processes. The resulting set of ODEs will generally have to be solved numerically. We do not take up this problem here but instead introduce two important system parameters that provide useful information for a preliminary assessment of these operations. These parameters are the pressure ratio ϕ and the membrane selectivity α.

The pressure ratio is an operational parameter defined as the ratio of feed pressure to that prevailing on the permeate side:

$$\phi = P_F/P_P \tag{8.18a}$$

where the subscripts refer to feed and permeate conditions, respectively. Selectivity α, on the other hand, is a property of the membrane and deals with the relation between compositions, usually expressed as mole fractions, y_1, on either side of the barrier. Its definition, at least in its form, is identical to that of relative volatility, which was seen in connection with vapor–liquid equilibria (see Equation 6.20a). For a two-component system, it is given by the relation

$$\alpha = \frac{y_{1P}(1-y_{1F})}{y_{1F}(1-y_{1P})} \tag{8.18b}$$

where the ratio $E = y_{1P}/y_{1F}$ can be viewed as an enrichment factor of the more permeable component 1 attainable by a particular type of membrane. Both α and ϕ are used, singly or in conjunction, to define two important regions of limiting operational behavior. When ϕ is small, and the selectivity is far in excess of the pressure ratio,

$$\alpha \gg \phi \tag{8.18c}$$

we speak of the process as being *pressure-ratio limited*. Conversely, if the pressure ratio substantially exceeds α,

$$\phi \gg \alpha \tag{8.18d}$$

the process is said to be in the *selectivity-limited region*. Let us consider these two cases and the information and lessons to be drawn from them.

PRESSURE-RATIO LIMITED REGION

We start by noting that gas permeation is driven by a concentration or partial pressure difference across the membrane — that is, it is a process based on diffusion and not on D'Arcy-type bulk flow. For diffusion to take place, the feed partial pressure must be higher than that on the permeate side. We must have

$$p_F > p_P \tag{8.19a}$$

or, equivalently in terms of mole fraction y or total pressure P,

$$y_F P_F > y_P P_P \tag{8.19b}$$

As a consequence of this inequality, the enrichment attained will generally be smaller than the total pressure ratio and can at most equal it. We must have

$$E = \frac{y_P}{y_F} < \frac{P_F}{P_P} \tag{8.19c}$$

and in the limit

$$(E)_{\text{Max}} = \frac{y_P}{y_F} = \frac{P_F}{P_P} = \phi \tag{8.19d}$$

This limiting value corresponds to the situation in which the partial pressures on either side of the membrane have attained equality, and diffusion has ceased. Note that while partial pressures are now equal, the total pressures on either side are not, nor are the corresponding mole fractions.

The enrichment given by Equation 8.18d represents the maximum value of E attainable, irrespective of the selectivity of the membranes involved. In other words, the process is now dependent on ϕ only and is, consequently, in the pressure-ratio-limited region, with a resulting enrichment given by Equation 8.19d. This condition arises when α exceeds ϕ by a factor of more than 5.

SELECTIVITY-LIMITED REGION

When the pressure ratio is very large, it ceases to be a limiting factor, and the process becomes solely dependent on the membrane selectivity α. The enrichment E will in this case no longer be dictated by the value of ϕ but must instead be extracted from Equation 8.18a. We obtain the following by simple rearrangement:

$$E = \frac{y_{1P}}{y_{1F}} = \frac{\alpha}{1 + y_{1F}(\alpha - 1)} \tag{8.19e}$$

which is entirely analogous to the enrichment obtained in vapor–liquid equilibria. This condition holds when ϕ exceeds α by a factor of more than 5.

Comments

The limiting expressions (Equation 8.19d and Equation 8.19e) are invaluable in arriving at a first assessment of gas separation processes. Suppose, for example, that $\alpha = 5$, which is a typical membrane selectivity for the separation of nitrogen and oxygen. Then in any practical operation, the pressure ratio will be in excess of α, and the operation will consequently take place in the selectivity-limited region. The maximum enrichment of nitrogen attainable is then given by (Equation 8.19e):

$$(E_{\text{Max}}) = \frac{\alpha}{1 + y_{1F}(\alpha - 1)} = \frac{5}{1 + 0.79(5 - 1)} = 1.2 \tag{8.19f}$$

This is still a respectable value and has led to the large-scale adoption of the membrane process for the separation and enrichment of air.

The limiting expressions set out above are also helpful in setting limits on the usefulness of enhancing membrane selectivity. This comes about as a result of the high cost of compressing feed gas to very high pressures or drawing a hard vacuum on the permeate side, which limits practical pressure ratios ϕ to the range 10 to 50. Suppose, for example, that a value of $\phi = 10$ is chosen. Then a membrane selectivity of 50 will locate the process in the pressure-ratio-limited region with the enrichment factor given by $E = \phi$, independent of α (Equation 8.18d). Any further increases in membrane selectivity will therefore have no effect on the performance of the process under these conditions.

Practice Problems

8.1 The Operating Line in a Gas Scrubber

Explain the meaning and consequences of the following special cases of an operating line for a gas scrubber:

a. The line is vertical.
b. The line is horizontal.
c. The line is curved.
d. The line is displaced parallel to itself and away from the equilibrium curve.
e. The line lies below the equilibrium curve.

What is the effect on NTU in cases a, b, and d?

8.2 The Effect of Feed Flow Rate in Gas Absorption

Analyze how an increase in the gas flow rate would affect the performance of an existing absorber.

8.3 The Effect of Packing Size in Gas Absorption

Qualitatively plot the component and total costs affected by packing size in a gas absorber. (*Hint*: Use Figure 8.2 as a guide.)

8.4 The Operating Diagram for Gas Absorption with a Reactive Solvent

Suppose a reactive solvent (e.g., a solution of NaOH) is used in excess to remove a component (e.g., H_2S — see Illustration 8.2) from a gas stream using a packed column.

a. What will be the position of the "equilibrium curve"?
b. What concentration units are used for the liquid phase?
c. How is the operating line constructed, and what represents the driving force in the diagram?
d. Why is the HETS concept inappropriate for this system?

8.5 Nickel Stripping from Its Organic Chelated Complex

Illustration 8.3 outlined the two-step extraction procedure used in recovering nickel from an acid leach. Some pertinent laboratory data for the second step involving the reverse extraction from the organic phase with hydrochloric acid were shown in Figure 8.6.

a. Estimate the value of the distribution coefficient m. Is it reasonable to assume that m is constant?
b. Set a "reasonable" value for the extraction factor $E = mB/A$, and calculate the amount of solvent required per kilogram of feed.
c. If a recovery of 99% is stipulated, how many transfer units would a packed column require?

8.6 Flooding in an Extraction Column: Minimum Column Diameter

"Flooding" of an extraction column commonly occurs when the superficial velocity of the continuous phase exceeds the *terminal velocity* of the dispersed droplets, assumed to be the lighter phase introduced at the bottom. The droplets will then cease to rise and will ultimately coalesce into a pool at the bottom of the column or be swept out of it. If the dispersed phase is the heavier of the two, the same phenomena will take place at the top of the column. Countercurrent operation comes to a halt.

One way of obtaining a "quick fix" of the problem is to determine the terminal velocity of rise v_t of the droplets and set the superficial velocity of the continuous phase at a fraction of its value, say 30% of v_t.

Assume a droplet diameter of 3 mm, a density ratio of 0.8, and a raffinate flow rate of 0.01 m^3/s. What is the minimum column diameter if the continuous phase velocity is to be kept at 30% of flooding? (*Hint*: Consult Illustration 5.4.)

8.7 Choice of Column and Operating Conditions

Styrene, used in the manufacture of polystyrene plastics, is produced by the dehydrogenation of ethyl benzene. The product stream containing some unreacted ethyl benzene is to be purified by distillation. What would be your recommended choice of type of column and operating pressure?

8.8 Another Look at Isotope Distillation

An isotope pair with a separation factor of 1.01 is to be separated into overhead and bottoms product with compositions of 0.99 and 0.10 mole fraction, respectively, by distillation in a packed column. A particular high-efficiency packing has been touted by its manufacturer as having HTU values of no more than 3 cm. The facility considering the process is, for various reasons, constrained to column heights of no more than 10 m. Will the process meet this requirement?

8.9 Concentration Polarization, Peclet Number, and Enrichment Factor

Show that Brian's equation can be recast in the form

$$E_{Cp} = \frac{C_w}{C_b} = \frac{\exp(Pe)}{1 + E_e[\exp(Pe) - 1]}$$

where $Pe = J\Delta x/D$ is the flux Peclet number, and E_e is an enrichment factor defined by the ratio C_p/C_w. An idea of the magnitude of these parameters for various membrane processes is conveyed in Table 8.5.

8.10 The Effect of Pressure in Reverse Osmosis

Typical seawater desalination plants using reverse osmosis produce 30 to 50 gfd of water [gal (U.S.) per square ft per day] (1 gfd = 4.72×10^{-7} m^3/m^2s). Suppose it is decided to double the operating pressure from 50 to 100 atm.

Will the production rate increase by less or by more than a factor of 2? Assume 100% salt rejection.

8.11 The Effect of Feed Rate and Water Salinity in Reverse Osmosis

Consider the process described in Illustration 8.7. Using the numerical parameter values cited there, determine the following:

a. The effect of halving salinity on J_w.
b. The effect of doubling shell-side feed rate on the same quantity. (*Hint*: Consider the effect on v in tubular flow.)

Answer:

a. $J_w = 8.9 \times 10^{-5}$ m^3/m^{2}s

8.12 Processing of Whey by Ultrafiltration

Whey is a by-product of cheese manufacture containing about 1% by weight of proteins (average molecular weight 20,000 Dalton) and several percent of sugars, principally lactose (~ 200 Dalton). Apart from the incentive of recovering these valuable components, most jurisdictions now prohibit the discharge of whey to the environment.

The current technology involves a two-step membrane separation procedure. In the first step, lactose is recovered by ultrafiltration, and the feed concentrates to about 5% by ultrafiltration (Δp = 400 kPa). In the second, high-pressure, step, the proteins are further concentrated by reverse osmosis (ΔP = 7 MPa).

If the feed rate is 5 L/s of unprocessed whey, what would be the membrane area required for the utlrafiltration step? Assume Pe = 5, $D = 4 \times 10^{-11}$ m^3/m^{2}s Pa.

Answer:

440 m^2

8.13 Concentration of an Isotonic Saline Solution

Isotonic saline is an NaCl solution that has the same osmotic pressure as blood. It is used in medical treatments, for example, to make up fluid loss, and also as the extracting fluid on the dialysate side of an artificial kidney. Calculate the concentration of this solution. (*Hint*: Consult Table 8.3.)

Answer:

0.141 molar

8.14 Translation of Medical Terms

Venture a translation of the following terms used by physiologists:

a. Proximal and distal
b. Hypertonic and hypotonic
c. Transmural and peritubular
d. Osmolarity, given that isotonic saline has an osmolarity of 0.242

8.15 The Ascending Limb of the Loop of Henle

The principal event taking place in the ascending limb of the Loop of Henle is the active transport of Na^+ out of the "lumen" (i.e., the interior of the tubule). The limb has an overall length of 1 cm and an average diameter of 15 μm.

a. Given that the Reynolds number in the tubule is of the order 10^{-2} ($v \cong 1$ mm/s!), and Sc can be set equal to 10^3, calculate the entry length for mass transfer and compare with similar situations described elsewhere in the text.
b. Given that the measured *overall* transport coefficient for active sodium transport is 3×10^9 m/s, what is the fractional resistance to mass transfer in the lumen?

8.16 Performance of a Hemodialyzer

Use the following data to calculate the urea concentration exiting from a hemodialyzer with a total surface area of 1 m^2 and 100 hollow fibers of length $L = 17$ cm.

Urea diffusivity: $D = 2 \times 10^{-5}$ cm^2/s
Urea inlet concentration: $C_{Bi} = 100$ ppm
Blood flow rate: $Q_B = 1$ cm^3/s (total)

Answer:

560 ppm

8.17 The Hemodialyzer

Derive the solution (Equation 8.13b) describing the performance of a hemodialyzer. (*Hint*: Start with a differential balance for the dialyzer to obtain the effluent blood concentration, and use this value for the balance over the blood compartment.)

8.18 Performance of a Membrane Process in Gas Separation

The membrane selectivity for the removal of CO_2 from natural gas is of the order 20. For a feed containing 5% CO_2, give an approximate evaluation of the enrichment attainable using a pressure ratio of 20.

9

Simultaneous Heat and Mass Transfer

In our treatment so far, occasional reference has been made to heat transfer, primarily in order to draw the reader's attention to the analogies that exist between the transport of heat and mass. For example, in Chapter 1, a point was made of highlighting the similarities between the rate laws governing convective and diffusive heat and mass transfer. The analogy that exists between the two phenomena when dealing with compartmental or countercurrent operations was brought out on several occasions, notably in Illustration 8.8, which drew a comparison between the models governing heat exchange and hemodialysis.

We now turn our attention to processes where heat and mass transfer occur in unison. This is far from being an unusual event, but it raises the complexity of the underlying model, a fact that persuaded us to defer its consideration to the last chapter.

Simultaneous heat and mass transfer occur in a natural way whenever the transport of mass is accompanied by the evolution or consumption of heat. An important class of such operations, and one that will occupy a considerable portion of the present chapter, involves the condensation of water vapor from an airstream and the reverse process of evaporation of liquid water into air. There are a host of important operations in which this type of transfer occurs. The humidification and dehumidification of air, often identified with air conditioning, is practiced both within a domestic context and on a commercial or industrial scale. Process water that has been used in a plant for cooling purposes is often cycled through a cooling tower where it is contacted in countercurrent flow with air and undergoes evaporative cooling before being returned for reuse. The drying of solids, an important class of operations in its own right, also draws on the principles underlying the transport of water between its liquid phase and air. The air–water system, and the temperature and concentration changes that arise in air–water contact, are of such importance in the physical sciences that they have led to the construction of so-called psychometric or humidity charts. These charts summarize in convenient fashion the thermal and concentration variables relevant to operations involving the air–water system. We will, in the illustrations and practice problems that follow, have occasion to use these charts repeatedly to establish parameters of interest in various operations based on the air–water system.

Simultaneous heat and mass transfer also occur in exothermic or endothermic heterogeneous reacting systems and in the absorption or adsorption from concentrated gas streams. Some of these topics are addressed in separate illustrations, but we retain the air–water system as the central theme of this chapter.

9.1 The Air–Water System: Humidification and Dehumidification, Evaporative Cooling

9.1.1 The Wet-Bulb Temperature

We start by examining the events that occur when a flowing gas comes in contact with a liquid surface. From personal experience, you are aware that this process results in a drop in the temperature of the liquid, often referred to as evaporative cooling. The chill we experience when wind blows over our perspiring bodies is one manifestation of this effect.

Assume that both the water and the air are initially at the same temperature. During the first stage of evaporation, the energy required for the process (i.e., the latent heat ΔH_v) will come from the liquid, which consequently experiences a drop in temperature. That decline, once it is triggered, will cause a corresponding amount of heat transfer to take place from the air to the water. At this intermediate stage, the latent heat of vaporization is provided both by the liquid and by heat transfer from the warmer gas.

As the liquid temperature continues to drop, the rate of heat transfer accelerates until a stage is reached where the entire energy load is supplied by the air. A steady state is attained at which the rate of evaporation is exactly balanced by the rate at which heat is transferred from the gas to the liquid. The liquid is then said to be at its "wet-bulb temperature," T_{wb}, and the corresponding air temperature is referred to as the "dry-bulb temperature," T_{db}. The difference ($T_{db} - T_{wb}$) constitutes the driving force for the heat being transferred from the gas to the liquid. This is indicated in Figure 9.1, which also shows the associated humidities of the air, Y_{wb} (kg H_2O/kg air), the saturation humidity prevailing at the surface of the liquid, and Y_{db}, the humidity in the bulk air. The wet-bulb temperature and its associated saturation humidity play a central role in humidification and dehumidification, in water cooling operations, as well as in drying processes. These are taken up in subsequent illustrations.

The relation between T_{wb}, Y_{wb}, and the system parameters is established by equating the rate of heat transfer from air to water to the rate of evaporation (i.e., the rate at which moisture is transferred from the water surface to the air). Thus,

Rate of evaporation = Rate of heat transfer

$$k_Y A(Y_{wb} - Y_{db})\Delta H_v = hA(T_{db} - T_{wb}) \qquad (9.1a)$$

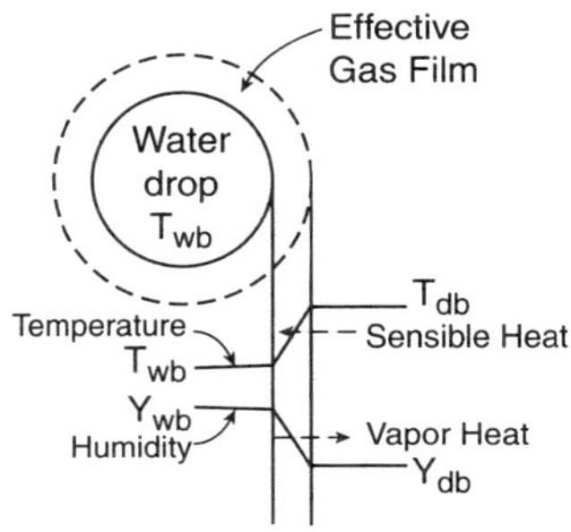

FIGURE 9.1
Temperature and humidity distribution around a water drop exposed to a flowing airstream.

where k_Y is the mass transfer coefficient (in units of kg $H_2O/m^2s\ \Delta Y$).

Canceling terms and rearranging, we obtain

$$\frac{Y_{wb} - Y_{db}}{T_{wb} - T_{db}} = -\frac{h}{k_Y \Delta H_v} \tag{9.1b}$$

where the difference $T_{db} - T_{wb}$ is referred to as the wet-bulb depression.

We note from Equation 9.1b that the humidity of the air Y_{db} can, in principle, be established from measured values of T_{db}, T_{wb}, and Y_{wb}, the latter being obtained from the relation

$$Y_{wb} = \frac{\left(P^o_{H_2O}\right)_{wb}}{P_{Tot} - \left(P^o_{H_2O}\right)_{wb}} \frac{M_{H_2O}}{M_{air}} \tag{9.1c}$$

where $P^o_{H_2O}$ is the vapor pressure of water, available from tables, and M = molar mass. T_{db} is measured by exposing a dry thermometer to the flowing air, while T_{wb} is obtained in similar fashion using a thermometer covered with a moist wick. More recent devices for measuring Y rely on changes in electrical properties of the sensor element with the moisture content of air. Both wet- and dry-bulb properties appear on the humidity charts, which will be discussed in this chapter.

9.1.2 The Adiabatic Saturation Temperature and the Psychrometric Ratio

Before addressing the properties and construction of the humidity charts, we consider a small variation in the simple contact of water with flowing air, which led to the wet-bulb conditions. In this modified arrangement, shown in Figure 9.2, a stream of air is humidified in contact with constantly recirculated water. Both the water and the exiting gas stream attain adiabatic saturation temperature, T_{as}, which is lower than the dry-bulb temperature of the entering air because of evaporative cooling.

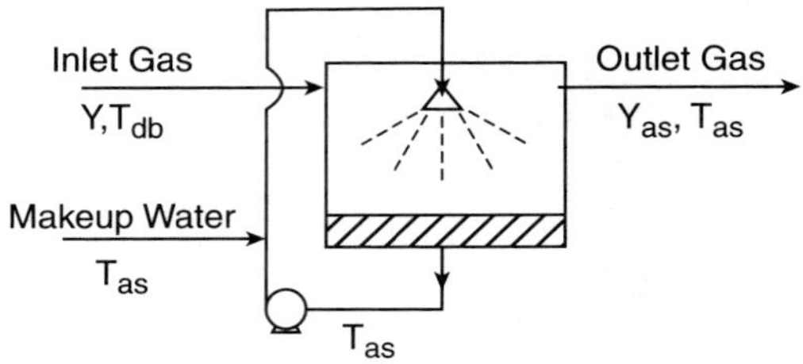

FIGURE 9.2
Flow sheet showing the attainment of adiabatic saturation conditions.

If care is taken to introduce the makeup water at the same adiabatic saturation temperature, and the datum temperature is set at T_{as}, a simple energy balance will yield

$$\text{Rate of energy in} - \text{Rate of energy out} = 0$$

$$[C_s(T_{db} - T_{as}) + Y_{db}\Delta H_v] - [C_s(T_{as} - T_{as}) + Y_{as}\Delta H_v] = 0 \tag{9.2a}$$

which upon rearrangement leads to the expression

$$\frac{Y_{as} - Y_{db}}{T_{as} - T_{db}} = -\frac{C_s}{\Delta H_v} \tag{9.2b}$$

where C_s is the specific heat of the air, also termed *humid heat*, in units of kilojoule per kilogram dry air. Plots of this equation appear in the humidity charts discussed in the next illustration.

The striking similarity between the adiabatic saturation and wet-bulb relations, Equation 9.2b and Equation 9.1b, led to a detailed examination of the ratio of the two slopes, $h/k_Y C_s$, also known as the psychrometric ratio. These studies culminated in the finding that for the water–air system, *and only for that system*, its value is approximately unity. Thus,

PSYCHROMETRIC RATIO

$$h/k_Y C_s \approx 1 \tag{9.2c}$$

This expression, known as the Lewis relation, when used to compare Equation 9.1b and Equation 9.2b, leads to the conclusion that the adiabatic saturation and wet-bulb temperatures are essentially identical. The Lewis relation has other important implications as well, as will become apparent in Illustration 9.4 dealing with the design of water-cooling towers. It is seen there that the underlying model equations can be enormously simplified by making use of the Lewis relation.

Illustration 9.1: The Humidity Chart and Its Application

The psychrometric or humidity charts to be taken up here are displayed in Figure 9.3 and Figure 9.4 for the low- and high-temperature ranges shown. To familiarize ourselves with the properties of these diagrams, we start by defining and deriving a set of variables, which appear implicitly or explicitly in the two figures.

ABSOLUTE HUMIDITY (Y)

This quantity was already referred to in connection with the wet-bulb temperature and is redefined here for convenience:

$$Y(\text{kg H}_2\text{O/kg dry air}) = \frac{18}{29}\frac{p_{\text{H}_2\text{O}}}{P_T - p_{\text{H}_2\text{O}}} \tag{9.3a}$$

where $p_{\text{H}_2\text{O}}$ is the partial pressure of water vapor.

Y appears as the right-hand side ordinate in the humidity charts.

RELATIVE HUMIDITY (RH)

To obtain a sense of the relative degree of saturation of the air, we define

$$\%RH = \frac{p_{\text{H}_2\text{O}}}{P^o_{\text{H}_2\text{O}}}100 \tag{9.3b}$$

where $P^o_{\text{H}_2\text{O}}$ is the saturation vapor pressure of water at the temperature.

RH varies over the range 0% (dry air) to 100% (fully saturated air) and appears in the humidity charts as a set of parametric curves that rise smoothly from left to right.

DEW POINT (T_{DP})

This is the temperature at which moist air, cooled at constant P_{Tot} and *Y*, becomes saturated (i.e., attains 100% relative humidity). Its value is established by moving from the initial defining point of a given air–water mixture on the humidity chart along a horizontal line to the eventual intersection with the curve of 100% relative humidity.

HUMID VOLUME (V_H)

The humid volume of moist air is the specific volume in cubic meter per kilogram dry air measured at P_{Tot} = 101.3 kPa (1 atm) and the temperature *T* of the mixture. Values of V_{H} appear in the humidity charts as a set of lines of negative slope.

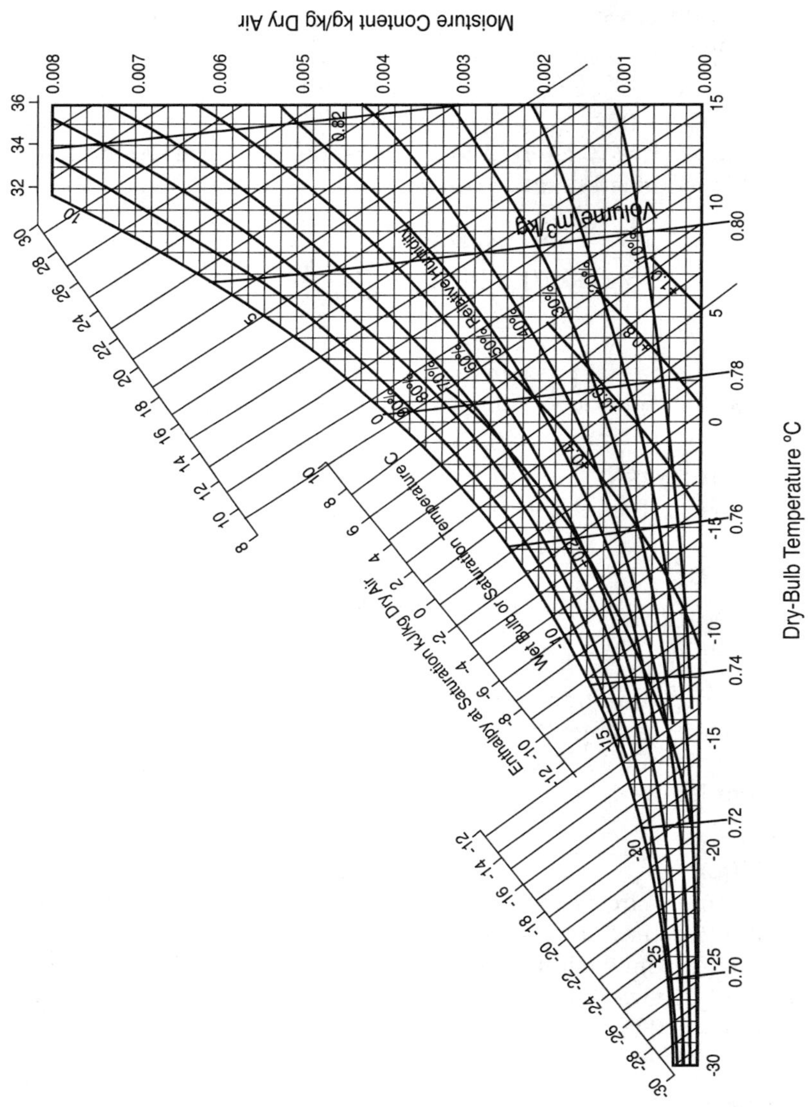

FIGURE 9.3
Humidity chart: low-temperature range.

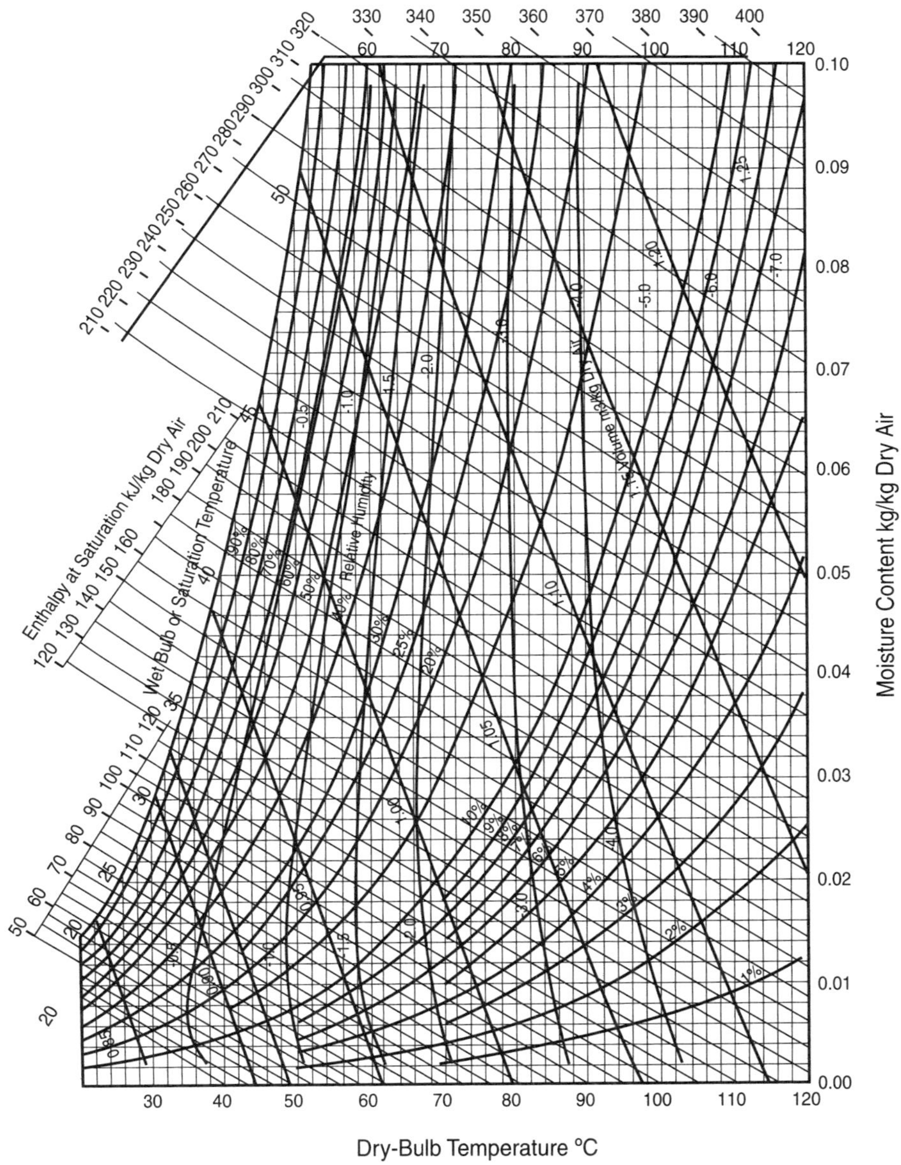

FIGURE 9.4
Humidity chart: high-temperature range.

HUMID HEAT (C_S)

This quantity, which was encountered in connection with the adiabatic saturation temperature, is the specific heat of moist air expressed in units of kilojoule per kilogram dry air.

Humid heat does not usually appear explicitly in the charts but is contained in the enthalpies shown there. It can be calculated from the following equation:

$$C_s(\text{kJ/kg dry air}) = 1.005 + 1.88\,Y \tag{9.3c}$$

ENTHALPY (H)

With the humid heat in hand, we are in a position to formulate the enthalpy of an air–water mixture. With $T°$ chosen as the datum temperature for both components and adding sensible and latent heats, we obtain

$$H(\text{kJ/kg dry air}) = C_s(T - T°) + Y\Delta H_v{}° \tag{9.3d}$$

Sensible heat – Latent heat

where the datum temperature is usually set equal to 0°C for both liquid water and dry air.

Values of the enthalpies of various air–water mixtures are read from the left-hand oblique ordinate.

ADIABATIC SATURATION TEMPERATURE (T_{AS}) AND WET-BULB TEMPERATURE (T_{WB})

Plots of the adiabatic saturation line, Equation 9.2b, appear in the humidity charts as lines extending from the abscissa to the 100% relative humidity curve. The point of intersection with that curve defines the wet-bulb temperature (T_{wb}), which is also the adiabatic saturation temperature (T_{as}).

EXAMPLE

This concrete example illustrates the various uses to which the humidity charts may be put. We choose moist air with a relative humidity of 25% and a (dry-bulb) temperature of 50°C and proceed to calculate various properties of interest using the chart shown in Figure 9.4.

ABSOLUTE HUMIDITY (Y)

This value is read from the right-side rectangular ordinate, which yields

$$Y = 0.0195 \text{ kg } H_2O/\text{kg dry air} \tag{9.4a}$$

DEW POINT (T_{DP})

We follow the horizontal line through the point $Y = 0.0195$, $T = 50°C$ to its intersection with the 100% relative humidity curve and obtain

$$T_{dp} = 24.5°C \tag{9.4b}$$

This corresponds to the temperature at which, on isobaric cooling of the moist air, the first condensation of water occurs.

WET-BULB TEMPERATURE (T_{WB})

Here the procedure is to follow the adiabatic saturation line to its intersection with the 100% relative humidity curve. We obtain

$$T_{wb} = 30.4°C \tag{9.4c}$$

Note that the wet-bulb temperature is not identical to the dew point.

WATER PARTIAL PRESSURE (p_{H_2O})

This quantity can be obtained directly from the absolute humidity and Equation 9.3a. Solving it for p_{H_2O} yields

$$p_{H_2O} = \frac{29\,Y\,P_{Tot}}{18+29\,Y} = \frac{29 \times 0.0195 \times 101.3}{18+29 \times 0.0195} = 2.09\text{kPa} \tag{9.4d}$$

HUMID VOLUME (V_H)

The plots for V_H are shown as steep lines of negative slope. The point $Y = 0.1095$, $T = 50°C$ is located between the lines for $V_H = 0.90$ and 0.95. Linear interpolation yields the value

$$V_H = 0.945 \text{ m}^3/\text{kg dry air} \tag{9.4e}$$

ENTHALPY (H)

This value is read from the oblique left-hand ordinate of Figure 9.4 and comes to

$$H = 103 \text{ kJ/kg dry air} \tag{9.4f}$$

WATER REMOVAL

Suppose the air mixture considered here is to be cooled and dehumidified to $T = 15°C$ and $RH = 20\%$. The water to be removed can then be calculated as follows:

$$(Y)_{initial} = 0.1095,\ (Y)_{final} = 0.0021 \tag{9.4g}$$

Water to be removed:

$$(Y)_{initial} - (Y)_{final} = 0.0195 - 0.0021 = 0.0174 \text{ kg H}_2\text{O/kg air} \tag{9.4h}$$

Alternatively, the result may be expressed in volumetric units by dividing by the humid volume of the original mixture: 0.0174/0.945 = 0.0184 kg H_2O/m^3 initial mixture.

WATER REMOVAL HEAT LOAD

In addition to the amount of water to be removed, an important parameter in the design of a dehumidification unit is the associated heat load. That quantity can be computed from the relevant enthalpies read from the humidity chart. We have, for the case cited,

$$(H)_{\text{initial}} = 103 \text{ kJ/kg dry air}, \ (H)_{\text{final}} = 20.3 \text{ kJ/kg dry air} \tag{9.4i}$$

$$\text{Heat removed} = (H)_{\text{initial}} - (H)_{\text{final}} = 103 - 20.3 = 82.7 \text{ kJ/kg dry air} \tag{9.4j}$$

Alternatively, using volumetric units,

$$\text{Heat removed} = [(H)_{\text{initial}} - (H)_{\text{final}}]/V_H = 82.7/0.945 = 87.5 \text{ kJ/m}^3 \tag{9.4k}$$

9.1.3 Model for Countercurrent Air–Water Contact: The Water Cooling Tower

As previously mentioned, warm process water that was used in a plant for cooling purposes can be restored to its original temperature by contacting it with an airstream, which causes it to undergo evaporative cooling. The operation is generally carried out in cooling towers containing stacked packings of large size and voidage to minimize pressure drop. We propose here to model the operation of such a tower and, in the course of the model development, introduce some ingenious simplifications based on the Lewis relation (Equation 9.2c).

As in all packed-column operations, the fundamental model equations consist of differential balances taken over each phase; the principal novelty here is the simultaneous use of mass and energy balances.

The pertinent variables and the differential elements around which the balances are taken are displayed in Figure 9.5a.

9.1.3.1 *Water Balance over Gas Phase (kg H_2O/mLs)*

This balance is no different from similar mass balances used in packed-gas absorbers and distillation columns (see Chapter 8), and it takes the following form:

$$\text{Rate of water vapor in} - \text{Rate of water vapor out} = 0$$

$$\begin{bmatrix} G_s Y\big|_z \\ +N_{\text{avg}} \end{bmatrix} - \left[G_S Y\big|_{z+\Delta z} \right] = 0 \tag{9.5a}$$

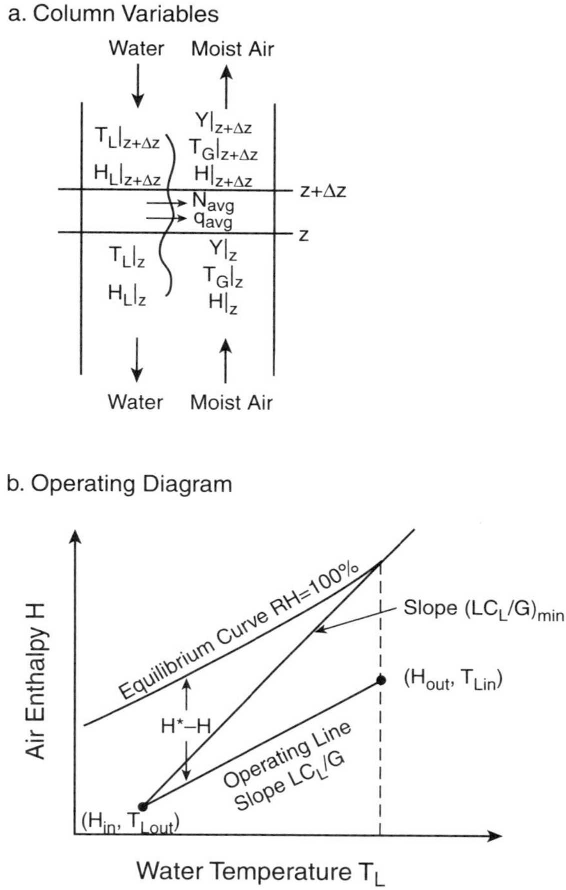

FIGURE 9.5
Variables and operating diagram for a packed cooling tower.

which upon introduction of the auxiliary mass transfer rate equation, division by Δz, and letting $\Delta z \rightarrow 0$, yields the usual form of ODE applicable to these cases:

$$G_s \frac{dY}{dz} - K_Y a(Y^* - Y) = 0 \tag{9.5b}$$

where $Y^* - Y$ is the humidity driving force.

9.1.3.2 *Water Balance over Water Phase*

This balance is omitted because the water losses are usually less than 1%.

9.1.3.3 Gas-Phase Energy Balance (kJ/m²s)

Here we must be careful to include both sensible heat transfer as well as the latent heat brought into the air by the water vapor. We obtain

$$\text{Rate of energy in} - \text{Rate of energy out} = 0$$

$$\begin{bmatrix} G_s H\big|_z + q_{avg} \\ +\Delta H_v N_{avg} \end{bmatrix} - \left[G_s H\big|_{z+\Delta z} \right] = 0 \quad (9.5c)$$

which, after applying the same procedure as before, yields

$$G_s \frac{dH}{dz} - Ua(T_L - T_G) - \Delta H_v K_Y a(Y^* - Y) = 0 \quad (9.5d)$$

Here T_L and T_G are the water and air temperatures, respectively, and H equals the enthalpy of the moist air at a given point in the tower.

9.1.3.4 Liquid-Phase Energy Balance (kJ/m²s)

A completely analogous derivation to the gas-phase energy balance yields

$$LCp_L \frac{dT_L}{dz} - Ua(T_L - T_G) - \Delta H_v K_Y a(Y^* - Y) = 0 \quad (9.5e)$$

where we have replaced the liquid enthalpy H_L by Cp_L ($T_L - T°$). The model is completed by adding the relevant equilibrium relation, which coincides with the 100% RH curve in the humidity charts and is also available in analytical form. Thus, for the equilibrium relation,

$$Y^* = f(T_L) \quad (9.5f)$$

Equation 9.5b, Equation 9.5d, Equation 9.5e, and Equation 9.5f together with the previously given expression for H (Equation 9.3d) constitute a set of five equations in the five state variables Y, Y^*, T_G, T_L, and H.

Although a numerical solution of these equations is today easily accomplished, early workers in the field had to cast about for alternative means of solving the model. To do this, they used the ingenious device of introducing the Lewis relation into the gas-phase energy balance, which has the effect of combining T_G and Y into a single variable, the air enthalpy H. We sketch the procedure below, using interfacial values in place of Y^* and T_L to accommodate the film coefficients h and k_Y used in the Lewis relation. We use that relation to replace h by $k_Y C_s$ and obtain in the first instance

$$G_s \frac{dH}{dz} - k_Y a\left[(C_s T_i + \Delta H_v Y_i^*) - (C_s T_G + \Delta H_v Y)\right] = 0 \tag{9.5g}$$

where, as seen from Equation 9.3d, the bracketed terms represent enthalpies of air–water mixtures. We therefore can write

$$G_s \frac{dH}{dz} - k_Y a(H_i - H) = 0 \tag{9.5h}$$

where $H_i - H$ can be considered an enthalpy driving force, which replaces and combines the temperature and humidity driving forces in the original model.

We now assume that the two-film theory can be applied to this system, with the result that Equation 9.5h can be cast in the form

$$G_s \frac{dH}{dz} - K_Y a(H^* - H) = 0 \tag{9.5i}$$

where $K_Y a$ is now the overall mass transfer coefficient (in units of kg/m^{3}s ΔY) and H^* the gas enthalpy in equilibrium with the bulk water temperature T_L.

This equation is of the same form as gas-phase differential balances encountered in gas absorption and distillation, so that the design procedures used there can be replicated, provided an appropriate operating line can be constructed. That line is obtained from an overall two-phase integral heat balance and takes the form

$$G_s(H_1 - H) = LC_L(T_{L1} - T_L) \tag{9.5j}$$

for part of the column, and for the entire tower,

$$G_s(H_1 - H_2) = LC_L(T_{L1} - T_{L2}) \tag{9.5k}$$

The gas-phase energy balance (Equation 9.5i) can, in turn, be formally integrated to yield the familiar HTU–NTU relation:

$$Z = \frac{G_s}{K_Y a}\int_{H_{in}}^{H_{out}} \frac{dH}{H^* - H} = \text{HTU} \times \text{NTU} \tag{9.5l}$$

The model is completed with the addition of the equilibrium relation:

$$H^* = f(T_L) \tag{9.5m}$$

which is constructed from the 100% *RH* curve of the psychrometric charts.

FIGURE 9.6
Cooling towers built for zinc refinery modernization (height: 18.3 m; diameter: 3.6 m). (Courtesy of Fabricated Plastics Ltd., Maple, Ontario.)

The original set of five equations, three of which were ODEs, have thus been reduced to three relations (Equation 9.5k, Equation 9.5l, and Equation 9.5m). Also, the set is now cast in the familiar form of a HTU–NTU expression, joined to an operating line and equilibrium relation. The procedure used to solve this much simpler set is outlined in Figure 9.5b and follows the usual routine of drawing an operating line, this one of slope LC_L/G_s, through the point (H_1, T_{L1}), and evaluating the NTU integral using the enthalpy driving force calculated from the operating diagram. Note that it is now G_{Min}, not L_{Min}, which corresponds to an infinitely high tower.

Illustration 9.2: Cooling of an Acid Leach Slurry

Evaporative cooling on a large scale is not limited to the treatment of process water. It can be applied equally well to aqueous solutions or even slurries that require cooling prior to further processing. The product that results from the high-temperature–high-pressure acid leaching of various ores (see Illustration 7.8) is a case in point. The system here has two unusual features. One is its high solids content (up to 40% by weight) that rules out the use of *packed* cooling towers. The other is the huge tonnage rates involved (~1 ton/s) that are an order of magnitude higher than those seen in conventional cooling towers. We would therefore expect to run several towers *in parallel* in order to accommodate the high loads. This is, in fact, the case, as seen in Figure 9.6. It will be assumed that these will be spray columns, with the particle diameter tentatively set at 3 mm, particle density ρ at 1500 kg/m³, and air density at 1 kg/m³.

The base case for our calculations will be the slurry that results from the acid leach of nickel ores addressed in Illustration 7.8. After depressurization, the slurry exiting the leach reactor stabilizes at a temperature of 90°C and a sulfuric acid concentration of 5% by weight. It now has to be cooled to a level acceptable for further processing, say 30°C. The tonnage rate is set at 1 ton/s, which corresponds roughly to the projected output of the Goro facility. These are the parameters to be used here.

We confine our calculations to a preliminary estimate of the number of transfer units, the number of spray columns required, and their diameter.

NTU

The first stumbling block here is to determine the saturation enthalpy H* of the vapor in equilibrium with the 5% H_2SO_4 solution. That vapor consists of pure steam (see Table 7.3), but its equilibrium pressure is not known and is not found in standard tabulations. Inspection of Table 7.3 shows that while acid concentrations in weight percentages may appear high, the corresponding mole percentages are only a *fraction* of those values, and those of the solvent are in turn inordinately high and close to those of pure water. For the 5% solution we are dealing with here, the numbers are as follows:

Weight fraction of acid $x_{H_2SO_4} = 0.05$
Mole fraction of acid $x_{H_2SO_4} = 0.010$
Mole fraction of water $x_{H_2O} = 0.990$

At those levels of water concentration, it is reasonable to assume that Raoult's law is obeyed:

$$p_{H_2O} = 0.990 \times P^o_{H_2O} \tag{9.6a}$$

and that, furthermore, for all practical purposes, H^* can be set equal to that of pure water shown in the humidity charts.

Calculation of the NTU starts in the usual way by first determining G_{Min}, setting an *actual* value for G, and using the resulting operating relation (Equation 9.5k) to calculate H for use in the NTU integral.

We assume air to be available at T = 20°C, Y = 0.01 kg H_2O/kg air, H = 25 kJ/kg. Using the humidity chart in Figure 9.4 for additional data, we obtain

$$G_{Min}(H^*_{90°C} - H_{In}) = LC_L(T_2 - T_1) \tag{9.6b}$$

$$G_{Min} = \frac{10^3 \times 4.18(90 - 30)}{360 - 25} = 752 \text{ kg/s}$$

where we set C_L equal to the heat capacity of water.

TABLE 9.1

NTU Calculation for a Cooling Tower

T_L (°C)	H^* (kJ/kg)	H (kJ/kg)	$1/H^*\ H$	$(1/H^*\ H)_{avg}\Delta H$
90	360	275	0.015	
80	345	235	0.0091	0.48
70	335	195	0.00071	0.32
60	320	150	0.0059	0.29
50	275	110	0.0061	0.24
40	165	70	0.011	0.32
30	100	25	0.013	0.50
				NTU = 2.21

Setting G_{actual} = 1000 kg/s and solving the operating relation (Equation 9.5k) for H yields

$$H = \frac{LC_L}{G_{actual}}(\Delta T) + H_{In} = \frac{10^3 \times 4.18}{10^3}\Delta T + 25 \tag{9.6c}$$

The values of H calculated from this equation are combined with H^* read from the humidity chart to evaluate the NTU integral. The results are summarized in Table 9.1.

NUMBER OF TOWERS AND TOWER DIAMETER

These two parameters hinge on the requirement that air velocity v_g must remain below the terminal settling velocity v_t of the spray. Guidelines for calculating v_t were given in Illustration 5.4, and the appropriate expression for the present case is seen to be Equation 5.10c:

$$v_t = 5.45\left[d_p(\rho_p \rho_f - 1)\right]^{1/2} \tag{5.10c}$$

which yields the following terminal velocity:

$$v_t = 5.45\left[3 \times 10^{-3}(1500 / 1 - 1)\right]^{1/2} = 11.6\,\text{m/s}$$

We fix the allowable air velocity below this value, at 7 m/s, and use it to explore various combinations of number of columns, N_C, and column diameter, d_C. The vehicle for this is the continuity equation which is solved for d_C:

$$d_C = \left(\frac{Q_{air} / N_C}{v_{air}}\frac{4}{\pi}\right)^{1/2} \tag{9.6d}$$

where $Q_{air} = 1000\ m^3/s$, and $v_{air} = 7\ m/s$. Some calculated results appear below.

Number of columns, N_C	5	10	15
Column diameter, d_C (m)	6.0	4.3	3.5

It will be a matter of an economic balance to choose the appropriate combination of N_C and d_C.

Comments

The inordinately low number of transfer units, typical of evaporative cooling operations, indicates that the *height* of the transfer unit G/K_Ya will be considerable, of the order of several meters. This does not come as a surprise, because the operation involves both high airflow rates G and low interfacial areas (a). A recently introduced technology attempts to rectify this through the use of *fluidized beds*, which are denser and have much higher interfacial areas than what one sees in spray columns (see cover). The operation has the added advantage of allowing the use of higher air velocities and smaller tower diameters. All these conclusions can be inferred from the simple calculations presented here.

9.2 Drying Operations

The drying of solids is a topic of considerable proportions that merits an entire monograph for its proper treatment. Our purpose here will be to provide a brief survey of the operation and some practice in carrying out simple calculations.

Drying can be carried out in a variety of physical configurations in which the solids can be stationary, conveyed on a moving belt, or allowed to tumble through an inclined rotary kiln. What these operations have in common is the use of heated air to assist in the drying process. When air passes through a stationary mass of solids, which may be contained in a set of perforated trays or in a fixed bed or column, the process is referred to as *through-flow drying*. *Cross-flow drying* occurs when air passes through a perforated conveyor belt at right angles to the solids being conveyed. It also refers to the drying of a stationary solid by a gas stream flowing over it. Rotary kilns, on the other hand, usually make use of a *countercurrent mode* of contact, with hot combustion gases flowing upward into the inclined kilns while the tumbling solids make their way downward in the opposite direction.

The drying of solids in general is a highly complex process involving both heat and mass transfer. If the solid is porous, moisture content and temperature will vary internally, as well as externally in the direction of airflow.

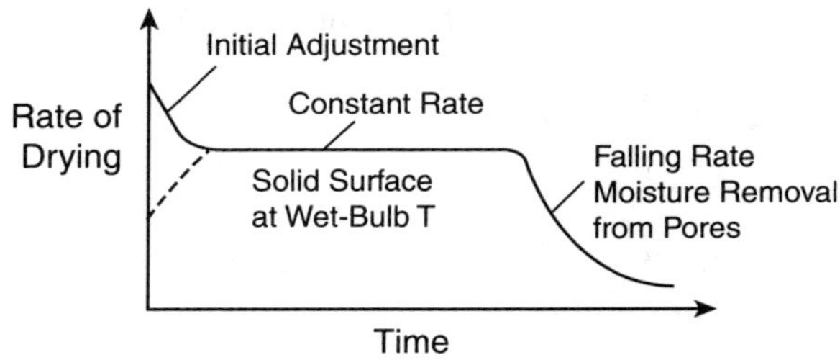

FIGURE 9.7
The various drying periods.

Thus, we could be dealing with at least two coupled PDEs (mass and energy balance) in time and two dimensions.

Early studies of drying processes revealed that considerable simplifications result by recognizing three distinct drying periods, shown in Figure 9.7. During an initial adjustment period, the surface moisture quickly drops to the *wet-bulb temperature*. If the moisture content in the air is either negligible or otherwise constant due to steady-state conditions, the humidity driving force will assume a constant value $Y^* - Y$, and the drying process will consequently proceed at a constant rate. During this constant-rate period, as it is called, the process can be modeled algebraically if we assume negligible change in air humidity in the direction of flow, or by an ODE if the latter varies. Thereafter, the drying process becomes more complex as moisture removal now must take place from the interior porous structure as well as remaining wet spots on the surface. A lengthy drying period may result as the moisture becomes increasingly inaccessible. This interval, called the *falling-rate period*, leads to the aforementioned PDEs.

Figure 9.8 translates these periods into temperature and humidity profiles that would result in a continuous countercurrent drying operation, taking place, for example, in a rotary kiln. The constant rate period predominates, usually occupying more than 50% of the kiln, or in excess of 95% in the case of nonporous solids. Very little evaporation takes place during the initial adjustment, and both the initial and constant-rate periods are mass transfer controlled. The falling-rate period becomes increasingly heat transfer controlled and is usually treated empirically.

The focus here will be on the dominant constant-rate period that is well described by conventional design equations for countercurrent gas–liquid operations. We can, in fact, adopt the gas scrubber equation (Equation 8.2) with only a slight change in form and write

Length of kiln
$$Z = \frac{G_s}{k_Y a} \int_{Y_1}^{Y_2} \frac{dY}{Y^* - Y} \tag{9.7a}$$

where Y^* is the saturation humidity corresponding to the wet-bulb temperature.

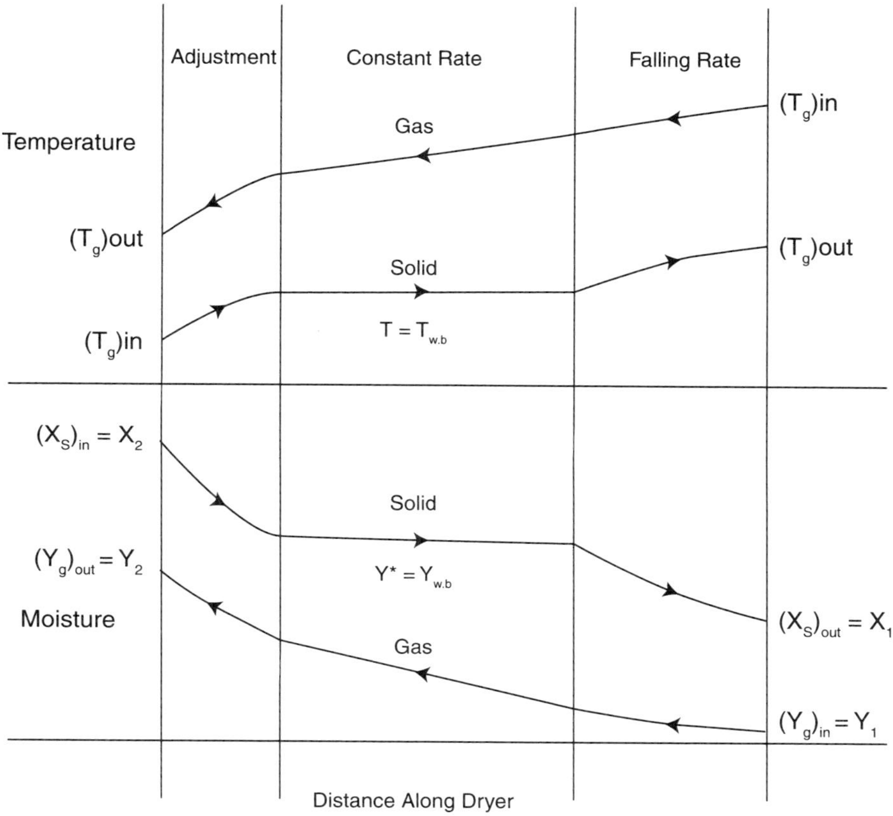

FIGURE 9.8
Temperature and moisture profiles in a countercurrent dryer. In the constant rate zone, the solid surface is a wet-bulb condition.

Because Y^* is constant, Equation 9.7a integrates immediately to the expression

$$Z = \mathrm{HTU} \times \mathrm{NTU} = \frac{G_s}{k_Y a} \ln \frac{Y^* - Y_1}{Y^* - Y_2} \tag{9.7b}$$

where Y_1 is the humidity of the incoming air. Y_2 is usually not specified and has to be determined from an overall water balance over the kiln. This is shown in Illustration 9.3.

Illustration 9.3: Rotary Kiln Drying of a Wet Granular Mass

A wet crystalline mass with a water content of $X_2 = 0.2$ kg H_2O/kg solid entering at a rate of $L_s = 0.25$ kg/s dry weight is to be dried to a water content of $X_1 = 0.01$ kg H_2O/kg solid. Air is available at a temperature of 110°C and

RH = 5%. The task will be to determine the important parameters for the design and operation of the dryer.

AIR PROPERTIES

Using the humidity chart (Figure 9.4), we obtain the following:

$T_1 = 110°C$, RH = 2%, $Y_1 = 0.019$ kg H_2O/kg air

$T_{wb} = 40°C$, $Y_{wb} = 0.05$ kg H_2O/kg air

MINIMUM AIR REQUIREMENTS

This quantity is obtained by setting $Y_2 = Y_{wb}$ (entering solid in equilibrium with the exiting air). A moisture balance over the constant rate zone, assumed to occupy the entire drying, then yields

$$L_s' (X_1 - X_2) = (G_s')_{Min} (Y_{wb} - Y_1) \tag{9.8a}$$

That is,

$$(G_s')_{Min} = \frac{0.5(0.2 - 0.01)}{0.05 - 0.019} = 6.23\,\text{kg/s} \tag{9.8b}$$

Set

$$(G_s')_{actual} = 8\,\text{kg/s} \tag{9.8c}$$

NUMBER OF TRANSFER UNITS

The exit humidity Y_2 required for the NTU is determined from a second moisture balance, this time using the actual airflow rate:

$$Y_2 = \frac{Y_s'}{G_s'}(X_1 - X_2) + Y_1 = \frac{0.5}{8} = 0.2 - 0.01) + 0.019 \tag{9.8d}$$

$$Y_2 = 0.031 \text{ kg } H_2O/\text{kg air} \tag{9.8e}$$

Hence,

$$\text{NTU} = \ln\frac{0.05 - 0.019}{0.05 - 0.031} = 0.49 \tag{9.8f}$$

DRYER DIAMETER

The diameter of the unit is largely determined, as it is in all other operations we have seen, by the allowable or upper velocity limit of one of the streams.

In rotary dryers, that limit is taken to be an air velocity of the order 1 m/s. The reason here is not the avoidance of "flooding" but rather the suppression of solids entrainment.

Using this value, we obtain for the minimum diameter d_{Min},

$$v = 1\,\text{m/s} = \frac{G_s'\rho_{\text{air}}}{\pi d^2 / 4} \tag{9.8g}$$

or

$$d_{\text{Min}} = \left(\frac{4 \times 8 \times 1}{1 \times \pi}\right)^{1/2} = 3.2\,\text{m} \tag{9.8h}$$

Comments

The striking feature here is the same as that seen in Illustration 9.2 — the low value of NTU. This inevitably entails a large value for the height (here length) of a transfer unit. $k_Y a$ values in these units are typically of the order of 0.05 kg/m^{3}s ΔY, mainly because the specific surface area a (m^2/m^3) is very low (order of 1). We obtain

$$\text{HTU} = \frac{G_s'}{k_Y a \times \pi d^2 / 4} = \frac{8}{0.05 \times \pi\, 3.2^2 / 4} \tag{9.9a}$$

$$\text{HTU} = 20\text{ m} \tag{9.9b}$$

and

$$\text{Length of dryer } L = \text{HTU} \times \text{NTU} = 9.8\text{ m} \tag{9.9c}$$

Both HTU and NTU (and d) can, of course, be reduced by raising the temperature of the air, which lowers the required airflow rate G_s. There is considerable room for manipulating these variables to achieve a more desirable result.

In the following illustration, which deals with the drying of a steamed activated carbon bed, we assume again that the operation takes place entirely in the constant-rate period. This is based on the fact that carbon is a hydrophobic substance that allows little penetration of its porous structure by the condensate produced during the steaming process. In other words, the condensate is assumed to be present entirely as surface moisture, which is removed by a constant-rate drying mechanism.

Illustration 9.4: Debugging of a Vinyl Chloride Recovery Unit

Vinyl chloride monomer (VCM) is a volatile substance (boiling point 14°C) used as a starting material for the production of polyvinyl chloride (PVC). It has been identified as a potential carcinogen, and occupational health regulations now call for an upper limit of 1 to 5 ppm VCM in factory air.

A preferred method of air purification is to pass VCM-laden air through beds of activated carbon. These beds operate on a four-step cycle:

1. Saturation with VCM to 1 ppm breakthrough
2. Stripping of the adsorbed VCM with steam, which is subsequently condensed, leaving essentially pure VCM that is dried and recycled
3. Drying of the carbon bed with hot air
4. Cooling of the regenerated bed with cold purified air

A dual-bed system is commonly employed, so that while one bed is "on-stream," the second bed can be regenerated and prepared for the adsorption step. The time period allowed for each step is typically as follows:

1. Saturation: 4 h
2. Steaming: 2 h
3. Drying: 1∫ h
4. Cooling: ∫ h

A schematic diagram of the cycle appears in Figure 9.9.

A major producer of PVC experienced difficulties with a newly installed adsorption system. The unit performed satisfactorily with fresh carbon, but during subsequent cycling, the VCM level in the effluent rose to unacceptable levels. Inspection of the beds after drying revealed considerable residual moisture. Inadequate drying during step 3 was therefore considered to be a possible reason for the malfunctioning of the bed. The total amount of condensate that needs to be evaporated was estimated at 100 kg.

When queried about their choice of airblower and its delivery rate (0.1 kg/s at 49°C), plant personnel responded that it was based on the assumption that the condensate was at or near 100°C. This was a major conceptual error. Basic knowledge of the psychrometric chart should have led to the realization that evaporative cooling would reduce the temperature of the condensate well below 100°C (i.e., to the wet-bulb level). This, in turn, would dramatically reduce the evaporation rate, because the vapor pressure and, hence, the driving saturation humidity are exponential functions of temperature.

If we assume that the entire bed attains the wet-bulb temperature after a brief start-up period, the model reduces to an ODE, which must be complemented by a cumulative mass balance in order to determine the drying time. We consider instead a limiting version of the full model in which the exiting airstream is assumed to be fully saturated. This case provides us with an

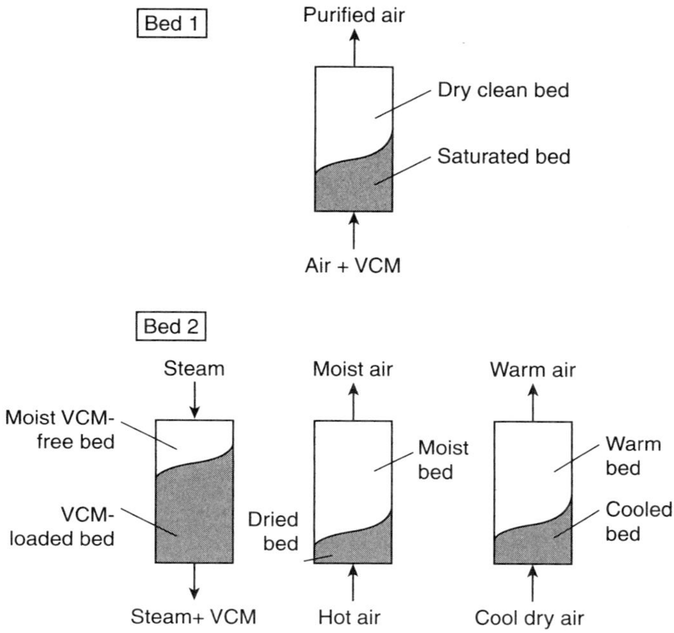

FIGURE 9.9
Removal of vinyl chloride monomer from air — operation of the adsorption purifier.

estimate of the *minimum* airflow rate required to evaporate the charge of 100 kg water in the assigned time of 1∫ h. To obtain that rate, we compose a cumulative water balance, which is given by

Water removed by exiting gas in $1\frac{1}{2}$ hr	=	Amount of water to be evaporated

$$Y_{wb}(G_s)_{Min}\, t = 100 \text{ kg} \tag{9.10a}$$

from which, with t = 1∫ h = 5400 s, we obtain

$$(G_s)_{Min} = \frac{100}{Y_{wb}\, 5400} = \frac{1}{54\, Y_{wb}} \tag{9.10b}$$

The humidity chart, Figure 9.4, is now used in Equation 9.10b to establish values of $(G_s)_{Min}$ for various levels of incoming air temperature, T_{air}. This is done in the usual fashion by first locating the incoming air temperature and humidity, assumed to be zero, on the chart, and then moving from that point upward and to the left along the adiabatic saturation line to a point of

TABLE 9.2

Minimum Airflow Rates for Drying a Carbon Bed Containing 100 kg Moisture

Air temperature (°C)	25	50	75	100	125
Y_{wb} (kg H_2O/kg air)	0.006	0.013	0.021	0.028	0.038
$(G_s)_{Min}$ (kg/s)	3.1	1.4	0.88	0.66	0.49

intersection with the 100% RH curve. The right-hand ordinate of that point yields the value of Y_{wb} to be used in Equation 9.6c. The results obtained are summarized in Table 9.2.

Examination of Table 9.2 shows that the air provided by the plant (T_{air} = 50°C, G_s = 0.1 kg/s) underestimated the minimum requirement by a factor of 14. Even at a temperature of 125°C, the minimum flow required was still five times that actually provided. Clearly, a combination of both higher temperature and greater blower capacity would be required to meet the drying specifications. A reasonable recommendation would be for a flow rate of 1 kg/s at 125°C at the point of delivery. This provides a safety factor of 2 over the tabulated $(G_s)_{Min}$ of 0.49 kg/s.

9.3 Heat Effects in a Catalyst Pellet: The Nonisothermal Effectiveness Factor

In Illustration 4.10, a model that describes the isothermal diffusion and reaction in a catalyst pellet was considered. Solution of that model yields the reactant concentration profile within the pellet, which is then converted by integration into the so-called catalyst effectiveness factor E. Such isothermal effectiveness factors apply to small particles with high thermal conductivities and relatively low reaction rates.

In general, the heat of reaction, which is of the order of 100 kJ/mol, cannot be ignored, and the mass balance must then be complemented by an appropriate shell energy balance. That balance must consider the heat conducted in and out of the shell, as well as the heat generated or consumed within the pellet.

We assume the same slab geometry and first-order reaction as before and consider the reaction to be exothermic, which is the more common case. The following formulation is then obtained:

$$\text{Rate of energy in} - \text{Rate of energy out} = 0$$

$$\left[-k_e A \left.\frac{dT}{dx}\right|_x\right] - \left[\begin{array}{c} -k_e A \left.\dfrac{dT}{dz}\right|_{x+\Delta x} \\ +k_r(T) C_A \Delta H_r A \Delta x \end{array}\right] = 0 \tag{9.11a}$$

Dividing by $A\Delta x$ and letting $\Delta x \rightarrow 0$, we obtain the second-order ODE:

$$\frac{d^2T}{dx^2} - k_r(T)C_A\Delta H_r / k_e = 0 \tag{9.11b}$$

where k_e is the effective thermal conductivity of the pellet.

This expression is supplemented by the mass balance given previously in Illustration 4.9 in which the rate constant k_r is now a function of temperature. We repeat it here for completeness:

$$\frac{d^2C}{dx^2} - k_r(T)C_A / D_e = 0 \tag{9.11c}$$

The two ODEs, which are coupled by the two variables C_A and T, generally have to be solved numerically. The resulting concentration profile $C_A(x)$ can then be integrated over the pellet volume V_p as was done in the isothermal case to obtain the nonisothermal effectiveness factor E_{ni}:

$$E_{ni} = \frac{\int_0^L k_r(T)C_A(x)dV_p}{k_r(T)C_AV_p} \tag{9.11d}$$

where $k_r(T)$ is given by the familiar Arrhenius relation:

$$k_r(T) = A\exp(-E_a/RT) \tag{9.11e}$$

Here the reference state is taken to be the surface concentration C_{AS} and the surface temperature T_s — the conditions that would prevail within the pellet in the absence of transport resistances.

A typical, unscaled plot of E_{ni} versus the nonisothermal Thiele modulus is shown in Figure 9.10. Two additional parameters that contain the thermal factors make their appearance here: the Arrhenius number E_a/RT_s, which contains the important activation energy E_a; and the dimensionless parameter β, which reflects the effect due to the heat of reaction and the transport resistances. For $\beta = 0$ (i.e., for a vanishing heat of reaction or infinite thermal conductivity), the effectiveness factor reduces to that of the isothermal case. $\beta > 0$ denotes an exothermic reaction, and here the rise in temperature in the interior of the pellet is seen to have a significant impact on E_{ni} which may rise above unity and reach values as high as 100. This means that the overall reaction rate in the pellet is up to 100 times faster than would be the case at the prevailing surface conditions. This is due to the strong exponential dependence of reaction rate on temperature, as expressed by the Arrhenius relation

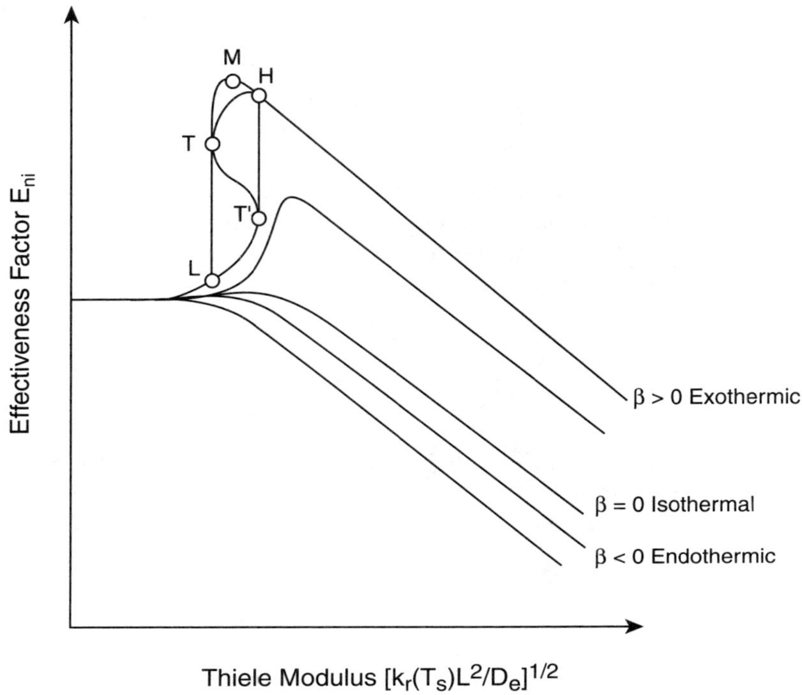

FIGURE 9.10
Nonisothermal effectiveness factor and its jump transitions (constant Arrhenius number).

(Equation 9.11e). As expected, the effect varies directly with the heat generated (ΔH_r) and inversely with the rate of heat removal. Thus, exothermicity, far from being undesirable, actually has a beneficial effect on catalytic conversion. We must guard, however, against an excessive rise in temperature that might adversely affect catalyst structure, causing a decline or even cessation of catalytic activity. The problem of estimating the maximum temperature that may occur in a catalyst pellet is addressed in Practice Problem 9.4.

For endothermic reactions, $\beta < 0$ applies; ΔH_r is now a positive number. The effectiveness factor is now below the value seen in the isothermal case, due once again to the dependence of k_r on temperature. The effect is shown in the plots of Figure 9.10.

An additional point needs to be noted in connection with the inflections shown by some of the effectiveness curves. It is seen that in this region of inflections, E becomes a multivalued function of the Thiele modulus. The question then arises as to which of the three values actually materializes in practice. A mathematical analysis of such systems, which goes beyond the scope of the present book, shows that of the three states, only the upper and lower ones are stable, and that a jump transition occurs from one to the other as one passes through this region. This transition occurs at different locations, depending on the direction from which the region is approached. Suppose, for example, that the Thiele modulus is gradually diminished, by reducing surface temperature T_s. Effectiveness then undergoes a gradual increase, reaches a maximum value M shown in Figure 9.10, and then begins a decline until the tangent point T is reached. Here the effectiveness factor experiences a sudden jump decrease to the lower value L, after which it continues a smooth decline with diminishing modulus value toward the limiting value of unity. A similar jump transition occurs when one approaches from the opposite direction, but this time it is a jump *increase*, and it occurs at a different location, from the tangent point T' to the location H. This phenomenon of obtaining different ordinate values, depending on the direction in which a curve is traversed, is referred to as a hysteresis effect, and the surface temperatures T_s at which the jump transition occurs are known as the *ignition* and *extinction* temperatures. In other words, when these temperatures are reached, the reaction rate either undergoes a sudden increase (ignition, point T') or it experiences a sudden drop (extinction, point T). These interesting features can be used to control the course of a particular catalytic reaction.

9.3.1 Comments

Although the system we considered here is relatively simple, involving a first-order reaction, it revealed the existence of some fascinating and exotic phenomena. Such phenomena are not limited to catalytic reactions but arise in other nonisothermal systems, for example, in continuous-flow stirred tank reactors, and even in isothermal systems. Their common feature is that the performance curve describing the system has to exhibit an *inflection*. Such inflections have also been observed in a biological context, where they play the role of a *biological switch*, which is activated in response to particular stimuli. This is taken up in the following illustration.

Illustration 9.5: A Biological Switch and the van Heerden Diagram

Blood coagulates when it comes in contact with an injured site or a foreign surface. The event triggers a cascade of enzyme-catalyzed reactions that culminate in the production of the clotting protein *thrombin*. Thrombin has the ability to form aggregates of platelets and strands of a high-molecular-weight protein, fibrin.

Examination of the coagulation cascade has shown that it gives rise to an S-shaped curve when the rate of production of thrombin is plotted against its concentration, C_w — the reaction is self-catalyzing or autocatalytic (Figure 9.11). The process is balanced by the rate at which thrombin is swept away from the vessel wall by mass transfer to the flowing blood. Because thrombin concentration in the bulk fluid, C_b, is nearly constant and low, we can write

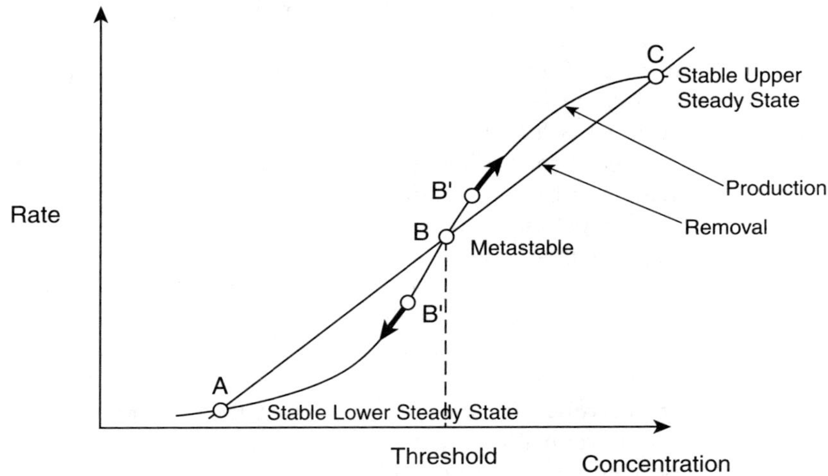

FIGURE 9.11
Example of a biological switch.

$$N/A \text{ (moles removed/m}^2\text{s)} = k_C\,(C_w - C_b) \qquad (9.12)$$

$$\cong k_C C_w$$

where k_C equals the slope of the line shown in Figure 9.10.

Suppose that thrombin is produced at a rate corresponding to point B' in Figure 9.11. Because the removal rate here is *higher* than the production rate, concentration will diminish until the stable lower steady state is reached at A. At this point, the rate of production equals the rate of removal. Any disturbance away from A, either upward or downward, will return the system to A, in line with the arguments just presented. By similar reasoning, a production rate B' will cause a continuous *increase* in concentration until the table upper state is reached at C. Using the same arguments, any disturbance away from the intermediate state B will lean away from it. That steady state is therefore metastable.

It follows that point B may be regarded as a *threshold* concentration. Any concentration above it will turn the switch ON to give the high concentration of thrombin needed for clotting to proceed. Conversely, concentrations below B will turn the switch OFF and cause thrombin concentration levels to subside to a "standby" level needed to initiate the cascade.

Comments

A diagram of the form shown in Figure 9.10 was first introduced by van Heerden to explain the behavior of exothermic, cold-stirred tank reactors. The reader is asked, in Practice Problem 9.8 to deduce the parameters involved in that analysis.

Practice Problems

9.1 Meteorology

a. Using the humidity charts, and given a daytime temperature and relative humidity, indicate how you would proceed in order to predict whether dew will form.
b. Again using the psychrometric charts, and assuming a strong wind to be blowing, indicate under what conditions frost may form, even though the temperature remains above 0°C. What is the maximum temperature that will allow this to happen?

9.2 Absolute and Relative Humidity

a. If absolute humidity changes, must relative humidity change as well?
b. Is it possible for relative humidity to stay constant if temperatures rise during the day?
c. Can perspiration evaporate if the surrounding air has a relative humidity of 100%?
d. Rationalize the following statement: "Here in Ottawa it may get really cold (–30°C), but it's always a *dry* cold. Toronto is warmer (0°C), but it feels much damper there."

9.3 Air Dehumidification and Cooling

This process, which is familiar to the users of air conditioners, can also be carried out on a large scale in packed towers.

a. Show how the basic model equations for water cooling (Section 9.1.3) should change to accommodate this case.
b. Does the operating diagram change?

9.4 Air Supply to a Drier

An air drier requires 1 kg/min (dry base) of air at 80°C and RH = 20%. The available air is at 25°C and RH = 50% and is to be brought to the desired conditions by direct injection of steam. What is the minimum rate at which steam must be supplied, given that its latent heat is 2450 kJ/kg?

Answer:

0.08 kg/min

9.5 Cross-Flow Drying of Plastic Sheets

Moist plastic sheets, 15 m long and 1.5 m wide, with an initial moisture content of 6 kg H_2O/sheet are to be dried by passing warm dry air ($T = 55°C$, $Y = 0$) over both sides of the supported sheets. Air is supplied at the rate of 1 kg (dry)/s.

a. What is the minimum time required to remove 90% of the moisture?
b. What is the *actual* time if $k_C = 0.0313$ kg/m²s ΔY?

Answer:

a. 5.6 min
b. 7.4 min

9.6 Design of a Cooling Tower

Water is to be cooled from 43.3 to 29.7°C in a packed column using air entering countercurrently at 29.5°C and a wet-bulb temperature of 23.3°C. The water flow rate is 2.71 kg/m²s, and airflow is to be set at 1.5 times the minimum value. The overall mass transfer coefficient for the column is estimated at $K_Y a = 0.90$ kg/m³s ΔY. Calculate the height of the tower.

Answer:

6.6 m

9.7 Maximum Temperature in a Catalyst Pellet

Derive an expression for the maximum possible temperature in a catalyst pellet. (*Hint*: Eliminate the reaction term by dividing mass and energy balances.)

9.8 The van Heerden Diagram and the Nonisothermal CSTR

A continuous stirred-tank reactor in which an exothermic reaction takes place, and which is cooled to remove a portion of the heat, can exhibit three steady states, as shown in Figure 9.10.

a. Identify "rates of production" and "rate of removal."
b. What quantity is plotted along the abscissa?
c. What is the slope of the "rate of removal" line?
d. What would cause the "rate of production" curve to inflect?

Selected References

General Mass Transfer Texts

The earliest major text on mass transfer processes is the following:

Treybal, R.E., *Mass Transfer Operations*, McGraw-Hill, New York, 1952. See also the third edition, 1979.

More recent texts on mass transfer, often presented in combination with heat transfer, include the following:

White, F.M., *Heat and Mass Transfer*, Addison-Wesley, Reading, MA, 1988.
Middleman, S., *Introduction to Mass and Heat Transfer*, John Wiley, New York, 1997.
Cussler, E.L., *Diffusion: Mass Transfer in Fluid Systems*, 2nd ed., Cambridge University Press, New York, 1997.
Mills, A.F., *Basic Heat and Mass Transfer*, 2nd ed., Prentice Hall, Englewood Cliffs, NJ, 1999.

An earlier, and largely ignored, monograph provides greater insight into mass transfer theory than more contemporary efforts:

Sherwood, T.K., Pigford, R.L., and Wilke, C.R., *Mass Transfer*, McGraw-Hill, New York, 1975.

Separation processes, which make extensive use of mass transfer concepts, are well presented in the recent text:

Seader, J.D. and Henley, E.J., *Separation Process Principles*, 2nd ed., John Wiley, New York, 2006.

Transport Phenomena

The classical text by Bird, Stewart, and Lightfoot (1960) was recently updated:

Bird, R.B., Stewart, W.R., and Lightfoot, E.N., *Transport Phenomena*, 2nd ed., John Wiley, New York, 2000.

A more easily comprehended version is by

Welty, J.R., Wicks, C.E., Wilson, R.E., and Dorner, G., *Fundamentals of Momentum, Heat and Mass Transfer*, 4th ed., John Wiley, New York, 2001.

Transport theory applied to biomedical and materials engineering is discussed in the following:

Poirier, D.R. and Geiger, G.H., *Transport Phenomena in Materials Processing*, Minerals, Metals and Materials Society, Warrendale, PA, 1994.
Fournier, R.L., *Transport Phenomena in Biomedical Engineering*, Taylor & Francis, London, 1999.
Truskey, G.A., Yuan, A., and Katz, D., *Transport Phenomena in Biological Systems*, Pearson–Prentice Hall, Englewood Cliffs, NJ, 2004.

A wealth of models related to transport in environmental systems are given in the following:

Thibodeaux, L.J., *Environmental Chemodynamics. Movement of Chemicals in Air, Water and Soils*, 3rd ed., McGraw-Hill, New York, 2007.

The fugacity concept pioneered by Mackay to model phase equilibria and transport in the environment is described in the following:

Mackay, D., *Multimedia Environmental Model*, 2nd ed., Lewis Publishers, Boca Raton, FL, 2002.

Diffusion

Authoritative compilations of solutions to Fick's and Fourier's equations can be found in the classical monographs:

Carslaw, H.S. and Jaeger, J.C., *Conduction of Heat in Solids*, Oxford University Press, Oxford, UK, 1959.
Crank, J., *Mathematics of Diffusion*, 2nd ed., Oxford University Press, Oxford, UK, 1978.

They include solutions of source problems.

Diffusivities, permeabilities, and solubilities in polymers are compiled in the following:

Brandrup, J., Immergut, E.H., and Grulke, E.A., Eds., *Polymer Handbook*, 4th ed., John Wiley, New York, 1999.

See also:

Crank, J. and Park, G.S., Eds., *Diffusion in Polymers*. Academic Press, New York, 1968.

Tables of diffusivities in various solids and liquids, including metals, molten salts, and semiconductors, appears in Poirier and Geiger cited above. Similar information on diffusion in porous catalysts can be found in the following:

Satterfield, C.N., *Mass Transfer in Heterogeneous Catalysts*, MIT Press, Cambridge, MA, 1970.

Diffusion and Reaction

The twin topics of diffusion and reaction are taken up in most texts on reactor engineering, including the following:

Missen, R.W., Mims, C.A., and Saville, B.A., *Introduction to Chemical Reaction Engineering and Kinetics*, John Wiley, New York, 1999.
Levenspiel, O., *Chemical Reactor Engineering*, John Wiley, New York, 1999.

Phase Equilibrium

There is a plethora of handbooks and other compilations of phase equilibria of relevance to mass transfer operations.

For gas, liquid, and solid solubilities in water, it is best to draw on environmental source books, among which the monumental treatise by Mackay et al. stands out:

Mackay, D., Shiu, W.-Y., and Ma, K.C., *Illustrated Handbook of Physico-Chemical Properties and Environmental Fate for Organic Chemicals*, Vol. 1–5, Lewis Publishers, Boca Raton, FL, 1991–1997.

The treatise also reports vapor pressures and bioconcentration factors of the cited substances.

For vapor–liquid and liquid–liquid equilibria, refer to the equally monumental compilations:

Behrens, D. and Eckermann, R., Eds., *Vapor-Liquid Equilibrium Collection* (25 volumes), Sorenson, J.M. and Arlt, W., Eds., *Liquid-Liquid Equilibrium Collection* (5 volumes), DECHEMA Chemistry Data Series, DECHEMA, Frankfurt, 1977–1982.

Equilibria of relevance to supercritical fluid extraction can be found in:

McHugh, M. and Krukonis, V., *Supercritical Fluid Extraction*, 2nd ed., Butterworths-Heinemann, Oxford, UK, 1994.

Compilations of adsorption equilibria appear in the following:

Valenzuela, D.P. and Myers, A.L., *Adsorption Equilibria Handbook*, Prentice Hall, Englewood Cliffs, NJ, 1989.

Equilibria involving metals and systems of metals can be found in the following:

Brandes, E. and Brooks, G.H., Eds., *Smithell's Metals Reference Book*, 7th ed., Butterworths-Heinemann, Oxford, UK, 1992.

Separation Processes

A somewhat dated, but still highly useful, reference for practicing engineers is by

Schweitzer, P.A., Ed., *Handbook of Separation Techniques for Chemical Engineers*, McGraw-Hill, New York, 1979.

Equilibrium stage separations with emphasis on distillation and gas absorption are exhaustively treated in the following:

Henley, E.J. and Seader, J.D., *Equilibrium Stage Separation Operations in Chemical Engineering*, John Wiley, New York, 1981.

See also the text on separation processes by the same authors cited above. Find an up-to-date treatment of distillation in the following:

Stichlmair, J.G. and Fair, J.R., *Distillation: Principles and Practice*, Wiley/VCH, New York, 1998.

Compilations of height equivalent to a theoretical plate (HETP) and height of a transfer unit (HTU) appear in the twin monographs by H.Z. Kister:

Kister, H.Z., *Distillation Operations. Distillation Design*, McGraw-Hill, New York, 1989 and 1992.

Treatments of liquid–liquid extraction are to be found in the following:

Treybal, R.E., *Liquid Extraction*, 2nd ed., McGraw-Hill, New York, 1963.
Lo, T.C., Baird, M.H.I., and Hanson, C., Eds., *Handbook of Solvent Extraction*, John Wiley, New York, 1983.

and in:

Thornton, J., *Science and Practice of Liquid-Liquid Extraction*, Vol. 1 and 2, Oxford University Press, Oxford, UK, 1992.

Packed tower design and operation can be found in the text:

Strigle, R.F., *Random Packings and Packed Towers. Design and Applications*, Gulf Publishing Co., Houston, 1987.
Billet, R., *Packed Towers in Processing and Environmental Technology*, VCH-Verlags-Gesellschaft, Weinheim, Germany, 1990.

The definitive and up-to-date monographs on membrane separation are by the following authors:

Matsunra, T., *Synthetic Membranes and Membrane Separation Processes*, CRC Press, Boca Raton, FL, 1994.
Baker, R.W., *Membrane Technology and Applications*, McGraw-Hill, New York, 2000.

Fundamentals of adsorption, separation, and purification are discussed in a slim volume by the author:

Basmadjian, D., *The Little Adsorption Book*, CRC Press, Boca Raton, FL, 1996.

Other

The illustration and practice problems in Chapter 3, which deal with transport in plants, used the following as a source:

Nobel, P.S., *Biophysical Plant Physiology and Ecology*, W.H. Freeman, San Francisco, 1987.

See also by the same author:

Nobel, P.S., *Physicochemical and Environmental Plant Physiology*, Academic Press, New York, 1999.

Appendix A1

The D-Operator Method

The basis of the D-Operator method consists of replacing the operational part of a derivative (i.e., d/dx) by the operator symbol D and treating that symbol as an *algebraic entity*. Thus, the second derivative is written in the form

$$\frac{d}{dx}\left(\frac{d}{dx}\right) + (D)(D) = D^2 \tag{A.1}$$

and in its full form

$$\frac{d}{dx}\left(\frac{dy}{dx}\right) = D^2 y \tag{A.2}$$

where D^2y is considered to be the algebraic product of D^2 and y. It follows that the quantity y can be separated from D^2y by factoring it out, just as one would an algebraic quantity. Thus, the ordinary differential equation (ODE)

$$\frac{d^2y}{dx^2} - y = 0 \tag{A.3}$$

can be written in the equivalent form

$$(D^2 - 1)y = 0 \tag{A.4}$$

from which it follows that

$$D^2 - 1 = 0 \tag{A.5}$$

with the solutions

$$D_1 = 1 \quad D_2 = 1 \tag{A.6}$$

Equation A.5 is termed the *characteristic equation* of the ODE (Equation A.3), and its solution (Equation A.6) is referred to as the characteristic roots of the ODE.

Consider now the general second-order ODE:

$$ay'' + by' + cy = 0 \tag{A.7}$$

Then it can be shown that its solution takes the form

$$y = C_1 \exp(D_1 x) + C_2 \exp(D_2 x) \tag{A.8}$$

where D_1 and D_2 are the characteristic roots of the ODE — the solutions of the characteristic equation

$$aD^2 + bD + c = 0 \tag{A.9}$$

When the roots are complex, the exponential terms in Equation A.8 are converted to a trigonometric form using Euler's formula:

$$e^{ix} = \cos x + i \sin x \tag{A.10}$$

We also note that the exponential terms can also be expressed in equivalent hyperbolic form, and that, when the roots are identical, one of the two solutions is premultiplied by x. This follows from the appropriate theory. The results are summarized in Appendix A2, Table A.1.

Appendix A2

Hyperbolic Functions and ODEs

TABLE A.1

Short Table of Hyperbolic Functions

$\sinh x = \dfrac{e^{x} - e^{-x}}{2}$	$\tanh x = \dfrac{\sinh x}{\cosh x}$
$\cosh x = \dfrac{e^{x} + e^{-x}}{2}$	$\coth x = \dfrac{\cosh x}{\sinh x}$

TABLE A.2

Solutions of the Second-Order ODE $ay'' + by' + cy = 0$

Characteristic Roots	Solutions
1. Distinct and real	$y = C_1 e^{D_1 x} + C_2 e^{D_2 x}$ or $y = C_1 \sinh D_1 x + C_2 \cosh D_2 x$
2. Identical and real	$y = C_1 e^{Dx} + C_2 x e^{Dx}$
3. Imaginary $D_{1,2} = \pm bi$	$y = C_1 \cos bx + C_2 \sin bx$
4. Complex conjugate $D_{1,2} = a \pm bi$	$y = e^{ax}(C_1 \cos bx + C_2 \sin bx)$

Index

A

Absorption, see Gas–liquid absorption
Accidental spills
  and ocean response to, 187
Acid rain
  and Biot number, 231
  calculation of pH, 231
Active transport, 10
Activity coefficients
  calculation, from low solubility substances, 292
  variation with concentration, 281
Additivity of resistances, principle of, 16, 18, 19
  at lung–blood interface, 19
  in carbon dioxide uptake by a leaf, 139
  in diffusion through composite cylinders, 17
  in heat transfer, 36
  in mass transfer, 36
  two film theory and, 33
Adsorption (see also Percolation processes)
  batch, of trace substance, 189
  crosscurrent cascade for, 320, 387
  Freundlich isotherm for, 273
  Henry's constants for, 256
  Hill's isotherm (O_2-hemoglobin), 262
  Langmuir isotherm for, 256
  minimum bed size in, 259
  moisture isotherms in, 255
Aeration trench
  design of, 213
Agitated vessels, 221
  dissolution time in, 223, 234, 235
  mass transfer correlations for, 223
Air-water system
  enthalpy of, 446
  humidity charts for, 443–445
  in drying operations, 455
  in slurry cooling, 452
  in water cooling, 448
Antoine equation
  table of constants for, 243
Artificial kidney, see Hemodialyzer
Azeotropes
  breaking of, 290
  diagrams for, 285
  table of, 287

B

Barometric formula, 12, 44
Batch crystallization
  McCabe ΔL law for, 236
  working equation for, 236
Batch distillation
  at constant overhead composition, 360–364
  at constant reflux, 364
  differential, and Rayleigh equation, 307
  multicomponent, 366
Biology, Biomedical engineering, and Biotechnology
  bioconcentration (BCF), 226
  biofilters, 410
    design of, criteria for success, 185–186
    Ottengraf model for, 184, 193
  biological oxygen demand (BOD)
    in wastewater aeration, 213
  blood coagulation, 190, 232, 233
  controlled-release devices, 43, 97, 104, 105
  diffusion in living cell, 153
  diffusivities of biological substances in water, 111
  drug delivery
    by implants, 97
    by nanoparticles, 190
    by transdermal patches, 45, 104, 105
  effective therapeutic concentration (ETC), 99
  hemodialysis, 416–418
  mass transfer in blood, 45
  mass transfer in kidney, 233
  mass transfer in leaf, 139
  partition coefficients in toxin transfer, 274
  plasma volume, 53
  toxin uptake and elimination in animals, 226, 274

Biot number
  and acid rain, 231
  as a measure of transport resistance, 199
  in heat transfer, 200
  in mass transfer, 198
Blood and blood flow
  absorption of O_2 in, 262
  anticoagulant release into, 233
  coagulation trigger in, 48
  critical vessel diameter in, 233
  determination by dye dilution, method of, 53
  hemodialysis of, 416–418
  mass transfer between lung and blood, 19
  mass transfer regimes in, 233
Buckingham's π Theorem, 206

C

Carbon dioxide
  and global warming, 153
  caffeine equilibrium in supercritical, 268
  compensation of emissions by plant life, 153
  emission from car, 140
  in carbonation of soft drink, 248
  in supercritical extraction of caffeine, 268
  in upper atmosphere, 44
  net global emissions of, 153
  sequestering of, 16
  uptake by leaves, 140
  uptake by water basin, 184
Catalyst pellet
  design of, 178, 191
  diffusivity in, 131
  effectiveness factor for, 177
  plot of Thiele modulus, 178
  Raschig Ring form of, 191
  reaction and diffusion in, 173
  temperature effect on performance of, 191
Coffee decaffeination (supercritical extraction), 268
Clearance of toxins
  in aquatic life, 236
Concentration polarization
  Brian's equation for, 420
  in alloy casting, 84
  in electorefining, 114
  in membrane separation, 420
  in reverse osmosis, 420
Continuous-contact operations
  biofilters, 410
  distillation, 402–403
  minimum solvent requirement in, 389
  supercritical fluid extraction, 408
  water cooling, 448
Controlled-release drug delivery, 45, 97, 104, 105
Cooling
  of slurries, 452–455
  of water, 448
Cooling tower
  design equation for, 448–452, 468
  operating diagram for, 449
Crystallization (see Batch crystallization)

D

D'Arcy's law, 2
Dialysis, 416
Diffusion
  and reaction, 170
  equimolar counter, 26
  from sources, 145, 148
  Fourier number in, 162, 166
  in animal tissue, 164
  in catalysts, 131
  in cylinder, 161
  in gases, 107
  in leaves, 128, 138, 139
  in liquids, 111
  in making of transformer steel, 163
  in metals, 122
  in plane sheet or slab, 161
  in polymers (gases), 118
  in porous media, 126
  in semi-infinite medium, 157
  in sphere, 161
  in sphere in finite volume solution, 103, 106
  in variable area geometries, 43
  mechanisms of, 108
  of dopant in silicon chip, 136, 188
  of neutrons, and critical mass, 48
  of solids in solids, 134, 140
  steady-state multidimensio, 93nal
  through stagnant film, 26
  transient, 143
Diffusion equation (see Fick's equation)
Diffusivities, 107–141
  effective, in porous media, 127, 132
  equations for, 108, 111, 127
  in air, 108
  in biofilms, 184
  in liquids, 111
  in metals, 114
  in molten salts and metals, 138
  in polymers, 118
  in water, 113
  Knudsen, 127, 137

tables of, 108, 110, 113, 114, 135
Dimensional analysis, 206
Dissolution of particulate solids
calculation of time of, 221
of $CaCO_4$ in water, 235
of sugar in coffee, 235
working equation for, 225
Distillation
batch (differential), see batch distillation
construction of trays for, 348
continuous fractional, 337
cryogenic, 381
effect of feed and reflux on, see McCabe–Thiele diagram
Fenske equation for, 352
isotope, 383
McCabe–Thiele diagrams for, 348
minimum number of trays for, 346
minimum reflux for, 346
packed-column, packing for, 216–218
Drug delivery, 45, 97, 99, 104, 105
Drying
periods, 456
time of, 102
with air blower, 48

E

Eddy diffusivity
effect of wind and flow in, 154
Effective therapeutic concentration (ETC), 99
Electrodeposition
Model, 114
size of plant, 138
voltage cycling and, 189
Emissions
accident response to, 187
concentration histories and profiles, 146
continuous, 146, 187
effect of wind on, 154
from chimney, 187
from embedded sources, 95
from point source, 147, 151, 186
from plane source, 146
from solvent spill, 188
from twin sources, and superposition principle, 187
instantaneous, 148
net global carbon dioxide and Kyoto Accord, 153
table of solutions for concentrations of, 148
Environmental topics (see also Emissions)
acid rain, 231
adsorption of pollutants in carbon bed, 257–260
aeration ditch, design of, 213
bioconcentration factors for toxins, 274
biofiltration of VOC's, 184
carbon dioxide uptake by p, 140lant life
clearance after pollution of river bed and soils, see Streeter–Phelps equation
DDT uptake by fish, 274
global warming, 153
mass transfer between oceans and atmosphere, 43
mass transfer in leaf, 139
ocean spill, analysis of, 33
octanol-water partition coefficient as measure of partitioning, 275
partitioning of toxin between animal and environment, 274
pollutant release from buried dumps, 95
pollutant release from groundwater onto, 260 soils
reaeration of river, 60, 100
Streeter-Phelps oxygen sag curve in polluted river, 100
uptake and clearance of toxins in animals, 226, 236
Eigen equation for pressure drop in packed beds, 383
Error function
graph of, 149
table of numerical values of, 149
table of properties of, 150
ETC, see Effective therapeutic concentration

F

Fenske equation, 352
Fermi problems, 48
Fick's equation, 141
general solutions of, 161
long-time solution of, 166, 190
short-time solution of, 166
Fick's law, 2
simple integrations of, 14, 16
Film theory, 21
Film thickness
Effective, 21
estimation of, 30
in entry region, 201
Fish
bioconcentration in (BCF), 225
uptake of toxin by, 226
Flooding
in liquid extraction, 434

in packed columns, 397
Food engineering
food packaging design, 124, 138, 139
freeze-drying, 192
vegetable oil extraction, 165
Fourier number, 166
Fourier's law, 2
Freezing of underground water mass, 103

G

Gas centrifuge, 10
Cascade, 13, 44
enrichment of U-isotopes by, 13
equation for separation factor, 12, 14
Gas–liquid absorption, Gas scrubbing
countercurrent, continuous contact, 66
diameter for packed column, 358
Henry's constants for, 248
HETP for, 396
Kremser equation for staged and linear equilibrium, 396
NTU's for linear equilibrium, 391–394
operating diagram for countercurrent, 349
trays for, 331
use of reactive solvent in, 182, 193
Gas-solid reactions and diffusion, 127, 132
Gradient-driven processes, 2, 4, 43
Graetz problem for mass transfer, 81

H

Hatta number (gas–liquid reaction and diffusion), 181
Heat exchangers, 102
Heat transfer
additivity of resistances in, 35, 102
analogy to mass transfer, 24, 32, 35, 81, 104, 231
Helium, underground storage of, 14
Henry's constants
in absorption of gas mixtures (landfill gases), 250
for adsorption from water onto carbon, 257
for gas-water equilibria, 248
in Langmuir isotherm, 256
liquid-liquid, 267
temperature dependence of, 242
HETP (height equivalent to a theoretical plate)
estimation of, 409
HETS (height equivalent to a theoretical stage)
in coffee decaffeination, 409
Higbie equation, 31, 159
estimation of transient mass transfer coefficients and film thickness, 160
HTU–HETP relation, 396
Humidity
Absolute, 446
and humid heat, 443
and humid volume, 443
charts, 444–445
Hypsometric formula (see Barometric formula)

I

Ion-exchange (see also Percolation processes)
efficiency of column, for linear equilibrium, 375–376
equilibrium isotherm for, 377
minimum bed size for, 383
Isotopes
uranium, separation of, 10, 132
use of Fenske equation in separation of, 132

K

Kidney
active transport in, 428
structure of, 429
Kremser or Kremser–Soudere Brown equation
table of parameters for various operations, 323

L

Laminar boundary layer, 201
Laminar flow
fully developed region for mass transfer in, 201–203
mass transfer coefficients for, 216–222
release of a substance into, 226
Landfill gases, 249
Laplace's equation, 93
solution of, using shape factors, 95
Leaching
in mineral processing, 183
Kremser equation for stage calculations in, 323
of oil-bearing seeds, 333–335
Linear phase equilibria
countercurrent cascades of systems with, 320
Kremser equation for staged operations with, 323

Liquid–liquid extraction
countercurrent cascade for, 320, 387
crosscurrent cascade for, 313
distribution coefficients for, 406
Kremser equation, use in, 323
operating diagrams for, 434
pulsed column in, 405
Log mean differences, 26
Loop of Henle, 233

M

Mass balances
classification of, 51
cumulative, 63, 74, 102
differential (gas absorber), 66
integral (compartment), 52
setting up of, 51–53, 78
table for, 64, 77
unsteady differential, 78
Mass transfer
analogies with heat transfer, 24, 32, 93
and the Biot number, 198–200
by diffusion, see Diffusion
in agitated vessels, 211
release of solute from wall to flowing fluid, 101, 226
simultaneous with chemical reaction, 102, 103
simultaneous with heat transfer, 439–468
through membranes, 47
uptake of toxins in animals, the bioconcentration factor BCF, 226
Mass transfer coefficients
conversion of, 25
definitions of, 23
estimation of, 30
film, 21
for column packings, 219
in agitated vessels, 211
in laminar flow around simple geometries, 263
in laminar tubular flow, 201
in turbulent flow around simple geometries, 206, 232
in turbulent tubular flow, 206, 232
overall, 36
tables for, 25, 29, 31, 92, 168, 203, 233
units of, 25
Materials science topics
binary liquid–solid equilibria, 84
casting of alloys, microsegregation in, 84
diffusion in metals and molten salts, 122
diffusion in polymers, 118
membranes, separation by, 416–417
membranes, structure of, 413
Sievert's law, 122
McCabe–Thiele diagram, 341, 348, 351, 361
Membranes
Asymmetric, 413
Ceramic, 412
hollow-fiber, 419
Loeb–Sourirajan, 413
spiral-wound, 414
structure of, 414
Membrane gas separation
pressure ratio, 431
pressure ratio limited, 432
selectivity, 431
selectivity limited, 432
Membrane processes
hollow-fiber dimensions for, 419
mass transfer coefficients for, 420
table of, 421
Michaelis–Menten equation
use in biofilter design, 410
Models
equivalence of, 103
information from, 76
Momentum transport, 3
Moving boundary problems
and freeze-drying of food, 192
and reacting particle, 171
shrinking core model for, 171

N

Norplant contraceptive implant, 97
Newton's viscosity law, 2
NTU (number of transfer units), 403
plot for calculation of, 393

O

Ocean spill, analysis of, 33
Oinology, 43
Operating diagrams
for continuous contact operations, 434
for percolation processes, 373
for supercritical extraction, 408
Osmotic pressures, table of, 417
Oxygen
content in blood, 262–264
production of by cryogenic air distillation, 381

P

Packaging materials, design of, 124
Packed columns
comparison with tray columns, 398
estimation of diameter, 398
flooding in, 397
preliminary design of, 398
pressure drop in, 397
Packings for packed-column operations, 216
estimation of mass transfer coefficient, 219
standard and modern types, 218
Partial differential equations
how to avoid, 70, 84, 93, 171
setting up of, 78, 82
vectorial form of, 90
Partitioning, see under "environmental topics"
in biology and environment, 226
Percolation processes
as staged operation, 373
efficiency of, 375
in adsorption from groundwater onto soil, 260–261
Permeability of polymers to gases, 120
Pharmacokinetics, 99
Phase equilibria
binary liquid–solid, 84
binary vapor–liquid, 227
effect of total pressure on, 288
ideal solutions of, 284
table of separation factors for isotopes, 288
gas–solid, 253
gas–liquid, 246
in supercritical CO_2, 270
liquid–liquid, 268
of water vapor on adsorbents, 255
table of azeotropes, 284
Photosynthesis, 138
Poiseuille equation, 2

Q

Quasi-steady-state assumption, 70, 171

R

Rayleigh equation
for constant α, 377
in ultrafiltration, 378
Reactive solvent, selection of, 182
Reverse osmosis (RO)
concentration polarization in, 420
design equation for, 424
effect of pressure in, 435
effect of salinity in, 436
flux Peclet number for, 422
hollow-fiber modules, 415
mass transfer coefficients for, 425
production rates of water by, 436

S

Schmidt number, 196
Separation factors (relative volatility)
and Rayleigh distillation, 379
Settling of particles
terminal velocity of, 271
Shape factors in 3-D diffusion, 95
Sherwood number, 196, 198, 203, 210, 233
Shrinking-core model for gas–solid reactions, 178
Sievert's law, 122
Solubilities of liquids and solids in water, table of, 252
Solubility of gases in water, table of, 248
Solution mining, 226
Spray columns
for slurry cooling, 452, 455
Stage efficiencies
in adsorption, 372
in distillation, 360–366
in gas–liquid absorption, 331
in liquid–liquid extraction, 332
in percolation processes, 373
Staged operations
co-current, 313
countercurrent, 320
crosscurrent, 313
differential, 307
in the real world, 330
single-stage, 301
with linear equilibria, 323
Stanton number, 196
Streeter–Phelps equation 100
Supercritical fluid extraction (SCE), 268
applications, 270
decaffeination by, 270
size of extraction vessel, 272
plant, at Houston, Texas, 272

T

Terminal velocity of rising or falling particles, 211
Thiele modulus for catalyst pellet, 177
Transdermal patch, 45

Transport diffusivities, 5, 6
Tray columns
limiting flow rates in, 358–360
preliminary design of, 356
Triangular diagram, 268
Turbulent flow
mass transfer coefficients in, 196
Two-film theory, 33

U

Uranium isotopes
separation by gas centrifuge, 10
separation by gaseous diffusion, 132
separation factors for, 10, 132

V

Vector operators, 90
in formulation of conservation laws, 90
table of, 92

W

Weather balloon, design of, 140

Z

Zero gradients, 8
forced transport against, 10

Other Related Titles of Interest

The Art of Modeling in Science and Engineering with Mathematica, Second Edition
Diran Basmadjian
ISBN: 1584884606

Basic Concepts of Environmental Chemistry, Second Edition
Des W. Connell
ISBN: 1566706769

Biological and Bioenvironmental Heat and Mass Transfer
Ashim K. Datta
ISBN: 0824707753

Chemical Process Performance Evaluation
Ali Cinar, Ahmet Palazoglu, and Ferhan Kayihan
ISBN: 0849338069

Conservation Equations and Modeling of Chemical and Biochemical Processes
Edited by Said S.E.H. Elnashaie and Parag Garhyan
ISBN: 0824709578

Industrial Gas Handbook: Gas Separation and Purification
Frank G. Kerry
ISBN: 0849390052

Phase Behavior of Petroleum Reservoir Fluids
Karen Schou Pedersen and Peter L. Christensen
ISBN: 0824706943

Thermodynamics in Materials Science, Second Edition
Robert DeHoff
ISBN: 0849340659